Sándor L. Erdö (Ed.)

GABA Outside the CNS

With 73 Figures

Springer-Verlag
Berlin Heidelberg New York
London Paris Tokyo
Hong Kong Barcelona
Budapest

Sándor L. Erdö, Ph. D.
Department of Anatomy
University of Göttingen
Kreuzbergring 36
W-3400 Göttingen, FRG

Proceedings of the International Symposium
on Peripheral GABAergic Mechanisms, Rome, July 19 – 21, 1990

ISBN-13: 978-3-642-76917-7 e-ISBN-13: 978-3-642-76915-3
DOI: 10.1007/978-3-642-76915-3

Library of Congress Cataloging-in-Publication Data. International Symposium on Peripheral GABAergic Mechanisms (1990: Rome, Italy). GABA outside the CNS / Sándor L. Erdö (ed.). p. cm. "Proceedings of the International Symposium on Peripheral GABAergic Mechanisms, Rome, July 19-21, 1990" – T.p. verso. Includes bibliographical references and index.
ISBN-13: 978-3-642-76917-7
1. GABA–Physiological effect–Congresses. 2. GABA–Receptors-Congresses. 3. Nerves, Peripheral–Congresses. I. Erdö, Sándor L. II. Title. QP364.7.I595 1990 599'.019245–dc20 91-38963 CIP

Softcover reprint of the hardcover 1st edition 1992

Typesetting: Camera ready by author
25/3130-5 4 3 2 1 0 – Printed on acid-free paper

Acknowledgements

I am grateful for financial support from the following pharmaceutical and chemical companies: Schering-Plough Ltd., Bloomfield, NJ, USA; Ciba-Geigy, AG Basel, Switzerland; Bayer AG, Frankfurt, Germany; and Research Biochemicals Inc., Natick, MA, USA. Without their generous help it would have been impossible to hold this symposium. The excellent secretarial assistance of Louise Müller is greatly appreciated.

Preface

The first comprehensive review of the possible role of γ-aminobutyric acid (GABA) outside the mammalian brain appeared 5 years ago[1]. Since that time, a large body of new evidence has accumulated to suggest that GABA may have a functional significance outside the central nervous system. The recent important findings obtained using biochemical, pharmacological, histochemical, physiological and other approaches made it necessary to update, summarize, and resynthesize the knowledge of peripheral GABAergic mechanisms. These were the major goals of the International Symposium on Peripheral GABAergic Mechanisms held in Rome in July 1990. Although this was a satellite symposium of the European Society for Neurochemistry meeting, the participants, and thus the contributors to this book, were also invited from overseas, from countries such as Australia, Canada, Japan and the United States. The meeting thus took on an international, or even intercontinental, character.

This book will provide the reader with state-of-the-art knowledge of the role of GABA in the mammalian periphery. Moreover, my hope is that the volume will serve not only as a comprehensive source of timely information for the GABA specialist, but also as thought-provoking reading for other life scientists with an interest in the biochemistry, physiology, pharmacology, or even morphology of the neuronal and nonneuronal tissues of mammals.

Now read on... The Editor, 1991

---

[1] Erdö SL, Bowery NG (eds) (1986) GABAergic Mechanisms in the Mammalian Periphery. Raven, New York.

Contents

List of Contributors

M. Ahonen
Department of Anatomy and Neurobiological Research Unit, University of Helsinki, Helsinki, Finland

F. Amenta
Dipartimento Di Scienze Neurologiche, Universita "La Sapienza", Via A. Borelli 50, I-00161 Rome, Italy

J. Barna
Research Centre for Animal Husbandry, H-2053 Herceghalom, Hungary

H.K. Boldizsár
Department of Physiology and Biochemistry, University of Veterinary Science, P. O. Box 2, H-1400 Budapest, Hungary

D.C. Bolser
Department of Allergy and Immunology, Schering-Plough Research 60 Orange Street, Bloomfield, NJ 07003, USA

N.G. Bowery
Department of Pharmacology, School of Pharmacy, University of London, 29/39 Brunswick Square, London WC1N 1AX, UK

M. Broekman
SmithKline Beecham Pharmaceuticals, The Frythe, Welwyn, AL6 9AR, UK

E. Castro
Instituto de Bioquimica, Centro mixto C.S.I.C.-U.C.M., Facultad de Farmacia, Ciudad Universitaria, E-28040 Madrid, Spain

M. Carlesimo
Dipartimento Di Scienze Neurologiche, Universita "La Sapienza", Via A. Borelli 50, I-00161 Rome, Italy

R.W. Chapman
Department of Allergy and Immunology, Schering-Plough Research, 60 Orange Street, Bloomfield, NJ 07003, USA

M. Ciuffi
Department of Preclinical and Clinical Pharmacology, "Mario Aiazzi Mancini", University of Florence, Viale G.B. Morgagni 65, I-50134 Florence, Italy

W.L. Collier
Dipartimento Di Scienze Neurologiche, Universita "La Sapienza", Via A. Borelli 50, I-00161 Rome, Italy

E. Dobó
Central Research Laboratory, Albert-Szentgyörgyi University, H-6720 Szeged, Hungary

R.W. Egan
Department of Allergy and Immunology, Schering-Plough Research, 60 Orange Street, Bloomfield, NJ 07003, USA

S.L. Erdö
Department of Anatomy, Georg-August-University, Kreuzbergring 36, W-3400 Göttingen, FRG

W.A. Fogel
Department of Biogenic Amines, Polish Academy of Sciences, Tylna 3, P.O. Box 225, PL-90-950 Lodz 1, Poland

S. Franchi-Micheli
Department of Preclinical and Clinical Pharmacology, "Mario Aiazzi Mancini", University of Florence, Viale G.B. Morgagni 65, I-50134 Florence, Italy

G. Gentilini
Department of Preclinical and Clinical Phramacology, "Mario Aiazzi Mancini", University of Florence, Viale G.B. Morgagni 65, I-50134 Florence, Italy

P. Gilon
Laboratory de Biologie Cellulaire, Batiment Claude Bernard, Place Croix du Sud 5, B-1348 Louvain-la-Neuve, Belgium
Present address: Laboratoire de Diabetologie et Nutrition, Tour Claude Bernard, UCL 54.74, Avenue Hippocrate 54, B-1200 Brussels, Belgium

M.P. Gonzalez
Instituto de Bioquimica, Centro mixto C.S.I.C.-U.C.M.,
Facultad de Farmacia, Ciudad Universitaria, E-28040 Madrid, Spain

O. Häppölä
Department of Anatomy and Neurobiological Research Unit,
University of Helsinki, Helsinki, Finland

J.A. Hey
Department of Allergy and Immunology, Schering-Plough Research,
60 Orange Street, Bloomfield, NJ 07003, USA

J.M. Hills
SmithKline Beecham Pharmaceuticals, The Frythe, Welwyn, AL6 9AR, UK

W. Howson
SmithKline Beecham Pharmaceuticals, The Frythe, Welwyn, AL6 9AR, UK

K. Jessen
Department of Anatomy, University College London, Gower Street, London
WC1E 6BT, UK

F. Joó
Institute of Biphysics, Biological Research Centre, Hungarian Academy of
Sciences, Szeged, Hungary

T. Karhula
Department of Anatomy and Neurobiological Research Unit,
University of Helsinki, Helsinki, Finland

P. Kása
Central Research Laboratory, Albert-Szentgyörgyi University,
H-6720 Szeged, Hungary

D.I.B. Kerr
Department of Anaesthesia and Intensive Care, The University of Adelaide,
South Australia 5000, Australia

W. Kreutner
Department of Allergy and Immunology, Schering-Plough Research,
60 Orange Street, Bloomfield, NJ 07003, USA

M.M. Larkin
SmithKline Beecham Pharmaceuticals, The Frythe, Welwyn, AL6 9AR, UK

Á. László
2nd Department of Gynecology, Semmelweis University Medical School, Budapest, Hungary
Present address: Medical College of Wisconsin, MFRC A 1000,
8701 Watertown Planck Road, Milwaukee, WI 53226, USA

S. Matsuyama
Department of Pharmacology, Kobe University School of Medicine, Kobe 650, Japan

G.Y. Minuk
Liver Diseases Unit, Rm RS-305, 820 Sherbrook St. HSC, Winnipeg, Manitoba, Canada R3A 1R9

J. Mistry
SmithKline Beecham Pharmaceuticals, The Frythe, Welwyn, AL6 9AR, UK

E. Monos
Experimental Research Department and 2nd Institute of Physiology, Semmelweis University Medical School, Budapest, Hungary

G.L. Nádasy
Experimental Research Department and 2nd Institute of Physiology, Semmelweis University Medical School, Budapest, Hungary

P. Napoleone
Dipartimento Di Scienze Neurologiche, Universita "La Sapienza", Via A. Borelli 50, I-00161 Rome, Italy

J. Ong
Department of Anaesthesia and Intensive Care, The University of Adelaide, South Australia 5000, Australia

M.J. Oset-Gasque
Instituto de Bioquimica, Centro mixto C.S.I.C.-U.C.M.,
Facultad de Farmacia, Ciudad Universitaria, E-28040 Madrid, Spain

H. Päivärinta
Department of Anatomy, University of Helsinki, Helsinki, Finland

Á. Párducz
Institute of Biophysics, Biological Research Centre, Hungarian Academy of Sciences, Szeged, Hungary

C. Remacle
Laboratoire de Biologie Cellulaire, Université Catholique de Louvain, Louvain-la-Neuve, Belgium

C.A. Rizzo
Department of Allergy and Immunology, Schering-Plough Research, 60 Orange Street, Bloomfield, NJ 07003, USA

N. Saito
Department of Pharmacology, Kobe University School of Medicine, Kobe 650, Japan

P. Sarlós
Research Centre for Animal Husbandry, H-2053 Herceghalom, Hungary

A.J. Sellers
SmithKline Beecham Pharmaceuticals, The Frythe, Welwyn, AL6 9AR, UK

S. Soinila
Department of Anatomy and Neurobiological Research Unit, University of Helsinki, Helsinki, Finland

K. Takeda
Department of Pharmacology, Kobe University School of Medicine, Kobe 650, Japan

C. Tanaka
Department of Pharmacology, Kobe University School of Medicine, Kobe 650, Japan

K. Taniyama
2nd Department of Pharmacology, Nagasaki University School of Medicine, Nagasaki 852, Japan

E. Vén
Research Centre for Animal Husbandry, H-2053, Herceghalom, Hungary

L. Wekerle
Research Centre for Animal Husbandry, H-2053, Herceghalom, Hungary

J.R. Wolff
Department of Anatomy, Georg-August-University, Kreuzbergring 36, W-3400 Göttingen, FRG

J.-Y. Wu
Department of Physiology and Cell Biology, University of Kansas, Lawrence, USA

L. Zilletti
Department of Preclinical, and Clinical Pharmacology, University of Florence, Viale G.B. Morgagni 65, I-50134 Florence, Italy

B. Zsolnai
2nd Department of Gynecology, Semmelweis University Medical School, Budapest, Hungary

GABA in the Peripheral Nervous System

The Role of GABA in the Peripheral Nervous System[1]

C. Tanaka and K. Taniyama

Introduction

It is universally accepted that γ-aminobutyric acid (GABA) is an important synaptic transmitter in the central nervous system. Significant amounts of GABA are also present in the peripheral tissues of mammals; however, it may be more difficult to elucidate the functional role of peripheral GABA. Evidence from physiological, neurochemical, and morphological studies suggests that GABA acts on peripheral tissues not only as a neurotransmitter in enteric and parasympathetic nervous systems, but additionally as a hormone [7, 14-16, 70, 72]. GABA receptors are present in the gut, gallbladder, urinary bladder, heart, blood vessels, platelets, lung, liver, kidney, knee joint, sympathic ganglia, uterus, oviduct, ovary, adrenal gland, thyroid, and pancreas. A physiological role for these receptors would depend on the presence or absence of GABA functioning as a neurotransmitter in these regions.

The criteria required to establish GABA as a neurotransmitter are: (a) the occurence of GABA and its forming enzyme within neurons, (b) the neuronal release of GABA, (c) the neuronal and glial uptake of GABA and its inactivation by GABA transaminase as mechanisms by which GABA is removed from the synaptic cleft and conserved for reuse, and (d) electrical stimulus-induced depolarization of the neuronal membrane which is mimicked by GABA and antagonized blockers. Most of these criteria are met in the enteric neurons. By analogy to the enteric nervous system, partial

---

[1] This work was partially supported by grants from the Ministry of Education, Science and Culture, Japan.

evidence is available for the presence of GABAergic neurons in the superior cervical ganglia, urinary bladder, gallbladder, synovial membrane of the knee joint, and sinus node of the heart.

Superior Cervical Ganglia

GABA was found to depress ganglionic transmission due to depolarization of related neurons [12]. This response to GABA was suggested to be mediated by an increase in the membrane Cl^- conductance of ganglion neurons, as in the case of central neurons. In the synaptic ganglia, it was thought that threr were no GABAergic neurons, although GABA is accumulated [2, 5, 89] and released [4].

Immunohistochemical studies using antisera against GABA revealed the presence of GABA-immunoreactive nerve fibers in the superior cervical ganglion, but the origin of the GABAergic nerve fibers was not elucidated [13, 31, 87]. In the superior cervical ganglion GABA containing nerve fibers surround the soma and the stem dendrites of a subpopulation of principal ganglion cells [31, 87]. Glutamate decarboxylase (GAD) [19, 32] and GABA transaminase [19, 23] have been found in neuronal structures of the sympathetic ganglia (Table 1). A specific presynaptic location for GABA and GAD within the sympathetic ganglia suggests GABAergic modulation of ganglion transmission in both sympathetic and parasympathetic ganglia.

Table 1. Evidence for GABAergic transmission in the superior cervical ganglia

	[1]
GAD	[19, 32]
Immunohistochemical demonstration of GABA	[13, 31, 87]
Inactivation of GABA	
Uptake of GABA	[2, 5, 89]
GABA-T	[19, 23]
Release of GABA	[4]
Response of $GABA_A$	
Depolarization	[3, 12]
Response of $GABA_B$	
Inhibition of ACh release	[17]

Myenteric Plexus of the Intestine

Since partial evidence of the presence of enteric GABAergic neurons was provided by Jessen et al. [27], the criteria required to establish GABA as a neurotransmitter have been nearly fulfilled in the enteric nervous system (Table 2) [28, 29, 34, 70]. GABA and its forming enzyme GAD are present within enteric neurons [27, 50, 73, 80]. Immunohistochemical studies using antisera against GABA revealed the presence of GABA immunoreactive neurons in the intestine [11, 21, 26, 58]. The appropriate stimulation of nerve fibers evokes release of GABA from various preparations [25, 33, 74, 75, 76, 82, 88]. The neuronal and glial uptake of GABA [25, 27, 33, 39, 57, 82] and its inactivation by GABA transaminase [38, 50] are mechanisms by which GABA is removed from the synaptic cleft and conserved for re-use. GABA includes a bicuculline-sensitive depolarization [8, 10] followed by a bicuculline-sensitive hyperpolarization [9] of the enteric neuronal membrane.

Table 2. Evidence for the GABAergic transmission in the myenteric plexus

Occurrence of GABA and GAD	[27, 50, 73, 80]
Immunohistochemical demonstration of GABA	[11, 21, 26, 58]
Inactivation of GABA	
Uptake of GABA	[25, 27, 33, 29, 57, 73, 82]
GABA-T	[38, 50]
Release of GABA	[25, 33, 51, 62, 66, 68, 69, 71, 74-76, 79, 82, 88]
Response of $GABA_A$	
Contraction	[18, 22, 24, 30, 37, 40, 44, 54-56, 85, 86]
ACh release	[36, 77]
Rapid depolarization	[8, 10]
Response of $GABA_B$	
Inhibition of contraction	[6, 18, 30, 35, 55]
Inhibition of ACh release	[36]
Non-desensitizing depolarization	[9]

There is evidence for GABA modulation of acetycholine (ACh) release from the intestine. GABA and muscimol evoked a transient contraction and ACh release via the $GABA_A$ receptor [36, 77], and GABA and baclofen inhibited stimulus-evoked contraction and ACh release via the $GABA_B$

receptor [36]. The $GABA_A$ receptor-mediated response was tetrotoxin-sensitive, while the $GABA_B$ receptor-mediated inhibition of ACh release was noted in the tetrotoxin-resistant high K^+-evoked release of ACh. Based on these findings, $GABA_A$ receptors may be located on the soma-dendritic regions of cholinergic neurons and $GABA_B$ receptors on nerve terminals in the myenteric plexus.

As shown in Table 3, 5-hydroxytryptamine (5HT) exerts an exitatory effect on the GABAergic neuron via the $5HT_3$ receptor and an inhibitory effect via the $5HT_1$ receptor in the guinea pig ileum. Biologically active peptides also effect the GABAergic neuron. Substance P, neurotensin, cholecystokinin, and bombesin, but no vasoactive intestinal polypeptide (VIP) and gastrin, evoke the release of GABA from the guinea pig ileum. The amount of GABA released by substance P was two to three times higher than that released by other peptides. These peptides have been shown to evoke the release of ACh from the same preparation. Bicuculline supresses the ACh release evoked by substance P, neurotensin, cholecystokinin, and bombesin thereby, indicating that this peptide-evoked release of ACh is mediated through the endogenous GABA released in response to these peptides. The peptidergic, GABAergic, and cholinergic neurons may interact to control motility of the intestine.

Table 3. Receptors on enteric GABAergic neurons

Stimulated by	
Serotonin ($5HT_3$)	[66]
Substance P	[71]
Neurotensin	[51]
Bombesin	[68]
Cholecystokinin	[62]
Somatostatin	[69]
Inhibited by	
GABA ($GABA_A$)	[74, 75]
Acetylcholine (muscarinic)	[20]
Norepinephrine (α_2)	[20]
Serotonin ($5HT_1$)	[66]

The release of GABA from the guinea pig ileum is subject to a presynaptic autoregulation via $GABA_A$ receptors which are coupled to benzodiazepine binding sites. The enteric GABA autoreceptor seems to differ from the central GABA autoreceptor, in which GABA has been reported to inhibit GABA release via the $GABA_B$ receptor. The tetrotoxin-resistant high K^+-evoked release of GABA was also inhibited either by ACh, via the muscarinic receptor, or norepinephrine (NE) via the $\alpha_2$adrenoceptor, thereby indicating that GABA release is regulated by cholinergic and adrenergic neurons.

As described above, there is evidence for the presence of GABAergic neurons in the myenteric plexus. Modulation of the neuronal control of the gut by GABA is probably of particular significance. With regard to the effect of systemically abministrated GABA agonists and antagonists of functions of the gut, these findings must be interpreted with caution.

Gallbladder

In addition to sympathetic and other nervous systems, the gallbladder is innervated by the parasympathetic enteric nervous system. The gallbladder is derived from the foregut, as is the intestine, and the motility of the gallbladder is physiologically related to that of the intestine.

GABA content and GAD activity measured in the gallbladder of guinea pig are similar to those in the intestine and the regional distribution closely relates to the number of ganglion cells [61].

[^{3}H]GABA in the gallbladder is accumulated by two different saturating processes with high and low affinities [61]. High affinity is sensitive to L-diaminobutyric acid (L-DABA) and is resistant to β-alanine, thereby suggesting the presence of neuronal uptake sites of GABA within the gallbladder wall.

Electrical stimulation evokes a tetrodotoxin-sensitive and Ca^{2+}-dependant release of GABA from gallbladder strips [61]. GABA and muscimol produce tetrodotoxin-sensitive contraction and release ACh from the gallbladder through $GABA_A$ receptors [60]. Hence the $GABA_A$ receptor seems to be located on the soma-dendritic of cholinergic neurons. Baclofen also produces contraction of gallbladder strips through $GABA_B$ receptors, which may be located on the non-cholinergic excitatory neurons or even on the smooth muscle itself [59].

These findings suggest the presence of GABA not only GABA receptors, but also GABAergic neurons in the gallbladders (Table 4). Immunohistochemical studies on GABA and GAD are needed to confirm or rule out the presence of the GABAergic neuron in the gallbladder.

Table 4. Evidence for GABAergic transmission in the gallbladder

Occurrence of GABA and GAD	[61]
Uptake and release of GABA	[61]
Response of $GABA_A$	
ACh release	[60]
Response of $GABA_B$	
Contraction	[59]

Urinary Bladder

Vesical ganglia within the urinary bladder receive input from both sympathetic and parasympathetic nervous systems and exhibit prominent inhibitory and exhibitory synaptic mechanisms that regulate the transfer of information from the central nervous system to the bladder.

Endogenous GABA and GAD are present in the urinary bladder [42] and their concentrations are about half those in the intestine. The distributions of GABA and GAD showed a positive correlation to the number of ganglion cells in the wall of the urinary bladder [42].

[^{3}H]GABA accumulates in both neuronal and non-neuronal elements of strips of the urinary bladder. When the detrusor strip is preloaded with [^{3}H]GABA in the presence of ß-alanine and superfused, the electrical field stimulation clearly evokes a Ca^{2+}-dependent, tetrodotoxin-sensitive release of [^{3}H]GABA [42]. In tissue treated with L-DABA, the stimulation failed to evoke the release of [^{3}H]GABA. Thus, [^{3}H]GABA may be taken up into neuronal elements and release from nerve terminals.

Electrophysiological studies have shown that GABA reacts with $GABA_A$ receptor sites on ganglion cells to cause a reduction in resting membrane potential and a depression of membrane excitability in vesical ganglia of the urinary bladder [12]. This action of GABA was supported by the finding that GABA inhibits the release of ACh from postganglionic cholinergic nerve terminals via the $GABA_A$ receptor [42]. The GABA action mediated by $GABA_A$ receptors in the vesical ganglia of the urinary bladder differs from that in the parasympathetic ganglia of intestine, in which GABA induces a release of ACh through $GABA_A$ receptors. There is evidence for $GABA_B$ receptors in the rabbit urinary bladder [63], newborn rat [47] and guinea pig [49].

GABA and baclofen inhibit the electrical stimulation-evoked contractions, and these effects are not antagonized by bicuculline or picrotoxin. These findings suggest that GABA receptors are present on neuronal elements within the urinary bladder and that $GABA_A$ and $GABA_B$ receptors are involved in the inhibition of mechanical activity of the urinary bladder (Table 5).

Table 5. Evidence for GABAergic transmission in the urinary bladder

Occurrence of GABA and GAD	[42]
Uptake and release of GABA	[42, 65, 67]
Response of $GABA_A$	
Inhibition of ganglion cell membrane excitability	[12]
Inhibition of contraction	[46, 48, 78]
Inhibition of ACh release	[42]
Response of $GABA_B$	
Inhibition of contraction	[49, 63]

In the urinary bladder of the guinea pig, substance P stimulates the cholinergic neuron directly and inhibits the cholinergic neuron indirectly by stimulation of the GABAergic neuron. The affinity of substance P for the cholinergic neuron was 5.3 times higher than that for the GABAergic neuron; therefore substance P may act predominantly on the cholinergic rather than on the GABAergic neuron [65].

GABA release from the urinary bladder is subject to presynaptic autoregulation via $GABA_A$ receptors [67]. The release of GABA was also inhibited by ACh via the muscarinic receptor and NE via the α_2 adrenoceptor [67], thereby indicating that the release of GABA is under the control of adrenergic and cholinergic neurons, as is the case for the intestine.

Sinus Node of the Heart

Central GABAergic mechanisms are involved in the regulation of cardiovascular function, whereas the peripheral GABA system has been shown to participate in the control of cardiovascular-related events. Administration of GABA into the sinus node artery of dogs produces an inconsistent augmentation of the sinus tachycardia due to right stellate stimulation and an attenuation of sinus bradycardia due to stimulation of the right vagosympathetic trunk [53]. In the guinea pig heart, GABA has been shown to inhibit the electrically stimulated release of NE from the atria of rats via the $GABA_B$ receptor [6]. In the case of $GABA_A$ receptors, which are presumably located on the preterminal region of capsaicin-sensitive sensory nerves and initiate a conducted impulse, GABA is involved in the release of transmitter [45]. We recently found a significant amount of GABA, GABA-accumulating cells, and a Ca^{2+}-dependent, tetrodotoxin-sensitive release of GABA in the sinus node of the guinea pig heart [81].

These findings suggest the presence of GABAergic neurons in the sinus node of guinea pigs and possibly GABA modulation of cardiac function (Table 4).

Table 6. Evidence for GABAergic transmission in the sinus node of the heart

Occurrence of GABA	[81]
Uptake and release of GABA	[81]
Response of $GABA_A$	
Sinus node, ACh release	[81]
Atrium, positive chronotropism	[45]
Response of $GABA_B$	
Atrium, inhibition of NE release	[6]

Synovial Membrane of the Knee Joint

Recent studies on the mammalian periphery revealed a possible GABAergic innervation of the knee joint via sciatic and femoral nerves [52]. The GABA content of the synovial membrane was 20.1 nmol/g tissue, a value much lower that in the myenteric neurons. Some 10 days after unilateral dissection of the sciatic nerve, femoral nerve, or both, the GABA content of the ipsilateral membrane was 13.8, 14.6, and 7.8 nmol/g tissue, respectively.

Depolarization of synovial membrane by high K^+ evoked the Ca^{2+}-dependent release of [^{3}H]GABA from intact membranes, but not from the membrane ipsilateral to the dissection of the sciatic and femoral nerves. The evoked release of [^{3}H]GABA was obtained in preparations preloaded with

[^{3}H]GABA in the presence of ß-alanine, but not in the presence of L-DABA (Table 7).

The functional role of GABA remains an open question. The possibility that GABA is present in neurons and modulates the release of ACh warrants further attention. GABAergic mechanisms are reported to regulate the release of substance P from primary afferent fibres and to exert an analgesic effect through $GABA_B$ receptors.

Table 7. Evidence for GABAergic transmission in the synovial membrane of knee joint

Occurrence of GABA	[52]
Release of GABA	[52]
Response of $GABA_A$	
ACh release	[41]

Subtypes of GABA Receptor

Activation of the $GABA_A$ receptor is known to open Cl^- channels both in the central and peripheral nervous systems. The central $GABA_A$ receptor sequences were determined and reconstituted in *Xenopus* oocytes [43, 64]. GABA has also been shown to inhibit neurotransmitter release from central and peripheral tissues via $GABA_B$ receptors, events which may involve Ca^{2+} and K^+ channels and not the Cl^- channel in the central nervous system. We reported that the $GABA_B$ receptor was expressed in the *Xenopus* oocytes following injection of RNA from rat cerebellum [83]. Under current-clamp conditions, the administration of GABA induced a hyperpolarization in the presence of bicuculline, isolating the $GABA_B$ response from the $GABA_A$ response. Under voltage-clamp conditions. When the holding potential was switched to -70 mV, GABA failed to induce any current. Baclofen mimicked the effect of GABA and the response to baclofen was antagonized by phaclofen. Thus, the properties of this $GABA_B$ receptor seem to be similar to those of postsynaptic $GABA_B$ receptors in the hippocampus.

Conclusion

Much progress has been made not only in elucidating the neurotransmitter role of GABA in the mammalian periphery, but also in isolating cDNAs encoding central $GABA_A$ receptor proteins. GABA receptors are present in a variety of peripheral tissues, including gut, gallbladder, urinary bladder, heart, blood vessels, platelets, lung, liver, kidney, knee joint, sympathetic ganglia, uterus, oviduct, ovary, adrenal gland, thyroid, and pancreas. There may be different properties of $GABA_B$ receptors in the peripheral tissue and the central nervous system. Examination of the nature of both peripheral $GABA_B$ receptors and $GABA_A$ receptors and whether $GABA_B$ receptors are divided into some subtypes are subjects of ongoing studies.

References

1. Bertisson L, Suria A, Costa E (1976) γ-Aminobutyric acid in rat superior cervical ganglion. Nature 260: 540-541
2. Bowery NG, Brown DA (1972) γ-Aminobutyric acid uptake by sympathetic ganglia. Nature 238: 89-91
3. Bowery NG, Brown DA (1974) Depolarizing actions of γ-aminobutyric acid and related compounds on rat superior cervical ganglia in vitro. Br J Pharmacol 50: 205-218
4. Bowery NG, Brown DA, Marsh S (1979) γ-Aminobutyric acid efflux from sympathetic glial cells: effect of depolarizing agents. J Physiol 293: 75-90
5. Bowery NG, Brown DA, White RD, Yamini G (1979) [^{3}H]γ-Aminobutyric acid uptake into neuroglial cells of rat superior cervical sympathetic ganglia. J Physiol 293: 51-74
6. Bowery NG, Doble A, Hill DR, Hudson AL, Shaw JS, Turnbull MJ, Warrington R (1981) Bicuculline-insensitive GABA receptors on peripheral autonomic nerve terminals. Eur J Pharmacol 71: 53-70
7. Bowery NG, Hill D (1986) GABA mechanism in autonomic ganglia. In: Erdö SL, Bowery NG (eds) GABAergic mechanisms in the mammalian periphery. Raven, New York, pp 135-152
8. Cherubini E, North RA (1984) Action of γ-aminobutyric acid on neurones of guinea pig myenteric plexus. Br J Pharmacol 82: 93-100
9. Cherubini E, North RA (1984) Inhibition of calcium spikes and tranmitter releaseby γ-aminobutyric acid in the guinea pig myenteric plexus. Br J Pharmacol 82: 101-105
10. Cherubini E, North RA (1985) Benzodiazepines both enhance γ-aminobutyric acid responses and decrease calcium action potentials in guinea pig myenteric neurones. Neuroscience 14: 309-315

11. Davanger S, Ottersen OP, Storm-Mathisen J (1987) Immunocytochemical localization of GABA in cat myenteric plexus. Neurosci Lett 73: 27-32
12. DeGroat WC (1970) The actions of γ-aminobutyric acid and related amino acids on mammalian autonomic ganglia. J Pharmacol Exp Ther 172: 384-396
13. Dobó E, Kása P, Wenthold RJ, Jóo F, Wolff JR (1989) Evidence for GABAergic fibers entering the superior cervical ganglion of rat from the preganglionic nerve trunk. Histo-chemistry 92: 133-136
14. Erdö SL (1985) Peripheral GABAergic mechanisms. Trends Pharmacol Sci 6: 205-208
15. Erdö SL, Kiss B (1986) Presence of GABA, glutamate, decarboxylase, and GABA transaminase in peripheral tissues: a collection of quantitative data. In: Erdö SL, Bowery NG (eds) GABAergic mechanisms in the mammalian periphery. Raven, New York, pp 5-17
16. Erdö SL, Wolff JR (1990) γ-Aminobutyric acid outside the mammalian brain. J Neurochem 54: 363-372
17. Farkas Z, Kása P, Balcar VJ, Joo F, Wolff JR (1986) Type A and type B GABA recep-tors mediate inhibition of acetylcholine release from cholinergic nerve terminals in the superior cervical ganglion of rat. Neurochem Int 8: 565-572
18. Giotti A, Luzzi S, Spagnesi S, Zilletti L (1983) $GABA_A$ and $GABA_B$ receptor-mediated effects in guinea pig ileum. Br J Pharmacol 78: 469-478
19. Happola O, Paivarinta H, Soinila S, Wu JY, Panula P (1987) Localization of L-glutamate decarboxylase and GABA transaminyase immunoreactivity in the sympathetic ganglia of the rat. Neuroscience 21: 271-281
20. Hashimoto S, Tanaka C, Taniyama K (1986) Presynaptic muscarinic and α-adrenoceptor-mediated regulation of GABA release from myenteric neurons of the guinea pig small intestine. Br J Pharmacol 89: 787-792
21. Hills JM, Jessen KR, Mirsky R (1987) an immunohistochemical study of the distribution of enteric GABA-containing neurons in the rat and guinea pig intestine. Neuroscience 22: 301-312
22. Hobbiger F (1958) Effect of γ-aminobutyric acid on the isolated mammalian ileum. J Physiol 142: 147-164
23. Hyde JC, Robinson N (1974) Histochemistry of gamma aminobutyrate transaminase in spinal cord, spinal and sympathetic ganglia. J Anat 118: 581-587
24. Inouye A, Fukuya M, Tsuchiya K, Tsujioka T (1960) Studies on the effect of γ-amino-butyric acid on the isolated guinea pig ileum. Jpn J Physiol 10: 167-182
25. Jessen KR, Hills JM, Dennison ME, Mirsky R (1983) γ-Aminobutyric acid as an autonomic neurotransmitter: release and uptake of $[^3H]$-γ-aminobutyrate in guinea pig large intestine and cultured enteric neurons using physiological method and electron microscopic autoradiography. Neuroscience 10: 1427-1442
26. Jessen KR, Hills JM, Saffrey MJ (1986) Immunohistochemical demonstration of GABAergic neurons in the enteric nervous system. J Neurosci 6: 1628-1634
27. Jessen KR, Mirsky R, Dennison ME, Burnstock G (1979) GABA may be a neurotransmitter in the vertebrate peripheral nervous system. Nature 281: 71-74
28. Jessen KR, Mirsky R, Hills JM (1986) GABAergic neurons in the vertebrate peripheral nervous system. In: Erdö SL, Bowery NG (eds) GABAergic mechanisms in the mammalian periphery. Raven, New York, pp 117-134
29. Jessen KR, Mirsky R, Hills JM (1987) GABA as an autonomic neurotransmitter: studies on intrinsic GABAergic neurons in the myenteric plexus of the gut. Trend Neurosci 10: 255-262

30. Kaplita PV, Waters DH, Triggle DJ (1982) γ-Aminobutyric acid action in guinea pig ileal myenteric plexus. Eur J Pharmacol 79: 43-51
31. Kása P, Jóo F, Dobó E, Wenthold RJ, Ottersen OP, Storm-Mathisen J, Wolff JR (1988) Heterogenous distribution of GABA-immunoreactive nerve fibers and axon terminals in the superior cervical ganglion of adult rat. Neuroscience 26: 635-644
32. Kenny SL, Ariano MA (1986) The immunofluorescence localization of glutamate decar-boxylase in the rat superior cervical ganglion. J Auton Nerv Sys 17: 211-215
33. Kerr DIB, Krantis A (1983) Uptake and stimulus-evoked release of [^{3}H]γ-aminobutyric acid by myenteric nerves of guinea pig intestine. Br J Pharmacol 78: 271-276
34. Kerr DIB, Ong J (1986) GABAergic mechanisms in the gut: their role in the regulation of gut motility. In: Erdö SL, Bowery NG (eds) GABAergic mechanisms in the mammalian periphery. Raven, New York, pp 153-174
35. Kerr DIB, Ong J, Prager RH, Gynther BD, Curtis DR (1987) Phaclofen: a peripheral and central baclofen antagonist. Brain Res 405: 150-154
36. Kleinrok A, Kilbinger H (1983) γ-Aminobutyric acid and cholinergic transmission in the guinea pig ileum. Naunyn Schmiedbergs Arch Pharmacol 322: 216-220
37. Krantis A, Costa M, Furness JB, Orbach J (1980) γ-Aminobutyric acid stimulates intrin-sic inhibitory and excitators nerves in the guinea pig intestine. Eur J Pharmacol 67: 461-468
38. Krantis A, Harding RK (1986) The distribution of GABA-transaminase-hydrogenase activity in the myenteric plexus of rat small intestine: a histochemical analyses. Neurosci Lett 64: 85-90
39. Krantis A, Kerr DIB (1981) Autoradiographic localization of [^{3}H]gamma-aminobutyric-acid in the myenteric plexus of the guinea pig small intestine. Neurosci Lett 23: 263-268
40. Krantis A, Kerr DIB (1981) GABA induced excitatory responses in the guinea pig small intestine are antagonized by bicuculline, picrotoxin and chloride ion blockers. Naunyn Schmiedbergs Arch Pharmacol 317: 257-261
41. Kubota S, Nakano A, Taniyama K, Tanaka C (1990) High affinity uptake of choline and release of acetycholine in rat synovial membrane (to be published)
42. Kusunoki M, Taniyama K, Tanaka C (1984) Neuronal GABA release and GABA inhibition of ACh in guinea pig urinary bladder. Am J Physiol 246: R502-R509
43. Levitan ES, Schofield PR, Burt DR, Rhee LM, Wisden W, Kohler M, Fujita N, Rodriguez HF, Stephenson A, Darlison MG, Barnard EA, Seeburg PH (1988) Structural and functional basis for $GABA_A$ receptor heterogeneity. Nature 335: 76-79
44. Luzzi S, Spagnesi S, Franchi-Micheli S, Rosi E, Ciuffi M, Zilletti L (1986) Diazepam potentiates GABA-contraction in guinea pig ileum. Arch Int Pharmacodyn Ther 279: 29-39
45. Maggi CA, Giuliani S, Manizini S, Meli A (1989) GABA-A receptor-mediated positive inotropism in guinea pig isolated left atria: evidence for the involvement of capsaicin-sensitive nerves. Br J Pharmacol 97: 103-110
46. Maggi CA, Santicioli P, Grimaldi G, Meli A (1983) The effect of peripherally administered GABA on spontaneous contractions of rat urinary bladder in vivo. Gen Pharmacol 14: 455-458
47. Maggi CA, Santicioli P, Meli A (1984) $GABA_B$ receptor mediated inhibition of field stimulation induced contraction of detrusor muscle strips from newborn rats. J Auton Pharmacol 4: 45-51

48. Maggi CA, Santicioli P, Meli A (1984) GABAinhibits excitatory neurotransmission in rat pelvic ganglia. J Pharm Pharmacol 37: 349-351
49. Maggi CA, Santicioli P, Meli A (1985) $GABA_A$ and $GABA_B$ receptors in detrusor strips from guinea pig bladder dome. J Auton Pharmacol 5: 55-64
50. Miki Y, Taniyama K, Tanaka C, Tobe T (1983) GABA, glutamic acid decarboxylase, and GABA transaminase levels in the myenteric plexus in the intestine of humans and other mammals. J Neurochem 40: 861-865
51. Nakamoto M, Tanaka C, Taniyama K (1987) Release of γ-aminobutyric acid and acetylcholine by neurotensin in guinea pig ileum. Br J Pharmacol 90: 545-551
52. Nakano A, Kondo M, Taniyama K, Tanaka S (1988) γ-Aminobutyric acid in synovial membrane of rat knee joint. Life Sci 42: 2657-2662
53. Neely BH, Hageman GR, James TN (1983) Effects of γ-aminobutyric acid on neural regulation of the canine sinus node. Am J Physiol 244: H266-H272
54. Ong J, Kerr DIB (1984) Potentiation of $GABA_A$-receptor-mediated responses by barbiturates in guinea pig ileum. Eur J Pharmacol 103: 327-332
55. Ong J, Kerr DIB (1987) Comparison of GABA-induced responses in various segments of the guinea pig intestine. Eur J Pharmacol 134: 349-353
56. Ong J, Kerr DIB (1989) Modulation of spontaneous motility by $GABA_A$ receptor antagonism in the guinea pig isolated ileum. Neurosci Lett 101: 203-298
57. Saffrey MJ, Marcus N, Jessen KR, Burnstock G (1983) Distribution of neurons with high-affinity uptake sites for GABA in the myenteric plexus of the guinea pig, rat and chicken. Cell Tissue Res 234: 231-235
58. Saito N, Tanaka C (1986) Immunohistochemical demonstration of GABA containing neurons in the guinea pig ileum using purified GABA antiserum. Brain Res 376: 78-84
59. Saito N, Tanaka C (1986) GABAergic mechanisms and their importance in the gallbladder. In: Erdö SL, Bowery NG (eds) GABAergic mechanisms in the mammalian periphery. Raven, New York, pp 185-191
60. Saito N, Taniyama K, Tanaka C (1984) ^{3}H-ACh release from guinea pig gallbladder evoked by GABA through the bicuculline-sensitive GABA receptor. Naunyn Schmiedbergs Arch Pharmacol 326: 45-48
61. Saito N, Taniyama K, Tanaka C (1985) Uptake and release of GABA in guinea pig gallbladder. Am J Physiol 249: G192-G196
62. Sano N, Taniyama K, Tanaka C (1989) Cholecystokinin, but not gastrin, induces γ-aminobutyric acid release from myenteric neurons of the guinea pig ileum. J Pharmacol Exp Ther 248: 378-383
63. Santicioli P, Maggi CA, Meli A (1984) $GABA_B$ receptor mediated inhibition on field stimulation-induced contractions of rabbit bladder muscle in-vitro. J Pharm Pharmacol 36: 378-381
64. Schofield PR, Darlison MG, Fujita N, Burt DR, Stephenson FA, Rodriguez H, Rhee LM, Ramachandran J, Reale V, Glencorse TA, Seeburg PH, Barnard EA (1987) Sequence and functional expression of the $GABA_A$ receptor shows a ligand-gated receptor superfamily. Nature 328: 221-227
65. Shirakawa J, Nakanisshi T, Taniyama K, Kamidono S, Tanaka C (1989) Regulation of the substance P-induced contraction via the release of acetylcholine and γ-aminobutyric acid in the guinea pig urinary bladder. Br J Pharmacol 98: 437-444
66. Shirakawa J, Takeda K, Taniyama K, Tanaka C (1989) Dual effects of 5-hydroxytryptamine on the release of γ-aminobutyric acid from myenteric neurones of the guinea pig ileum. Br J Pharmacol 98: 339-341

67. Shirakawa J, Taniyama K, Iwai S, Tanaka C (1988) Regulation of [^{3}H]GABA release from strips of guinea pig urinary bladder. Am J Physiol 255: R888-R893
68. Takeda K, Taniyama K, Saitoh Y, Tanaka C (1990) Mechanism of contractile response to bombesin of the longitudinal muscle from guinea pig ileum. J Pharmacol Exp Ther 252: 339-343
69. Takeda T, Taniyama K, Baba S, Tanaka C (1989) Putative mechanisms involved in excitatory and inhibitory effects of somatostatin on intestinal motility. Am J Physiol 257: G532-G538
70. Tanaka C (1985) γ-Aminobutyric acid in peripheral tissues. Life Sci 37: 2221-2235
71. Tanaka C, Taniyama K (1985) Substance P provoked γ-aminobutyric acid release from the myenteric plexus of the guinea pig intestine. J Physiol 362: 319-329
72. Tanaka C, Taniyama K (1985) GABA transport in peripheral tissues: uptake and efflux. In: Erdö SL, Bowery NG (eds) GABAergic mechanisms in the mammalian periphery. Raven, New York, pp 57-75
73. Taniguchi H, Hosoya Y, Okada Y, Baba S (1982) Presence of GABA and GAD and high uptake of [^{3}H]-GABA in Auerbach's plexus. In: Okada Y, Roberts E (eds) Problems in GABA research from brain to bacteria. Excerpta Medica, Amsterdam, pp 113-118.
74. Taniyama K, Hanada S, Tanaka C (1985) Autoreceptors regulate γ-[^{3}H]aminobutyric acid release from the guinea pig small intestine. Neurosci Lett 55: 245-248
75. Taniyama K, Hashimoto S, Hanada S, Tanaka C (1988) Benzodiazepines and barbiturate potentiate the pre- and postsyaptic γ-aminobutyric acid (GABA)A receptor-mediated response in the enteric nervous system of guinea pig small intestine. J Pharmacol Exp Ther 245: 250-256
76. Taniyama K, Kusunoki M, Saito N, Tanaka C (1982) Release of γ-aminobutyric acid from cat colon. Science 217: 1038-1040
77. Taniyama K, Kusunoki M, Saito N, Tanaka C (1983) GABA evoked ACh release from isolated guinea pig ileum. Life Sci 32: 2349-2353
78. Taniyama K, Kusunoki M, Tanaka C (1983) γ-Aminobutyric acid inhibits motility of the isolated guinea pig urinary bladder. Eur J Pharmacol 89: 163-166
79. Taniyama K, Miki Y, Kusunoki M, Saito N, Tanaka C (1983) Release of endogenous and labeled GABA from isolated guinea pig ileum. Am J Physiol 245: G717-721
80. Taniyama K, Miki Y, Tanaka C (1982) Presence of γ-aminobutyric acid and glutamic acid decarboxylase in Auerbach's plexus of cat colon. Neurosci Lett 29: 53-56
81. Taniyama K, Saito N, Matsuyama S, Takeda K, Tanaka C (1990) GABAergic mechanisms and cardiovascular function. In: Kito S, Segawa T, Olsen R (eds) Neurotransmitter receptors: neuroreceptor mechanisms in brain. Plenum, New York (to be published).
82. Taniyama K, Saito N, Miki Y, Tanaka C (1987) Enteric γ-aminobutyric acid-containing neurons and relevance to motility of the cat colon Gastroenterol 93: 519-525
83. Taniyama K, Takeda K, Ando H, Tanaka C (1990) Expression of the $GABA_B$ receptor in *Xenopus* oocytes and desensitization by activation of protein kinase C. In: Kito S, Segawa T, Olsen R (eds) Neurotransmitter receptors: neuroreceptor mechanisms in brain. Plenum, New York (to be published).
84. Taniyama K, Tanaka C (1986) GABAergic mechanisms and their functional relevance in the urinary bladder. In: Erdö SL, Bowery NG (eds) GABAergic mechanisms in the mammalian periphery. Raven, New York, pp 175-183

85. Tonini M, Onori L, Rizzi CA, DePonti F, Perucca E, Budassi P, Crema A (1986) Inhibitory action of capsaicin on cholinergic responses induced by $GABA_A$ agonists in the guinea pig ileum. Eur J Pharmacol 128: 273-276
86. Tonini M, Onori L, Rizzi CA, Perucca E, Manzo L, Crema A (1987) Involvement of substance P in the excitatory action of GABAA agonists on cholinergic neurons in the guinea pig ileum. Naunyn Schmiedbergs Arch Pharmacol 335: 629-635
87. Wolff JR, Kása P, Dobó E, Wenthold RJ, Jóo F (1989) Qualitative analysis of the number and distribution of neurons richly innervated by GABA-immunoreactive axons in the rat superior cervical ganglion. J Comp Neurol 282: 264-273
88. Yau W, Verdun PR (1983) Release of γ-aminobutyric acid from guinea pig myenteric plexus synaptosomes. Brain Res 278: 271-273
89. Young JAC, Brown DA, Kelly JS, Schon F (1973) Autoradiographic localization of sites of [^{3}H]γ-aminobutyric acid accumulation in peripheral autonomic ganglia. Brain Res 63: 479-486

GABAergic Neurons in the Myenteric Plexus

K.R. Jessen

Introduction

Some 20 years ago, Hobbiger [8] speculated on the basis of pharmacological studies that GABA, then recently discovered as a constituent of nervous tissue, might have a role in neurotransmission in the enteric nervous system of the vertebrate gut. Subsequent work, however, soon focussed on GABA in relation to the vertebrate central nervous system, partly because early measurements of endogenous GABA levels failed to detect significant levels of GABA in peripheral nervous tissue, suggesting that in vertebrates GABA was only present in the brain and spinal cord. It has since become accepted that in vertebrates GABA is a major neurotransmitter in the central nervous system (CNS). It has also been established that the peripheral nervous system (PNS) shows several GABA-related properties such as GABA receptors, high affinity glial uptake of GABA, the GABA synthesizing enzyme GAD and endogenous GABA. Many of these properties have come to be regarded as ubiquitous features of neural cells that are unrelated to the existence of conventional GABAergic synaptic transmission (for reviews see [9,25]). More recently, however, firm evidence has emerged that conventional GABAergic neurons are indeed present in some parts of the PNS, most notably in the enteric nervous system. The enteric nervous system constitutes the third division of the autonomic nervous system, the other two being the sympathetic and parasympathetic divisions. In addition to efferent neurons innervating intestinal muscle, blood vessels and glands, it contains both sensory neurons and interneurons. It is now believed that this complex system employs a number of different neurotransmitters in addition to acetylcholine and norepinephrine, the classical neurotransmitters of the vertebrate periphery. Thus GABA is the latest member in a group of putative enteric neurotransmitters, which at present consists of purine nucleotides, 5-HT or a 5-HT-like substance and several peptides including vasoactive intestinal polypeptide (VIP), substance P and enkephalin [3].

Biochemical Measurements

Inevitably, the GABA contained in a population of myenteric neurons will be diluted by surrounding tissues when endogenous GABA levels are measured in samples from the gut wall. The amount of GABA in a longitudinal muscle-myenteric plexus preparation from the guinea pig caecum, nevertheless, was about three times the amount of GABA found to be present in a control muscle tissue [10]. This artefact was avoided by measuring endogenous GABA in thin layers of the gut wall. The layer containing the myenteric plexus contained 4.6 nmol mg protein which was 15-20 times more than that obtained from surrounding layers that did not include the plexus [32]. A sharp peak in GABA levels in the whole wall of the rat small intestine was found as postnatal days 7-8 in a recent developmental study. The significance of this is not clear [6]. The myenteric plexus can be freed from the gut wall using a microdissection procedure, and it has been shown that the isolated plexus can synthesize [^{3}H] GABA from its precursor [^{3}H] glutamic acid and accumulate the newly synthesized product [10]. The amount of [^{3}H] GABA obtained in the myenteric plexus in these experiments was about 50 % of that present in cerebellar slices, but in the cerebellum GABAergic neurons are relatively numerous. In homogenates of the guinea pig myenteric plexus the specific activity of GAD (EC 4.1.1.15) was 21.8 nmol/mg protein per hour [10]. In other experiments on homogenates of layers of cat colon containing the myenteric plexus the GAD activity was 45.7 nmol/mg protein per hour, a value ten times that obtained from surrounding plexus-free layers [23,32].

Enteric GABAergic Neurons can be Revealed by Autoradiography

Most or all GABAergic neurons possess high affinity GABA uptake sites by which they are able to selectively accumulate [^{3}H] GABA following incubation in low [^{3}H] GABA medium. This has been widely used to identify GABAergic cells in the CNS and in the PNS of invertebrates, since the [^{3}H] GABA taken up in this way can be visualized by autoradiographic methods. A population of [^{3}H] GABA-labeled nerve fibres and neuronal cell bodies in the myenteric plexus of guinea pig, rat and chicken has been observed in several studies using cell cultures, whole mounts of the myenteric plexus or sections from the gut wall [10, 11, 19, 20, 26]. In the whole mounts it can be seen that the [^{3}H] GABA-labeled fibres lie both inside myenteric ganglia and in the primary, secondary and tertiary nerve plexuses [18]. The target of some of these labeled fibres can be deduced from examination of [^{3}H] GABA autoradiography in tissue sections [11, 20]. This revealed autoradiographic labeling in the intramuscular nerves outside the myenteric plexus. By electron microscopic autoradiography it could be observed that the labeling was only over some of the axons and that other axons were unlabeled. This suggests that only a subpopulation of the intramuscular axons have high affinity [^{3}H] GABA uptake sites and these may be expected to be GABAergic. Since GABA does not have any direct effects on smooth muscle, the role of these fibres in intramuscular nerves is probably the prejunctional regulation of transmitter release from their neighbours (see below).

GABA Release

Several studies have revealed that gut neurons loaded with [^{3}H] GABA release it in response to depolarization, and that this release is Ca^{2+}-dependent and tetrodotoxin-sensitive. Release of preloaded [^{3}H] GABA has been obtained from whole segments of cat colon [30], from strips of the guinea pig taenia coli containing the myenteric plexus [11] from the guenea pig ileal longitudinal muscle-myenteric plexus preparation [16] and from synaptosomes prepared from guinea pig small intestines [33]. Clear evidence that [^{3}H] GABA is released from neurons intrinsic to the enteric ganglia has come from experiments on cultured enteric neurons devoid of extrinsic innervation [11]. In these studies elevated potassium caused release of [^{3}H] GABA in a Ca^{2+} -dependent fashion from cultures containing the guinea pig myenteric plexus from the taenia coli. In other experiments it has been shown that release of endogenous GABA can also be obtained from the gut. Electrical stimulation [31], pharmacological challenge by CCK, neurotensin and substance P [24,28,29] and exposure to ethylene diamine (EDA) [21], which is known to release GABA from central nerve terminals, all release endogenous GABA from guinea pig gut preparations. There ist also evidence that GABA released from gut nerves causes release of acetylcholine and other transmitters from enteric terminals [7,17,18]

Immunohistochemistry

Recently it has been possible to show directly high levels of endogenous GABA in individual cells in tissue sections by the use of antibodies that selectively recognise GABA. In the mammalian gut GABA-immunolabeled neurons were first described in whole mount preparations of the guinea pig and rat intestine [12]. This observation has been confirmed and extended with the help of several different anti-GABA antibodies. GABA-immunolabeled enteric neurons have now been described in several species, including frog, cat and chicken, in addition to guinea pig and rat [1, 2, 4, 14, 27]. The distribution of GABA-immunopositive cells and fibres agrees very well with the previous autoradiographic descriptions of putative GABAergic cells previously obtained with autoradiographic methods. The immunolabeled cells have been considered to be the same population of neurons as those possessing high affinity GABA uptake sites, although this has not been proved directly. In the guinea pig, GABA-positive cell bodies are present in the myenteric ganglia, and they also contain some labeled fibres. Most of the immunolabeled fibres are, however, seen outside the ganglia in the secondary and tertiary plexuses and, most prominently, in the intramuscular nerves. Thus in the guinea pig immunolabeling is most extensive outside the myenteric ganglia. In the rat, the reverse is true and a dense network of fibres is seen inside the ganglia. At least in the guinea pig and rat, GABA-immunolabeled fibres extend throughout the entire length of the gastrointestinal tract. They are essentially confined to the extrinsic musculature, very few fibres being found in the submucosa or mucosa. In both species GABA-immunopositive neurons are essentially absent from the submucous plexus. The morphology and projections of GABA-immunolabeled neurons have been studied in detail [4] in the guinea pig small intestine. In these experiments the preparations had been pre-incubated with GABA to enhance the level of immunolabeling. This pre-incubation did not affect the distribution of immunoreactive structures. This is an important observation since it supports the idea that neurons with high affinity GABA uptake sites and GABA-immunoreactive neurons are the same cell population. The GABA-positive cells had a characteristic morphology which was similar to the morphology of Dogiel type I neurons. Some of these cells projected to the tertiary plexus while others projected to the circular muscle. The axon of these cells ran distally in most cases, extending for approximately 1-5 mm before splitting into branches which joined the circumferentially oriented intramuscular nerves in the circular muscle.

The possibility has been raised that some of the myenteric GABAergic neurons project away from the gut to other autonomic ganglia (15). In this study GABA-immunolabeled nerve fibres were found in two pre-vertebral sympathetic ganglia in guinea pigs, the inferior mensenteric and the coeliac ganglia. No labeled neuronal cell bodies were seen in these ganglia while labeled fibres were present in nerves peripheral, but not central, to the ganglia. Thus, the inferior mesenteric and coeliac ganglia appear to receive GABAergic innervation from some source peripheral to the ganglia, the GABAergic cells in the myenteric plexus of the colon being obvious candidates. It was also shown that exogenous GABA suppressed excitatory postsynaptic potentials and action potentials in the neurons of these ganglia [8].

GABAergic Neurons as Pre-juctional Modulators of Neuromuscular Transmission

Enteric neurons possess both $GABA_A$ and $GABA_B$ receptors. The effects of GABA in the gut must chiefly be mediated via these neuronal receptors, since GABA receptors have not been found on gut muscle [22]. We have previously suggested that GABAergic neurons act as interneurons within the myenteric plexus and take part in pre-junctional modulation of transmitter release within the intramuscular nerves of the circular muscle [11, 13]. Comparison of the relative density of GABA-positive fibres in myenteric ganglia and intramuscular nerves in the rat and guinea pig indicates that action within ganglia may be more important in the rat and that action within intramuscular nerves may be the predominant role in the guinea pig [12, 14].

To appreciate how GABAergic fibres might act within intramuscular nerves, it is necessary to bear in mind the complex structure of the pre-junctional element in gut muscle innervation [5]. When axons from myenteric neurons enter the circular muscle they branch extensively and become organized into fine intramuscular nerves each of which consists of several varicose axons generally accompanied by an irregular Schwann cell element. In these nerves varicosities containing synaptic vesicles lie facing the muscle, well within the axon bundles and in close proximity to other axons. These axon bundles constitute the pre-junctional element in the innervation of the circular muscle layer.

Although the complex nature of the pre-junctional part of the gut neuromuscular junction is well documented in electron microscopic studies, the functional role of this arrangement remains poorly analysed in pharmacological experiments. One possibility is that transmitter that is released from one fibre can modulate release from neighbouring axons. The gut pre-junctional element might then be envisaged as being composed of two functionally different classes of fibre (1) neuro-effector fibres - those axons directly involved in neuromuscular transmission by secretion of substance that relax or contract surrounding muscle (e.g. acetylcholine) and (2) regulatory fibres - accompanying axons whose chief function is to regulate transmitter release from the former class, i.e. to modulate pre-junctionally enteric neuromuscular transmission. Transmitter released from any fibre in a bundle is, however, likely to diffuse to surrounding muscle. A complete separation of pre-junctional control from direct neuromuscular transmission could therefore only be achieved if the transmitters secreted by the modulatory fibres did not influence the contractile state of gut muscle.

Some of the axons within the intramuscular nerves are GABAergic. Although GABA does not act directly on gut smooth muscle, which does not possess GABA receptors, GABA has been shown to affect release of other enteric transmitters including acetylcholine. We suggest that the role of the GABAergic fibres is to modulate the release of transmitter from neighbouring axons (Fig. 1). In this way GABA would be indirectly involved in the neuronal control of gut motility. GABA seems particularly suitable for this type of function by virtue of its inability to affect smooth muscle. Other substances present in the intramuscular nerves, such as substance P and enkephalin, may, however, also act partly in this manner. Thus it seems likely that the functional relevance of the innervation of gut muscle by axon bundles, instead of single fibres, lies in the opportunity this arrangement provides for such pre-junctional control.

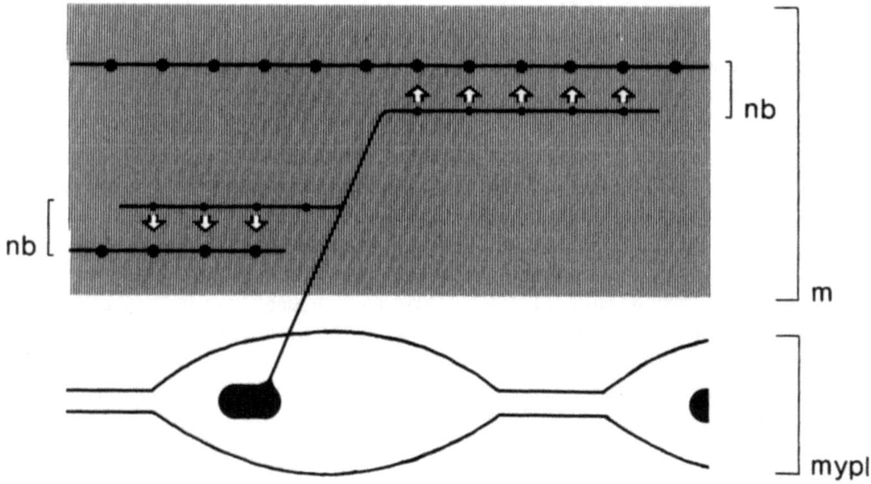

Fig. 1. Possible role of GABAergic neurons in the enteric nervous system. The drawing shows ganglia of the myenteric plexus (mypl) and adjoining musculature (m) innervated by bundles of varicose axons (nb). The cell body of a GABAergic neuron sits in a myenteric ganglion, sending its axon into the intramuscular nerve bundles, where GABA, released from varicosities, exerts pre-synaptic action on neighbouring fibres. (From [11] with permission)

References

1. Baetge G, Gershon MD (1986) GABA in the PNS: demonstration in enteric neurons. Brain Res Bull 16: 421-424
2. Davanger S, Ottersen OP, Storm-Mathisen J (1987) Immunocytochemical localization of GABA in cat myenteric plexus. Neurosci Lett 73: 27-32
3. Furness JB, Costa M (1987) The enteric nervous system. Churchill Livingstone, Edinburg
4. Furness J, Trussel DC, Pompolo S, Bornstein JC, Maley BE, Storm-Mathisen J (1989) Shapes and projections of neurons with immunoreactivity for gamma-aminobutyric acid in the guinea pig small intestine. Cell Tissue Res 256: 293-301
5. Gabella G (1972) Innervation of the intestinal muscular coat. J Neurocytol 1: 341-362
6. Gilon P, Reusens-Billen B, Remacle C, Janssens de Varebeke Ph, Pauwels G, Hoet JJ (1987) Localization of high-affinity GABA uptake and GABA content in the rat duodenum during development. Cell Tiss Res 249: 593-600
7. Giotti A, Luzzi S, Spagnesi S, Zilletti L (1983) GABAa and GABAb receptor mediated effects in guinea-pig ileum. Br J Pharmacol 78: 469-478
8. Hobbiger F (1958) Effects of gamma-aminobutyric acid on the isolated mammalian ileum. J Physiol (Lond) 142: 147-164

9. Iversen LL, Kelly JS (1975) Uptake and metabolism of gamma-aminobutyric acid by neurons an glial cells. Biochem Pharmacol 24: 933-938
10. Jessen KR, Mirsky R, Dennison ME, Burnstock G (1979) GABA may be a neurotransmitter in the vertebrate peripheral nervous system. Nature 281: 71-74
11. Jessen KR, Hills JM, Dennison ME, Mirsky R (1983) γ-Aminobutyrate as an autonomic neurotransmitter: release and uptake of ^{3}Hγ-aminobutyrate in guinea-pig large intestine and cultured enteric neurons using physiological methods and electron microscopic autoradiography. Neuroscience 10: 1427-1442
12. Jessen KR, Hills JM, Saffrey MJ (1986) Immunohistochemical demonstration of GABA-ergic neurons in the enteric nervous system. J Neurosci 6: 1628-1634
13. Jessen KR, Mirsky R, Hills JM (1987) GABA as an autonomic neurotransmitter: studies on intrinsic GABAergic neurons in the myenteric plexus of the gut. Trends Neurosci 10: 255-262
14. Hills JM, Jessen KR (1987) An immunohistochemical study of the distribution of enteric GABA-containing neurons in the rat and guinea-pig intestine. Neuroscience 22: 301-312
15. Hills JM, King BF, Mirsky R, Jessen KR (1988) Immunohistochemical localization and electrophysiological actions of GABA in prevertebral ganglia in guinea-pig. J Auton Nerv Syst 22: 129-140
16. Kerr, DIB, Krantis A (1983) Uptake and stimulus evoked release of 3Hγ-aminobutyric acid by myenteric nerves of guinea-pig intestine. Br J Pharmacol 78: 271-276
17. Kleinrok A, Kilbinger H (1983) γ-Aminobutyric acid and cholinergic transmission in the guinea-pig ileum. Naunyn Schmiedebergs Arch Pharmacol 332: 216-220
18. Krantis A, Costa M Furness JB, Orbach J (1980) γ-Aminobutyric acid stimulates intrinsic inhibitory and excitatory nerves in the guinea-pig intestine. Eur J Pharmacol 67: 461-468
19. Krantis A, Kerr DIB (1981) Autoradiographic localisation of ^{3}H-gamma aminobutyric acid in the myenteric plexus of the guinea-pig small intestine. Neurosci Lett 23: 263-268
20. Krantis A, Kerr DIB, Dennis BJ (1986) Autoradiographic study of the distribution of (^{3}H) gamma aminobutyrate-accumulating neural elements in guinea-pig intestine: evidence for a transmitter function for gamma-aminobutyre. Neuroscience 17: 1243-1255
21. Kerr DIB, Ong J (1984) Evidence that ethylene diamine acts in the isolated ileum of the guinea-pig by releasing endogenous GABA. Br J Pharmacol 83: 169-178
22. Kerr DIB, Ong J (1986) GABAergic mechanisms in the gut: their role in the reglation of gut motility. In: Erdö SL, Bowery NG (eds) GABAergic mechanisms in the mammalian periphery. Raven, New York
23. Miki Y, Taniyama K, Tanaka C, Tobe T (1983) GABA, glutamic acid decarboxylase and GABA transaminase levels in the myenteric plexus in the intestine of humans and other mammals. J Neurochem 40: 861-865
24. Nakemoto M, Tanaka C, Taniyama K (1987) Release of gamma-aminobutyric acid and acetylocholine by neurotensin in the guinea pig ileum. Br J Pharmacol 90: 545-551
25. Roberts E, Chase TN, Tower DB (1976) GABA in nervous system function. Raven, New York
26. Saffrey MJ, Marcus N, Jessen KR, Burnstock G (1983) Distribution of neurons with high affinity uptake sites for GABA in the myenteric plexus of the guinea-pig, rat and chicken. Cell Tissue Res 234: 231-235

27. Saito N, Tanaka C (1986) Immunohistochemical demonstration of GABA containing neurons in the guinea pig ileum using purified GABA antiserum. Brain Res 372: 78-84
28. Tanaka C, Taniyama K (1985) Substance P provoked γ-aminobutyric acid release from the myenteric plexus of the guinea-pig small intestine. J Physiol (Lond) 362: 319-329
29. Sano I, Taniyama K, Tanaka C (1989) Cholecystokinin, but not gastrin, induces γ-aminobutyric acid release from myenteric neurons of the guinea pig ileum. J Pharmacol Ther 248: 378-383
30. Taniyama K, Kusonoki M, Saito N, Tanaka C (1982) Release of gamma-aminobutyric acid from cat colon. Science 217: 1038-1040
31. Taniyama K, Miki Y, Kusonoki M. Satio N, Tanaka C (1983) Release of endogenous and labelled GABA from isolated guinea-pig ileum. Am J Physiol 245: G717-G721
32. Taniyama K, Miki Y, Tanaka C (1982) Presence of γ-aminobutyric acid and glutamic acid decarboxlase in Auerbach's plexus of cat colon. Neurosci Lett 29: 53-56
33. Yau WM, Verdun PR (1983) Release of γ-aminobutyric acid from guinea-pig myenteric plexus synaptosomes. Brain Res 278: 271-273

GABA and Gut Motility[1]

D.I.B. Kerr and J. Ong

Introduction

Motility in the gut is a complex series of movements with an underlying orderliness that ensures descending propulsion of the intestinal contents. Prominent amongst these are rhythmic contractile activity and coordinated descending waves of contraction and relaxation. Rhythmic activity can occur in the absence of obvious stimulation but probably has a mixing function, whereas propulsive activity is triggered by distension or by chemical stimulation of the mucosa. There is an obvious polarity in the propulsive activity; contractions occur above the stimulated region and relaxation below. This is the basis of the peristaltic reflex [3] which persists in extrinsically denervated gut segments having no connections with the central nervous system. Indeed, isolated segments continue to exhibit a range of motility, including the peristaltic reflex, even when totally removed and maintained in artificial media. Such isolated preparations lend themselves to pharmacological investigation and have formed the basis of our studies on GABA involvement in motility patterns of the gut.

[1] We are grateful to the Australian National Fellowship Advisory Committee for the award of a Queen Elizabeth II Research Fellowship to Jennifer Ong.

Enteric Nervous System

Structure

The autonomic nervous system can be classified into three main divisions, namely, sympathetic, parasympathetic, and enteric [44]. The enteric nervous system functions autonomously but under the combined influence of the other two systems. It contains as many neurons as does the spinal cord and consists of the myenteric and submucosal plexuses, which are intrinsic neural plexuses within the wall of the gastrointestinal tract. The myenteric plexus is situated between the outer longitudinal and the inner circular smooth muscle layers ot the muscularis externa, whilst the submucosal plexus supplies innervation to the glands an villi of the innermost mucosa. Myenteric neurons are found in the ganglia and primary meshwork of the plexus, the latter giving rise to a finer secondary meshwork that branches further to form the teritary and deep muscular plexuses innervating the intestinal smooth muscle; interstitial cells of cajal (ICC) are in turn found at intersections of the tertiary meshwork. Each myenteric ganglion contains a variety of neurons with differing morphology. Fundamentally, these provide the sensory elements, interneurons and motor neurons of an integrative circuitry that generates coordinated excitatory and inhibitory activity. Ultimately, efferent (motor) fibers from the myenteric ganglia innervate the outer longitudinal muscle layer and inner circular muscle, providing both excitatory and inhibitory input to these muscle layers.

There is a surprising range of transmitters and modulators to be found in the enteric nervous system, including many peptides present as co-transmitters. In the present context, notable amongst these are neurons containing γ-aminobutyric acid (GABA). On average, about five such GABA-ergic neurons are found per ganglion, constituting some 10 % of the total neurons present. Thus, GABA is likely to be a significant transmitter in the enteric nervous system.

Neurophysiology

Based on electrophysiological recordings, enteric neurons with different electrophysiological and membrane properties have been classified as either S or AH cell types [78]. Of these, S neurons have a low resting membrane potential with high input resistance, exhibit tetrodotoxin (TTX)-sensitive repetitive spike discharges and commonly show fast excitatory post-synaptic potentials (EPSPs) and inhibitory post-synaptic potentials (ISPSs). AH neurons are distinguished by a higher resting membrane potential with lower input resistance, exhibit mainly TTX-resistant calcium-dependent spikes of very limited train length and show prolonged post-spike hyperpolarising potentials. In 1899 Dogiel [10] described three morphologically different types of enteric cells: type I possess short, club-shaped processes, type II cells have many long smooth processes, and type III cells are multipolar. To date, despite significant advances in various research techniques, it has not yet been possible to develop a complete and definite classification of enteric neurons based on correlations between structure, electrophysiological behaviour and transmitter neurochemistry. However GABA-containing neurons uniformly belong in the Dogiel type I category.

Myenteric GABAergic Neurons

GABA-containing neurons in the myenteric plexus, as defined by GABA-immunoreactivity [15], have numerous short, flattened dendritic processes within the ganglia, and a single emergent process (Dogiel type I). Some of the latter run in the tertiary plexus on the inner face of the longitudinal muscle, whilst other neurons provide a dense innervation of the circular muscle layer where GABA-containing fibres appear to be in bundles running in parallel with the circular muscle fibers. In the guinea-pig, few GABAergic fibers run between or within the ganglia, but spines are common on these processes as they leave the cell body. These spines may provide synaptic contacts within the ganglia, otherwise virtually the entire GABAergic projection system is essentially confined to the muscle layer innervation. This description confirms the original autoradiographic studies of Krantis and Kerr [39] and Krantis et al. [42], as well as the immunohistochemical demonstration of enteric GABAergic neurons by Jessen et al. [24]. However, the latter stresses that inter- and intraganglionic GABAergic fibers are to be found in the rat. Nevertheless, these studies agree that the most dense projections run into the muscle layers. Since GABA has no direct action on the smooth muscle of the longitudinal or circular muscle layers, this implies that the principal function of these emergent GABAergic neurons is to provide some form of modulation, probably presynaptic, on the motor innervation of the smooth muscle. As will emerge, this could be mediated through either $GABA_A$ or $GABA_B$ receptors, but most probably the latter. Interestingly, the nerve fiber bundles containing myenteric motor fibers also carry an opioid-peptidergic projection that might well also contribute to such modulation.

In the autoradiographic studies, [^{3}H]GABA accumulation was also found in the region of the ICC, located at junctions of the GABAergic meshwork in the tertiary plexus, implying that some ICC may be GABAergic or receive a dense innervation of GABAergic fibers [42]. At any event, ICC are densely innervated and are connected to each other and to associated smooth muscle cells by gap junctions. They generate pacemaker activity [2] which is directly responsible for contractile activity of colonic smooth muscle cells. ICC receive an inhibitory innervation mediated through hyperpolarisation caused by increased potassium conductance [22]. The latter could well be induced through $GABA_B$ receptors activated through a GABAergic input.

Myenteric GABAergic Neurophysiology

When GABA was applied to myenteric neurons in neurophysiological studies [7-9], two major responses were described: 1) a $GABA_A$ receptor-mediated depolarisation of AH (and possibly S) cells that is bicuculline-sensitive and potentiated by benzodiazepones and 2) a bicuculline-insensitive shortening of calcium-dependent spikes in AH cells. The latter effect is typical of $GABA_B$ receptor-mediated action as seen also in dorsal root ganglia of the spinal cord. In addition, there is a bicuculline-insensitive reduction of transmitter release, leading to a depression of EPSPs that is mimicked by baclofen. This too is a classical $GABA_B$ receptor-mediated action and, being presynaptic, would have no associated measurable IPSP. Slow IPSPs with an equilibrium potential close to that of potassium have been described in myenteric neurons. The latter, in turn, could very well be $GABA_B$ receptor-mediated IPSPs but do not appear to have been challenged with $GABA_B$ receptor antagonists. By definition, such IPSPs would be insensitive to bicuculline, a property that could explain the inability so far [78] to demonstrate bicuculline-sensitive IPSPs. Also, $GABA_B$ receptor antagonists shorten AG cell calcium aciton potentials in myenteric neurons, again as in dorsal root ganglia.

The excitatory depolarising action of GABA on AH cells might be responsible for the excitation seen when GABA is applied to isolated intestinal segments. These cells have a higher resting potential and lower input resistance than S cells. The latter are generally motor neurons of the Dogiel type I morpohology, whereas AH cells, so called because they exhibit long lasting after-hyperpolarisations, are Dogiel type II cells, possibly afferent. If this is so, then GABA and $GABA_A$ receptor agonists would depolarise these afferent cells just as with dorsal root ganglia. Consequently GABA may activate AH cells which would then relay onto (motor) neurons in the ganglia. This implies the release of some transmitter other than acetylcholine from AH cells, since ganglionic blockers do not alter GABA-induced responses in the gut; however, the responses do involve cell firing as they are TTX-sensitive.

Myenteric GABA and GABA Receptors

Originally, applied GABA was shown both to excite and inhibit in the guinea-pig isolated ileum [20,21]. This preparation remains the best test object for such studies, segments from [38] and rabbit being considerably less reliable due to high tone and spontaneous activiy. In the intervening years, aspects of Hobbiger's observations were occationally confirmed and extended but remained little more than a pharmacological curiosity, since the possibility that GABA might be an enteric transmitter was generally discounted. However, it was the demonstration by Jessen et al.(1979) [25] of glutamic acid decarboxylase (GAD) activity and the synthesis of GABA in the myenteric plexus, toghether with GABA uptake by cultured neurons from the plexus, that sparked more general interest in the possibility that GABA might be a transmitter in the gut. This was reinforced by the further demonstration of Krantis et al. (1980) [43] that GABA could stimulate both cholinergic excitatory and NANC (nonadrenergic noncholinergic) inhibitory responses in segments of guinea-pig isolated intestine. At that time our own studies on GABA as an enteric neurotransmitter had already begun, and received a considerable impetus from the demontration by Bowery and Hudson [4] and Bowery er al. [5] of a novel GABA receptor on peripheral autonomic motor fibers.

This new class of GABA receptors proved to be insensitive to the classical GABA antagonists bicuculline and picrotoxin, but activated by the central depressant baclofen. In the field stimulated isolated ileum, both GABA and baclofen depress the cholinergic twitch response, as they do in the stimulated istolated vas deferens of the mouse [5] and guinea-pig [32]. These quite different inhibitory actions immediately recall aspects of Hobbiger's original studies on GABA-induced responses in isolated intestinal segments. In retrospect, he had already described $GABA_B$ receptor-mediated actions in the intestine in 1958, as can be seen in Figs. 8 and 9 of Hobbiger [20], in which GABA depresses spontaneous cholinergic activity and nicotine-induced contractions in the rabbit isolated ileum. Interestingly, a prolonged relaxant response to GABA was found in preparations stored overnight in the cold. Such preparations do not show

$GABA_A$ receptor-induced actions, as we have confirmed (unpublished results). Unfortunately Hobbiger did not use picrotoxinin, which was already known to antagonise some actions of GABA, and so it was not until Bowery's work some 20 years later that the true significance of these observations became apparent.

Myenteric $GABA_A$ Receptors

In isolated segments of intestine GABA stimulates intrinsic neurons leading to transient cholinergic contractions and NANC relaxations. Such effects are mediated through typical $GABA_A$ receptors coupled to chloride channels [40] and are not to be confused with the delayed prolonged relaxation of $GABA_B$ receptor origin that follows them and which can be elicited in the presence of $GABA_A$ receptor antagonists. Ileal $GABA_A$ receptors are sensitive to antagonism by picrotoxinin or bicuculline and are potentiated by neuronal and glial GABA uptake inhibitors [50] by barbiturates which reverse picrotoxinin [33,54], and by benzodiazepines [45,76]. Using intracellular recording techniques in guinea-pig myenteric neurons, benzodiazapines were shown to potentiate the bicuculline- and chloride-sensitive depolarising responses to GABA [9]. In addition, benzodiazapines and barbiturates potentiate the pre- and post-synaptic $GABA_A$ receptor-mediated modification of the release of [^{3}H]GABA and [^{3}H]acetylcholine in the longitudinal muscle-myenteric plexus preparations of the guinea-pig [70]. These properties all closely resemble those of the classical oligomeric $GABA_A$ receptor-induced ileal contraction is a useful pharmacological tool for investigating such actions.

More recently, in addition to barbiturates and benzodiazapines, a diverse number of synthetic and natural steroids have also been found to modulate $GABA_A$ receptors. For example, the anaesthetic steroid alphaxalone enhances the inhibitory actions of GABA in brain tissue [17] and the $GABA_A$ receptor-mediated contractile responses in guinea-pig isolated ileum [62] at lower concentrations; at higher doses, it has direct GABA-mimetic actions sensitive to bicuculline and picrotoxinin. Other steroids such as cortisol and pregnenolone are biphasic modulators of $GABA_A$-induced responses in the periphery, potentiating at low doses and antagonising at high doses [1, 26, 60, 61]. Other modulatory agents also include alcohols [71] and insecticidal agents such as avermectin B1a, a macrolide anthelmintic [64]. The latter induce neurally mediated rhythmic activity in the isolated ileum of the guinea-pig, presumably by releasing endogenous GABA which then modulates the $GABA_A$ receptor chloride-ionophore complexes, since the avermectin-induced responses are sensitive to bicuculline and 3-mercaptopropionic acid [29]. However, although alcohol facilitates activation of $GABA_A$ receptor-linked chloride channel by GABA in the spinal cord [71], it does not appear to directly activate or enhance $GABA_A$ receptor function in the guinea-pig ileal myenteric plexus. [14].

It might be thought that such $GABA_A$ receptor-mediated effects merely represent some form of pharmacological artefact in the intestine, since $GABA_A$ receptors are ubiquitously found on autonomic terminals [6]. Nevertheless, spontaneous activity that is sensitive to $GABA_A$ receptor antagonists is often seen in isolated intestinal segments in the absence of exogenous GABA. Furthermore, in different preparations, this activity can consist of either contractions or relaxations [31,58]. Thus endogenous GABA is involved in the generation of two distinct types of activity that are ultimately dependent on the release of totally different transmitters from motor and inhibitory fibers at the level of the smooth muscle. The anatomical organisation of the innervation through the plexus at the level of the smooth muscle makes it unlikely that these two patterns could arise as an artefact due to spillover of GABA onto extra-synaptic $GABA_A$ receptors. To do so, GABA, presumably being released at axonal $GABA_B$ receptors, would have to diffuse preferentially to $GABA_A$ receptors on motor fibers at one time yet onto inhibitory fibers at another, which is quite improbable given the tight proximity of all fibers within the tertiary bundles. Much more likely, GABA is in fact a transmitter within the ganglia, where it is involved in the generation of spontaneous or induced rhythmic activity, and may even be involved in coordinating reflex patterns. In all probability, the GABAergic projections in the tertiary plexus simply inhibit motor activity by depressing terminal transmitter release.

Myenteric $GABA_B$ Receptors

Once Bowery's work on the new $GABA_B$ receptor type was detailed [5], it was quickly confirmed by Kaplita et al.[27] and by Ong and Kerr [51]. Indeed, we have demonstrated that in the guinea-pig ileum, the $GABA_B$ receptor-mediated depression of cholinergic neurotransmission is sensitive to extracellular calcium concentrations, with an increased sensitivity and sinistral shift of the baclofen dose-response curve at lower calcium concentractions [59]. In the meantime, the designation "$GABA_B$ receptors" was introduced by Hill and Bowery in 1981 [19], based on binding studies with synaptic membranes from rat brain. The first antagonist for $GABA_B$ receptors to be introduced was δ-aminovaleric acid (DAVA), suggested by

Muhyaddin et al. (1982) [47] using the rat anococcygeus muscle as a test object. Baclofen remains the best selective agonist for $GABA_B$ receptors, although 3-aminopropylphosphinic acid is a potent agonist for pre-synaptic $GABA_B$ receptors [63]. Phaclofen, the phosponic analogue of baclofen, was the first selective antagonist that we introduced [34], followed by 2-hydroxy-saclofen, a sulphonic analogue of baclofen [35,36]. Neither DAVA nor 3-aminopropanesulphonic acid (3-APS) are selective antagonists because they exhibit $GABA_A$ receptor agonist activity together with the $GABA_B$ receptor antagonism [16,52].

GABA and Myenteric Inhibition

Whilst rhythmic activity in the plexus could arise simply by virtue of spontanously discharging pacemaker neurons, such activity is clearly coordinated into longer lasting and/or localised patterns, which implies inhibitory interactions between and within ganglia of the plexus. Indeed, IPSPs have been recorded from both S and AH cells, which form the major classes of myenteric neurons. This being so, what might be the transmitter for these myenteric IPSPs? In the central nervous system, glycine and GABA are the principal inhibitory transmitters, with the latter predominating in the brain but also present in the spinal cord. Analogies between the central and enteric nervous systems immediately suggest that GABA would be a prime candidate for myenteric IPSPs. Against this, however, it is persistently claimed that no bicuculline-sensitive potentials can be elicited in the myenteric plexus [78], a claim that for many years was used to deny that GABA could be an enteric neurotransmitter. This failure may reflect an inability to record from or stimulate the relevant neurons or, as it now seems, may suggest that $GABA_B$ receptor-mediated slow IPSPs predominate in the ganglia, rather than fast $GABA_A$ receptor-mediated potentials; certainly it suggests that a reappraisal of this question with the new $GABA_B$ receptor antagonists would be in order.

GABA-Induced Excitation

At one time 5-hydroxytryptamine (5-HT) was suggested [73], but later retracted [77] as an intermediary between GABA stimulation in the ganglia and the responses at the level of the smooth muscle. We have discounted this possibility, although 5-HT does stimulate myenteric neurons and the idea has some attractions [52,55]. Some 30 years ago, in 1958, Hobbiger

[21] described a weak antagonism of GABA to the stimulatory actions of nicotine and 5-HT in various mammalian isolated intestinal preparations, but through a mechanism which was not operating through their rather individual receptor sites. Rather, the antagonism appeared to be the result of an action of GABA on those neuronal stuctures which are involved in the transmission of impulses generated by both the stimulant drugs. Subsequently, in later years, we re-evaluated these interactions between GABA, 5-HT, and nicotinic agents in the guinea-pig ileum and showed that the depression of contractile responses to 5-HT and nicotine by GABA is due to a common depressive $GABA_B$ receptor-mediated effect on cholinergic transmission [52]. However, a brief $GABA_A$ receptor-mediated action was also demonstrated to arise from stimulation of NANC inhibition.

Whilst neurochemical studies showed that substance P produces a dose-dependent increase in the neuronal release of radiolabeled and endogenous GABA in the guinea-pig small intestine, indicating a possible excitatory role of the peptide on the GABAergic neurons [68], isolated organ bath studies have shown that capsaicin (a substance P releaser and depleter) depresses cholinergic responses to $GABA_A$ receptor activation [74, 75]. Curiously, it has been suggested that $GABA_A$ receptor activation evokes both direct and indirect stimulation of enteric cholinergic neurons and that substance P may play and important role in mediating the indirect component of the cholinergic response [75, 76]. Neurotensin also modulates enteric GABAergic neurons [48, 69], as does 5-HT [66], as well as muscarinic M_1 receptors and the α_2-adrenoreceptor [18, 46].

GABA and Motility

Motility can be best described as varied movements of the gastrointestinal tract involving propulsion, mixing and accomodative changes in volume. The ways in which these movements are manifested vary from one part of the tract to the other. So far, studies on the influence of GABA and GABAergic drugs on motility have largely concentrated on the ileum and large intestine, mainly in guinea-pig. Our own studies on the possible significance of GABA in the gut motility have centered on both spontaneous motility and peristaltic activity using isolated segments of small and large intestine from this animal. Peristaltic propulsion of solid faecal contents (pellets) in the distal colon is due to the peristaltic reflex [3]. The small bean shaped pellets within the colon elicit this reflex by distending the gut wall and by mechanically stimulating the mucosa. This results in a contraction of the circular muscle above and a relaxation below the pellet which is thus driven toward the rectal end of the colon. Freshly excised lenghts of distal colon from well-fed guinea-pigs contain a chain of such pellets that are driven at a rather uniform rate. This rate of emptying can then be used as a measure of peristaltic activity that can be subjected to pharmacological manipulation. Alternatively, pellets that have been collected, dried and coated with, e.g. Araldite, can be introduced into the oral end of the isolated segment of colon after its initial contents have been expelled and the rate of descending propulsion again measured [41].

Such experiments have been conducted by three different groups from our laboratory, each time with similar results [11, 41, 51]. In these, inactivation of GABA receptors, by desensitization or with antagonists of $GABA_A$ and/ or $GABA_B$ receptors, always causes significant slowing of the propulsive activity. Indeed, most often after a few minutes, in the continued presence of the drug, the propulsion ceases entirely, the time to cessation also being a measure of the drug influences on propulsion. These effects are seen with either bicuculline or picrotoxinin to block $GABA_A$ receptors and with DAVA to block $GABA_B$ receptors. Similar results are found with desensitization to 3-APS, a $GABA_A$ receptor antagonist. In general, the slowing of propulsion is more effective with combined $GABA_A$ and $GABA_B$ receptor antagonism or desensitization, which implies that both these GABA receptor types are in some way(s) involved in the generation or coordination of peristaltic reflexes.

GABA and Peristaltic Reflexes

Originally, Hobbiger [20] and Inouye et al. [23] found a transient influence of applied GABA on peristalsis in isolated segments of guinea-pig intestine but did not pursue the matter further. According to Krantis et al. [43], bicuculline has no influence on peristaltic reflexes elicited by mechanical stretch, although their method may not be an entirely satisfactory model for this reflex [12]. Flachsenberger [11] has shown that bicuculline, in low doses, slows the ascending excitatory component of the peristaltic reflex elicited by a small inflatable balloon. However, Frigo et al. [13] have shown, and we confirm this, that bicuculline at higher doses has a stimulatory effect on cholinergic transmission in the guinea-pig ileum. This action, however, does not explain our results on the altered velocity of peristaltic propulsion with $GABA_A$ receptor antagonism nor does it explain the interactions of bicuculline (at low doses) and pirenzepine on peristalsis using a modified Trendelenburg method [65]; GABA itself has no influence on the latter preparation.

There is no doubt that a range of $GABA_A$ receptor antagonists can markedly depress or abolish spontaneous activity initiated by avermectins that potentiate GABAergic responses and release GABA from GABAergic neurons [30]. Such blocking actions of $GABA_A$ receptor antagonists are found in preparations exhibiting spontaneous relaxations as well as those exhibiting spontaneous contractions. Furthermore, GABAergic responses (contractions and relaxations) are elicited by ethylenediaminemonocarbamate (EDAC), which presumably releases GABA by some homoexchange mechanism [28, 56]. These responses are also sensitive to 3-mercaptopropionic acid which prevents the release of [^{3}H]GABA from preloaded intestinal segments and thus presumably of endogeneous GABA from the myenteric plexus. In the guinea-pig, only the ileum exhibits the biphasic contractile-relaxant response to GABA and EDAC, whilst the duodenum, jejunum, and colon respond essentially by a $GABA_B$ receptor-mediated relaxation [56].

GABA and Junctional Modulation

In the tertiary plexus on the longitudinal muscle and the muscular-plexus of the circular muscle layer the anatomical arrangement suggests that $GABA_B$ receptor-mediated inhibition of transmitter output will predominate in these terminal regions of the myenteric plexus. Such junction modulation by GABAergic neurons suggests that a GABA-dependent reduction of excitation may contribute to the relaxation phase that precedes the passage of a faecal pellet. Commonly this relaxation is ascribed to a NANC inhibition exerted at the level of the smooth muscle, but it should be borne in mind that the $GABA_B$ receptor-mediated relaxant action could well contribute to or even be entirely responsible for the relaxation preceding the pellet. That this can occur is shown in Fig. 3a of Kerr and Ong [30], where it is shown that DAVA blocked the descending inhibition following inflation of the colon with a small balloon. Unfortunately, these experiments were done before phaclofen or 2-hydroxy-saclofen were readily available and should be repeated with these newer $GABA_B$ receptor antagonists.

Conclusion

There can be little or no doubt that GABA is a transmitter in the enteric nervous system where it elicits responses through both $GABA_A$ and $GABA_B$ receptors. The precise locus of the $GABA_A$ receptors on enteric neurons is unknown, but they closely resemble those found in the central nervous system. Antagonism of these receptors can modify both spontaneous and elicited responses, including the peristaltic reflex. $GABA_B$ receptors are found on motor terminals at all levels of the intestine, where they depress excitatory transmitter release leading to relaxations. In the colon, this relaxant action can mimic the descending inhibitory component of the peristaltic reflex, which may represent their physiological function. Clearly the subject of GABA and GABA receptor involvement in the regulation of gut motility is ripe for development.

References

1. Andres-Trelles F, Bibby V, Lustman S, Simmonds MA (1989) Effects of cortisol on $GABA_A$ receptor-mediated responses compared in guinea-pig ileum and rat cuneate nucleus. Neuropharmacology 28: 705-708
2. Barajas-Lopez C, Berezin I, Daniel EE, Huizinga JD (1989) Pacemaker activity recorded in interstitial cells of Cajal of the gastrointestinal tract. Am J Physiol 257: C830-C835
3. Bayliss WM, Starling EH (1899) The movements and innervation of the small intestine. J Physiol (Lond) 24: 99-143
4. Bowery NG, Hudson AL (1979) Gamma-aminobutyric acid reduces the evoked-release of [^{3}H]-noradrenaline from sympathetic nerve terminals. Br J Pharmacol 66: 108P
5. Bowery NG, Doble A, Hill DR, Hudson AL, Shaw JS, Turnbull MJ, Warrington R (1981) Bicuculline-insensitive GABA receptors on peripheral autonomic nerve terminals. Eur J Pharmacol 71: 53-70
6. Brown DA, Marsh S (1978) Axonal GABA-receptors in mammalian peripheral nerve trunks. Brain Res 156: 187-191
7. Cherubini E, North RA (1984) Actions of gamma-aminobutyric acid on neurones of guinea-pig myenteric plexus. Br J Pharmacol 82: 93-100
8. Cherubini E, North RA (1984) Inhibition of calcium spikes and transmitter release by gamma-aminobutyric acid in the guinea-pig myenteric plexus. Br J Pharmacol 82: 101-105
9. Cherubini E, North RA (1985) Benzodiazepines both enhance gamma-aminobutyrate responses and decrease calcium action potential in guinea-pig myenteric neurones. Neuroscience 14: 309-315

10. Dogiel AS (1899) Über den Bau der Ganglien in den Geflechten des Darmes und der Gallenblase des Menschen und der Säugetiere. Ach Anat Physiol 5: 130-159
11. Flachsenberger WA (1986) Studies on the peristaltic reflex. Thesis, University of Adelaide, Adelaide
12. Frigo GM, Lecchini S (1970) An improved method for studying the peristaltic reflex in the isolated colon. Br J Pharmacol 39: 346-356
13. Frigo GM, Galli A, Lecchini S, Marcoli M (1987) A facilitatory effect of bicuculline on the enteric neurones in the guinea-pig isolated colon. Br J Pharmacol 90: 31-41
14. Frye GD, Fincher AS, Lorenzetti M, Trzeciakowski J, Griffith WH (1990) Comparison of contractile responses of the guinea-pig ileum longitudinal muscle to ethanol and $GABA_A$ agonists. J Pharmacol Exp Ther 252: 474-481
15. Furness JB, Trussell DC, Pompolo S, Bornsterin JC, Maley BE, Storm-Mathisen J (1989) Shapes and projections of neurons with immunoreactivity for gamma-aminobutyric acid in the guinea-pig small intestine. Cell Tissue Res 256: 293-301
16. Giotti A, Luzzi S, Spagnesi S, Zilletti L (1983) Homotaurine: a $GABA_B$ antagonist in guinea-pig ileum. Br J Pharmacol 79: 855-862
17. Harrison NL, Simmonds MA (1984) Modulation of the GABA receptor complex by a steroid anaesthetic. Brain Res 323: 287-292
18. Hashimoto S, Tanaka C, Taniyama K (1986) Presynaptic muscarinic and alpha-adrenoceptor-mediated regulation of GABA release from myenteric neurones of the guinea-pig small intestine. Br J Pharmacol 89: 787-792
19. Hill DR, Bowery NG (1981) ^{3}H-Baclofen and ^{3}H-GABA bind to bicuculline-insensitive $GABA_B$ sites in rat brain. Nature 290: 149-152
20. Hobbiger F (1958) Effects of gamma-aminobutyric acid on the isolated mammalian ileum. J Phyiol (Lond) 142: 147-164
21. Hobbiger F (1958) Antagonism by gamma-aminobutyric acid to the actions of 5-hydroxytryptamine and nicotine in isolated organs. J Physiol (Lond) 144: 349-360
22. Huizinga JD, Berezin I, Daniel EE, Chow E (1990) Inhibitory innervation of colonic smooth muscle cells and interstitial cells of Cajal. Can J Physiol Pharmacol 68: 447-454
23. Inouye A, Fukuya M, Tsuchiya K, Tsujioka T (1960) Studies on the effects of gamma-aminobutyric acid on the isolated guinea-pig ielum. J Physio (Jap) 10: 167-182
24. Jesse KR, Hills JM, Saffrey MJ (1986) Immunohistochemical demonstration of GABAergic neurons in the enteric nervous system. J Neuroscience 6: 1628-1634
25. Jessen KR, Mirsky R, Dennison ME, Burnstock G (1979) GABA may be a neurotransmitter in the vertebrate peripheral nervous system. Nature 281: 71-74
26. Johnston GAR, Kerr DIB, Ong J (1987) Stress, steroids and GABA receptors. In: Rand MJ, Raper C (eds) Elsevier, Amsterdam, pp 121-124
27. Kaplita PV, Waters DM, Triggle DJ (1982) Gamma-aminobutyric acid action in guinea-pig ileal myenteric plexus. Eur J Pharmacol 79: 43-51
28. Kerr DIB, Ong J (1984) Evidence that ethylenediamine acts in the isolated ileum of the guinea-pig by releasing endogenous GABA. Br J Pharmacol 83: 169-177
29. Kerr DIB, Ong J (1986) Gamma-aminonutyric acid-dependent motility induced by avermectin B1a in the isolated intestine of the guinea-pig. Neurosci Lett 65: 7-10
30. Kerr DIB, Ong J (1986) GABAergic mechanisms in the gut: their role in the reglation of gut motility. In: Erdö SL, Bowery NG (eds) GABAergic mechanisms in the mammalian periphery. Raven, New York

31. Kerr DIB, Ong J (1989) $GABA_B$ receptors in peripheral function. In: Bowery NG, Nistico G (eds) GABA: basic research and clinical applications. Phytagora, Rome, pp 239-259
32. Kerr DIB, Ong J, Prager RH (1990) Antagonism of $GABA_B$ receptor-mediated responses in the guinea-pig isolated ileum and vas deferens by phosphono-analogues of GABA. Br J Pharmacol 99: 422-426
33. Kerr DIB, Ong J, Prager RH Ward DA (1986) Caprolactam-barbiturate interaction at the $GABA_A$ receptor complex in the guinea-pig intestine. Eur J Pharmacol 124: 203-206
34. Kerr DIB, Ong J, Prager RH, Gynther BD, Curtis DR (1987) Phaclofen: a peripheral and central baclofen antagonist. Brain Res 405: 150-154
35. Kerr DIB, Ong J, Johnston GAR, Abbenante J, Prager RH (1988) 2-Hydroxysaclofen: an improved antagonist at central and peripheral $GABA_B$ receptors. Neurosci Lett 92: 92-96
36. Kerr DIB, Ong J, Johnston GAR, Abbenante J, Prager RH (1989) Antagonism at $GABA_B$ receptors by saclofen and related sulphonic analogues of baclofen and GABA. Neurosci Lett 107: 239-244
37. Kerr DIB, Ong J, Johnston GAR, Berthelot P, Debaert M, Vaccher C (1989) Benzofuran analogues of baclofen: a new class of central and peripheral $GABA_B$ receptor antagonists. Eur J Pharmacol 164: 361-364
38. Krantis A, Harding RKH (1987) GABA-related actions in isolated in vitro preparations of the rat small intestine. Eur J Pharmacol 141: 291-298
39. Krantis A, Kerr DIB (1981) Autoradiographic localisation of ^{3}H-gamma aminobutyric acid in the myenteric plexus of the guinea-pig small intestine. Neurosci Lett 23: 263-268
40. Krantis A, Kerr DIB (1981) GABA induced excitatory responses in the guinea-pig small intestine are antagonized by bicuculline, picrotoxinin and chloride ion blockers. NaunynSchmiedebergs Arch Pharmacol 317: 257-262
41. Krantis A, Kerr DIB (1981) The effect of GABA antagonism on propulsive activity of the guinea-pig large intestine. Eur J Pharmacol 76: 111-114
42. Krantis A, Kerr DIB, Dennis BJ (1986) Autoradiographic study of the distribution of [^{3}H] gamma aminobutyrate-accumulating neural elements in guinea-pig intestine: evidence for a transmitter function for gamma-aminobutyre. Neuroscience 17: 1243-1255
43. Krantis A, Costa M Furness JB, Orbach J (1980) γ-Aminobutyric acid stimulates intrinsic inhibitory and excitatory nerves in the guinea-pig intestine. Eur J Pharmacol 67: 461-468
44. Langley JN (1921) The autonomic nervous system, Part I. Heffer, London
45. Luzzi S, Spagnesi S, Franchi-Micheli S, Rosi E, Ciuffi M, Zilletti L (1986) Diazepam potentiates GABA-contraction in guinea-pig ileum. Arch Int Pharmacodyn Ther 279: 29-39
46. Micheletti A, Schiavone A, Giachetti A (1988) Muscarinic M1 receptors stimulate a nonadrenergic noncholinergic inhibitory pathway in the isolated rat duodenum. J Pharmacol Exp Ther 244: 680-684
47. Muhyaddin M, Roberts PJ, Woodruff GN (1982) Presynaptic gamma-aminobutyric acid receptors in the rat anococcygeus muscle and their antagonism by 5-aminovaleric acid. Br J Pharmacol 77: 163-168
48. Nakamoto M, Tanaka C, Taniyama K (1987) Release of gamma-aminobutyric acid and acetylocholine by neurotensin in the guinea pig ileum. Br J Pharmacol 90: 545-551

49. Olsen RW (1981) GABA benzodiazapines-barbiturate receptor interactions. J Neurochem 37: 1-13
50. Ong J (1987) Uptake inhibitors potentiate gamma-aminobutyric acid-induced contractile responses in the isolated ileum of the guinea-pig. Br J Pharmacol 91: 9-15
51. Ong J, Kerr DIB (1983) $GABA_A$ and $GABA_B$ receptor-mediated modification of intestinal motility. Eur J Pharmacol 86: 9-17
52. Ong J, Kerr DIB (1983) Interaction between GABA and 5-hydroxytryptamine in the guinea-pig ileum. Eur J Pharmacol 94: 305-312
53. Ong J, Kerr DIB (1984) Evidence for a physiological role of GABA in the control of guinea-pig intestinal motility. Neurosci Lett 50: 339-343
54. Ong J, Kerr DIB (1984) Potentiation of $GABA_A$ receptor-mediated responses by barbiturates in the guinea-pig ileum. Eur J Pharmacol 103: 327-332
55. Ong J, Kerr DIB (1984) Evidence that 5-hydroxytryptamine does not mediate GABA-induces contractile responses in the guinea-pig proximal ileum. Eur J Pharmacol 106: 665-668
56. Ong J, Kerr DIB (1987) Comparison of GABA-induced responses in various segments of the guinea-pig intestine. Eur J Pharmacol 134: 349-353
57. Ong J, Kerr DIB (1989) GABAergic modulation of spontaneous motility in the guinea-pig isolated intestine. Neurosci Lett 101: 203-208
58. Ong J, Kerr DIB (1990) GABA receptors in peripheral tissues. Life Sci 46: 1489-1501
59. Ong J, Kerr DIB, Johnston GAR (1987) Calcium-dependence of baclofen- and GABA-induced depression of responses to transmural stimulation in the guinea-pig isolated ileum. Eur J Pharmacol 134: 369-372
60. Ong J, Kerr DIB, Johnston GAR (1987) Cortisol: a potent biphasic modulator at $GABA_A$ receptor complexes in the guinea-pig isolated ileum. Neurosci Lett 82: 101-106
61. Ong J, Kerr DIB, Johnston GAR (1987) Pregnenolone and its sulphate modulate $GABA_A$ receptor-mediated contractile responses in the guinea-pig isolated ileum. Eur J Pharmacol 142: 461-464
62. Ong J, Kerr DIB, Johnston (1988) Alfaxalone potentiates and mimics GABA-induced contractile responses in the guinea-pig isolated ileum. Br J Pharmacol 95: 33-38
63. Ong J, Harrison NL, Hall RG, Barker JL, Johnston GAR, Kerr DIB (1990) 3-Amino-propanephosphinic acid is a potent agonist at peripheral and central presynaptic $GABA_B$ receptors. Brain Res (to be published)
64. Pong SS, Wang CC, Fritz LC (1980) Studies on the mechanism of action of avermectin B1a: stimulation of release of gamma-aminobutyric acid from brain synaptosomes. J Neurochem 34: 351-358
65. Schworer H, Kilbinger H (1988) Enhancement of guinea-pig intestinal peristalsis by blockade of muscarinic M1-receptors. Br J Pharmacol 93: 715-720
66. Shirakawa J, Takeda K, Taniyama K, Tanaka C (1989) Dual effects of 5-hydroxytryptamine on the release of gamma-aminobutyric acid from myenteric neurones of the guinea-pig ileum. Br J Pharmacol 98: 339-341
67. Study RE, Barker JL (1981) Diazepam and (-)pentobarbital-fluctuation analysis reveals different mechanisms for potentiation of GABA responses in cultured central neurones. Proc Natl Acad Sci USA 78: 7180-7184
68. Tanaka C, Taniyama K (1985) Substance P provoked gamma-aminobutyric acid release from the myenteric plexus of the guinea-pig small intestine. J Physiol (Lond) 362: 319-329

69. Tanaka C, Taniyama K (1986) GABA transport in peripheral tissues; uptake and efflux. In: Erdö SL, Bowery NG (eds) GABAergic mechanisms in the mammalian periphery. Raven, New York, pp 57-75
70. Taniyama K, Hashimoto S, Hanada S, Tanaka C (1988) Benzodiazepines and barbiturate potentiate the pre- and postsynaptic gamma-aminobutyric acid (GABA)-A receptor-mediated response in the enteric nervous system of guinea-pig small intestine. J Pharmacol Exp Ther 245: 250-256
71. Ticku MK, Lowrimore P, Lehoullier P (1986) Ethanol enhances GABA-induced $^{36}Cl^-$ influx in primary spinal cord cultured neurons. Brain Res Bull 17: 123-126
72. Tonini M, De Petris G, Onori L, Manzo L, Rizzi CA, Crema A (1989) The role of $GABA_A$ receptor function in peristaltic activity of the guinea-pig ileum: a comparative study with bicuculline, SR 95531 and picrotoxinin. Br J Pharmacol 97: 556-562
73. Tonini M, Onori L, Lecchini S, Frigo GM, Perucca E, Saltarelli P, Crema A (1983) 5-HT mediated GABA excitatory responses in the guinea-pig proximal ileum. Naunyn-Schmiedebergs Arch Pharmacol 324: 180-184
74. Tonini M, Onori L, Rizzi CA, De Ponti F, Perucca E, Budassi P, Crema A (1986) Inhibitory action of capsaicin on cholinergic responses induced by $GABA_A$ agonists in the guinea-pig ileum. Eur J Pharmacol 128: 273-276
75. Tonini M, Onori L, Rizzi CA, De Ponti F, Perucca E, Budassi P, Crema A (1987) Involvement of substance P in the excitatory action of $GABA_A$ agonists on cholinergic neurons in the guinea-pig ileum. Naunyn Schmiedebergs Arch Pharmacol 335: 629-635
76. Tonini M, Onori L, Manzo L, Rizzi CA, Crema A (1988) Cholinergic contractions induced by $GABA_A$ receptor activation in the guinea-pig ileum are inhibited by alpha-chymotrypsin and potentiated by diazepam. Arch Int Pharmacodyn Ther 296: 173-183
77. Tonini M, Saltarelli P, Onori L, De Ponti F, Garlaschelli L, Rizzi CA, Crema A (1986) A re-appraisal of the mode of action of 5-HT in inhibiting GABA-induced cholinergic contractions in the guinea-pig ileum. Eur J Pharmacol 127: 267-270
78. Wood JD (1983) Neurophysiology of parasympathetic and enteric ganglia. In: Lars-Gosta (ed) Autonomic ganglia. John Wiley, Chichester, pp 367-398

The GABAergic Innervation of Paravertebral Sympathetic Ganglia[1]

J.R. Wolff, P. Kása, E. Dobó, Á. Párducz, and F. Joó

Introduction

According to the classical scheme, the sympathetic innervation of target organs is noradrenergic and mediated by postganglionic axons of principal ganglion cells. These principal neurons (PN) are in turn innervated by cholinergic axons originating from neurons in the intermediolateral column of the spinal cord [33, 29]. This view has been extended in several directions: Many but not all PN coexpress neuropeptide Y together with tyrosine hydroxylase which is responsible for norepinephrine production [33]. A considerable proportion of preganglionic axons and the corresponding spinal neurons contain certain peptides, such as met-enkephalin. Peptides are either colocalized with choline acetyltransferase and acetylcholine (ACh) or may occur separately in spinal afferents, e.g., to the superior cervial ganglion (SCG) [19]. In addition to PN, there are interneurons or neuroendocrine cells in sympathetic ganglia, so-called small intensively fluorescent (SIF) cells which produce catecholamines [18]. Some of the peptides (e.g., metenkephalin [19]) and catecholamines inhibit ganglionic transmission.

Until recently, little was known about the possible involvement of amino acids in the innervation of sympathetic ganglia. However, rapidly accumulating evidence indicates that γ-aminobutyric acid (GABA), a major inhibitory transmitter in the central nervous system, also plays a role in the sym-

[1] Our studies have been supported by grants from the German Science Foundation (Wo 279/8-2), the Hungarian Academy of Science and the Scientific Research Council, Ministry of Health 06/4-20/457.

pathetic ganglion system. In 1986, Bowery and Hill [5] concluded that "there is still insufficient evidence to proclaim it (GABA) as an endogenous agent, at least in sympathetic ganglia." According to new findings, GABA may be a transmitter, a neuromodulator, and/or a (neuro)trophic agent not only in sympathetic ganglia, but also in many other parts of the autonomous nervous system. Here we have collected new data strongly indicating the presence and neuronal location of an endogenous GABA system, although we are still far from understanding its functional role in these ganglia. Since this volume includes reports on GABA in the intestinal plexus, (Jessen et al., this volume), preaortal ganglia (see Häppölä et al., this volume), adrenal medulla (Oset-Gasque et al., this volume), we shall focus on the paravertebral sympathetic trunk, its ganglia and its connectives.

Is There an Endogenous GABA System?

On the basis of the pharmacological actions of GABA and agonistic agents (diazepam) on blood pressure regulation, Stanton and Woodhouse [36] first suggested that sympathetic ganglia may include an endogenous GABA system. Electrophysiological studies confirmed that there are both $GABA_A$ and $GABA_B$ receptors, which may suppress synaptic transmission via pre- and postsynaptic inhibitory mechanisms in the SCG if activated by exogenous GABA or the respective agonists (muscimol, baclofen) [5]. In addition, it has been shown that GABA is able to inhibit ACh release from preganglionic nerves via both receptor types [17]. Thus, the activity of PN might be effectively modulated by GABA if this amino acid were produced and released in sympathetic ganglia.

Several authors have found significant but variable concentrations of GABA in the SCG: [34] 15.0 pmol/mg tissue [7], 29.7 pmol/mg [2], and 153.0 pmol/mg [6]. These biochemical data have recently been confirmed by immunohistochemical techniques using antibodies against GABA (glutaraldehyde conjugates of GABA with serum albumin) [40]. The release of endogenous GABA has not yet been demonstrated, but preloaded $[^3H]$GABA is spontaneously released from sympathetic ganglia. This release can be enhanced by depolarizing agents and electrical stimulation [37]. Thus, there is at least a chance for neuronal activity to induce the release of endogenous GABA by increasing extracellular concentrations of potassium. The source of endogenous GABA is not yet clear. On the one hand, a considerable capacity for GABA uptake has been demonstrated in the SCG of various species. GABA seemed to be exclusively accumulated by glial cells [4, 45, 6, 39]. This was an interesting possibility, because the GABA content of serum appears to be high enough to allow for ganglionic accumulation via high affinity uptake (e.g., 1.5 μM GABA in bullfrog serum) [30]. On the other hand, GABA may be produced by glutamate decarboxylase (GAD) which is apparently present at least in the SCG. Biochemical evidence provided by Kanazawa et al. [25] has recently been supported by GAD immunohistochemistry [31, 18, 26, 14]. Since axons, axon terminals, and presynaptic elements containing GABA and GAD have been found in sympathetic ganglia [40, 44, 14] (Figs. 1 A, B), it is no longer clear how much GABA is accumulated by glial uptake from extraganglionic sources and/or locally provided by neuronal synthesis. There is evidence for some endogenous production of GABA in the SCG; however, it remains to be determined which proportion of the GABA pool in the SCG depends on accumulation from extraganglionic sources and whether data on the SCG are representative for other sympathetic ganglia.

Cellular Location of GABA and GAD

Until recently, GABA was thought to be exclusively located in glial cells [5]. This assumption was based on several lines of evidence: First, ganglionic GABA uptake can be inhibited by ß-alanine which is a specific competitor for GABA uptake sites on glial cells but not on neurons [4]. Second, the biochemical characteristics of induced GABA release differ from those found in brain tissue [37]. Third, in autoradiograms following accumulation of [^{3}H]GABA, labeling was restricted to what appeared to be glial sheaths around PN [45, 39]. In contrast, the immunocytochemical localization of endogenous GABA only occasionally visualized weak and irregular immunoreaction in glial processes (Fig. 1C), while certain neuronal elements were consistently and strongly stained in the same preparations of rat SCG. In some preparations, glial immunoreactivity for GABA was completely absent [40]. It is unclear whether the GABA content of ganglionic glial cells without preloading is too low to be detected by immunohistochemistry and/or whether endogenous GABA is selectively lost from glial cells during tissue fixation. Thus, glial cells may contain GABA, but the content seems to be low or labile and variable. GAD immunoreaction has not been found in ganglionic glial cells [18] (confirmed by unpublished own observations). A prominent immunoreactivity for GABA was regular-

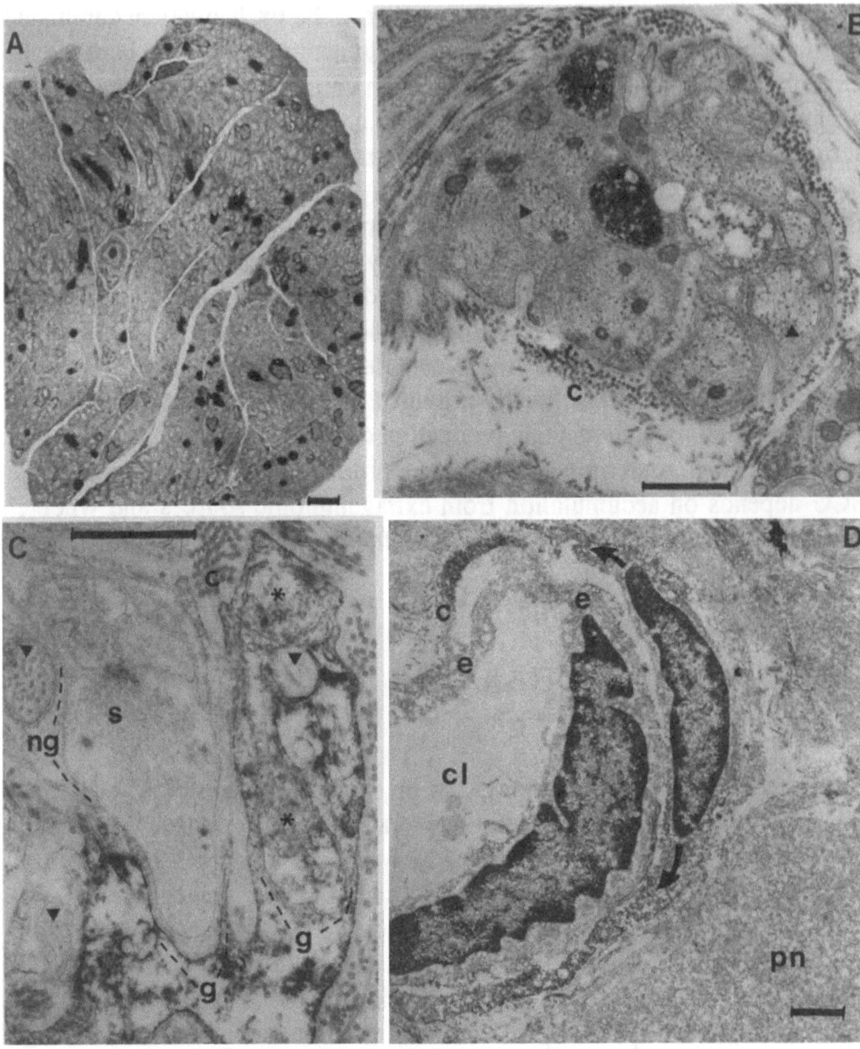

Fig. 1 A-D. Electron micrograph of GABA-immunoreactive (IR) structures. **A** IR axons in preganglion nerve of the rat SCG. **B** IR and nonlabeled axons resemble each other in size and glial relations. **C** IR glial process (*g*) in contact with weakly IR axon varicosities (*asterisk*). **D** Non-neural, mesenchymal (?) cell in perivascular position (*arrows*), GAD immunoreaction. *Arrows*, non-IR axon profiles; *c*, collagen fibrils; *cl*, capillary lumen; *ng*, non-IR glial processes; *pn*, non-IR principal neuron; *s*, non-IR synapse. *Bars* = 10 μm (**A**) and 1 μm (**B-D**)

lyseen in small neuron-like cells which varied in shape between oval, bipolar, and multipolar. Their dendrites were innervated by nonreactive (cholinergic) presynaptic elements forming asymmetric synaptic junctions [40] (Fig. 2A). GABA-positive, neuron-like cells occur as solitary cells in both paravertebral ganglia and the nerve trunks connecting them [13, 14]. Others form groups which are integrated in clusters of SIF-like neurons. GABA immunoreactivity was restricted to subpopulations of SIF-like cell clusters in all paravertebral ganglia studied (Fig. 3A-D). This is in agreement with findings of Karhula et al. [26] (see also Häppölä, this volume), who used GAD as a marker and found such GABAergic cells not only in the SCG, but also in the coeliac-superior mesenteric ganglion complex and the major pelvic ganglion. In addition, these authors reported that all GAD-positive neurons also reacted with tyrosine hydroxylase antibodies. Hence, at least subpopulations of GABAergic neurons are SIF cells in sympathetic ganglia. It is not yet clear whether this also holds true for solitary multipolar cells and those bipolar neurons located in the preganglionic nerve trunk of the SCG (Fig. 3I). It is the latter population of GABA-immunoreactive cells which seems to produce most of the GABA-positive axons, form axon varicosities and terminals, and make synaptic junctions with PN [28, 44, 14, Fig. 3E-I]. The vast majority of GABA-positive axons are unmyelinated with diameters varying between 0.08 and 0.8 μm (Fig. 1A, B). Only a few GABA-labeled axons were larger and myelinated among more than one thousand fibers checked. Axon varicosities and terminal axon branches were mostly filled with small clear vesicles, although some of them contained considerable numbers of dense cored vesicles and larger granules (Fig. 2B). Most synaptic junctions with spines had an asymmetric structure (Fig. 2B-D) sometimes supplemented with subsynaptic bodies and bands.

This synaptic structure was also seen on dendritic stems and somatic protrusions. Only GABA-positive axonal varicosities directly attached to the soma surface of PN made synaptic junctions of symmetric type showing inconspicuous postsynaptic densities [14].

Finally, another GABA-immunoreactive cell type has been observed which is apparently neither of glial nor of neuronal nature. These cells are located within the connective tissue of ganglia. They are covered neither by glial sheaths nor basal laminae and are often found in perivascular positions. Interestingly enough, these fibroblast-like cells also react with GAD antibodies (Fig. 1D). It is unclear whether these cells are identical with those described by Häppölä et al. [18], because these authors described SIF cells in a perivascular position while our observations revealed flattened or elongated fibroblast-like cells.

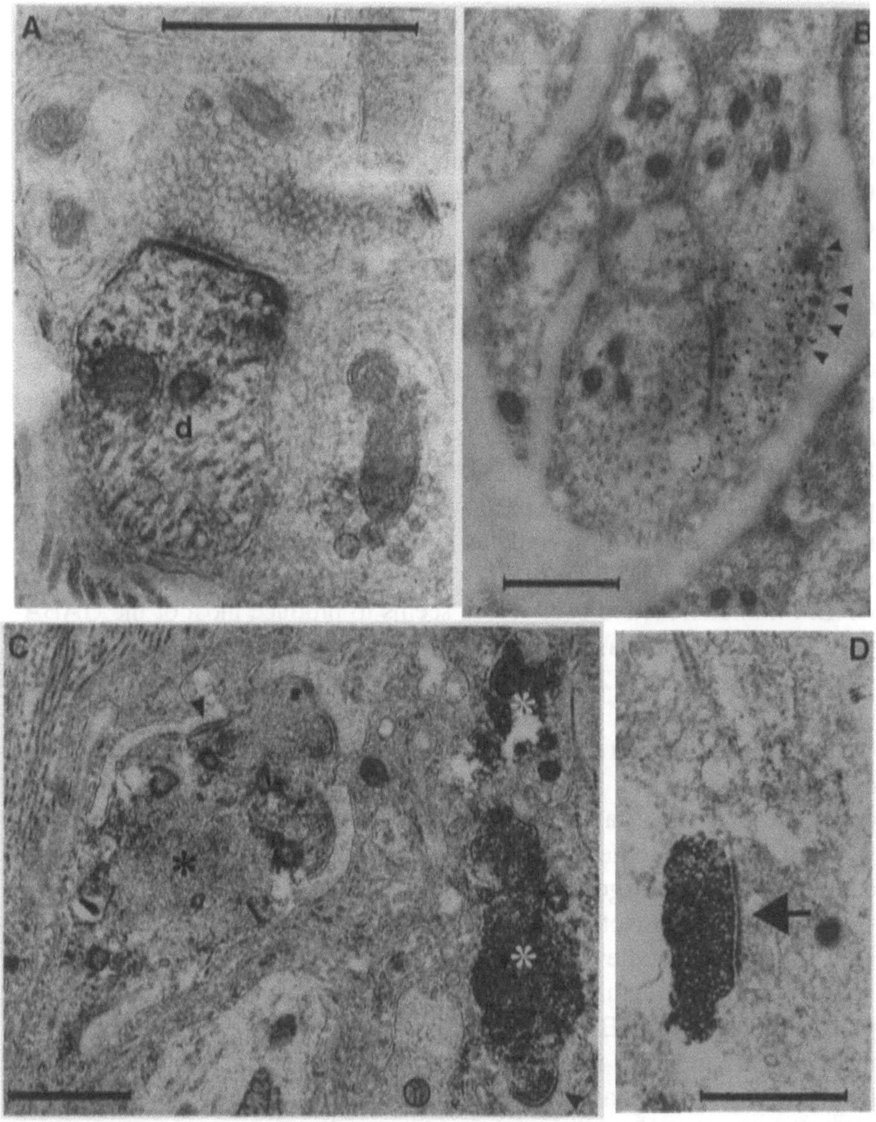

Fig. 2 A-D. Afferent and efferent synapses of GABA-positive neurons.
A GABA-immunoreactive (IR) dendrite *(d)* is postsynaptic to non-IR presynaptic element. **B** IR of an presynaptic element making an asymmetric synapse; note small and large dense granules *(arrowheads)* among a majority of clear vesicles (immunogold, postembedding IR). **C** Non-IR *(black asterisk)* and IR *(white asterisk)* axon varicosities make synapses with similar postsynaptic structures *(arrowheads)*. **D** GAD immunoreaction is restricted to presynaptic elements and shows asymmetric synapse *(arrows)*. *Bars* = 1 μm

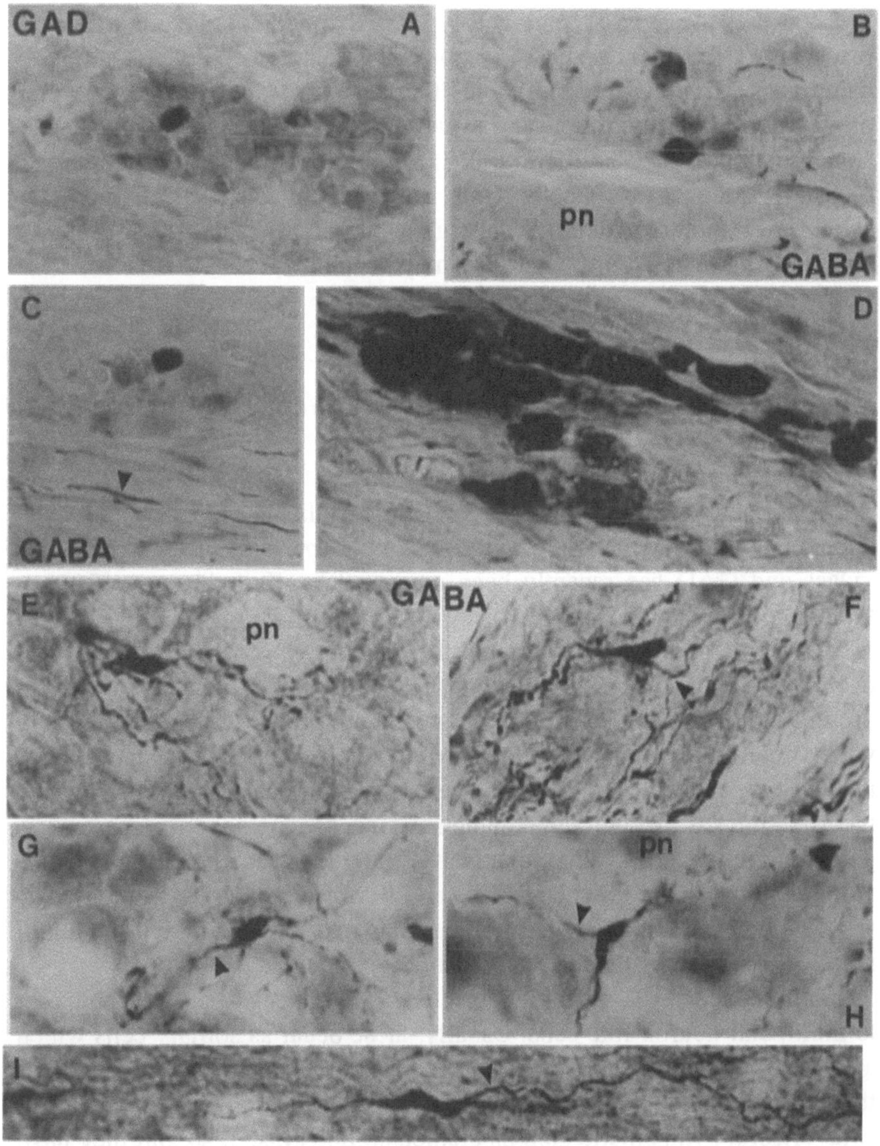

Fig. 3 A-I. GABA-immunoreactive (IR) neuron-like cells which appear SIF cell-like (**A-D**) or bi- to multipolar (**E-I**) in the SCG and stellate ganglion. The latter cells form varicose and axon-like processes *(arrowheads)*. **A** GAD immunoreactivity; **B-I** GABA-IR. Note that SIF-like IR cells are integrated in clusters non-IR SIF cells. *Bars* = 10 μm

Thus, at present, at least three cell types in sympathetic ganglia appear to react with GABA antibodies: (1) small neuron-like cells which probably comprise subtypes (SIF cell-like, multipolar, bipolar) at least one of which produces GABAergic axons and synapses; (2) glial cells which are occasionally stained, at least in parts of their processes, and (3) non-neuronal (possibly mesenchymal) cells. Only the first and the last cell type also show GAD immunoreactivity. This suggests that glial cells are not capable of producing GABA, although they may accumulate it.

Distribution of GABA-Immunoreactive Cells

As mentioned above, a subpopulation of SIF cells may produce both catecholamines and GABA. Direct evidence for coexpression of tyrosine hydroxylase and GAD has recently been presented by Karhula et al. [26], although colocalization was previously not found in the rat SCG [31]. Thus, the former authors ascertained that immunoreactivity for GAD and GABA occurs in SIF-like cells [31, 18, 14] (Fig. 2A-D). Evidence available at present indicates that SIF cells immunoreactive for GABA and GAD are widely distributed in paravertebral and preaortal sympathetic ganglia, e.g., in the SCG, the stellate ganglion, the superior thoracic ganglion down to the first lumbar paravertebral ganglion, the coeliac-superior mesenteric ganglion complex, and the major pelvic ganglion (see Häppölä et al. this volume, but see [21] for the coeliac and inferior mesenteric ganglion). The relation between these SIF cells and axons and axon terminals immunoreactive for GABA and GAD remains to be elucidated.

Another type of GABA-immunoreactive cell has been described as bipolar to multipolar neuron-like cells. Usually, these occur as solitary cells which are located in paravertebral ganglia and nerve trunks connecting ganglia of the cervical sympathetic complex. Such cells have been identified in all paravertebral sympathetic ganglia studied so far (Figs. 3, 4), but they have never been seen in post-ganglionic nerves [40, 41, 28, 13, 14]. It is not yet clear whether all of these cells belong to a uniform cell type, because only some of them clearly show axon-like processes, especially in lower thoracic ganglia.

Identification of GABA (or GAD) immunoreactivity in glial or non-neuronal (mesenchymal) cells requires the high resolution of electron microscopy. Both cell types are labeled by GABA antibodies in the SCG and stellate ganglion (unpublished observations; Fig. 2F-I), but corresponding studies in other paravertebral ganglia are still needed.

Distribution of GABAergic Terminal Axons and Synapses

Electron microscopical analysis of GABAergic synapses is at present restricted to the SCG [14] and stellate ganglion [41, 43]. In both ganglia, the vast majority of GABA- and GAD-positive axon varicosities form synaptic junctions with PN. Most of the GABAergic synapses are located on dendritic spines and spine-like protrusions from somata, while a minority make direct contacts with dendritic stems and somata of PN. This suggests that PN are the major target of GABAergic innervation, at least in cervical sympathetic ganglia.

In the SCG of rat [28, 44] and mouse (unpublished observations) GABA-positive axon varicosities are unevenly distributed. Terminal axon branches immunoreactive for GABA form a loose network in the whole ganglion. Around some PN, however, GABA-positive axon varicosities aggregate and form dense, basket-like arrays enclosing somata and stem dendrites (Fig. 5). PN, richly innervated by GABAergic axons (so called "RIG cells"), are unevenly distributed in the paravertebral ganglion chain and follow a rostral-to-caudal gradient. In the SCG, the relative frequency of RIG cells among all PN varied between 9 % in the rostral part to 3 % in the caudal part (average about 5 %) [44]. In the cervical sympathetic ganglion complex their number is smaller and they are virtually absent below the 4th thoracic ganglion.

As a consequence of the uneven distribution of GABA-positive axons and axon varicosities, probably not all PN receive GABAergic innervation. This is especially the case in lower levels of the sympathetic trunk, where GABA-immunoreactive axons are scarce. In contrast, in the cervical and upper thoracic ganglia, many PN receive at least some GABAergic synapses and a minority of PN are strongly innervated by GABAergic axons. The distribution of RIG cells follows neither the distribution pattern of GABA-positive cells nor that of GABAergic preterminal axons. Heavy innervation of RIG neurons rather formation of numerous varicosities near innervated dendrites and somata. Neighbouring PN may not be approached by these GABAergic axons. This suggests depends on a local increase in terminal branching of GABAergic axons and that GABAergic innervation

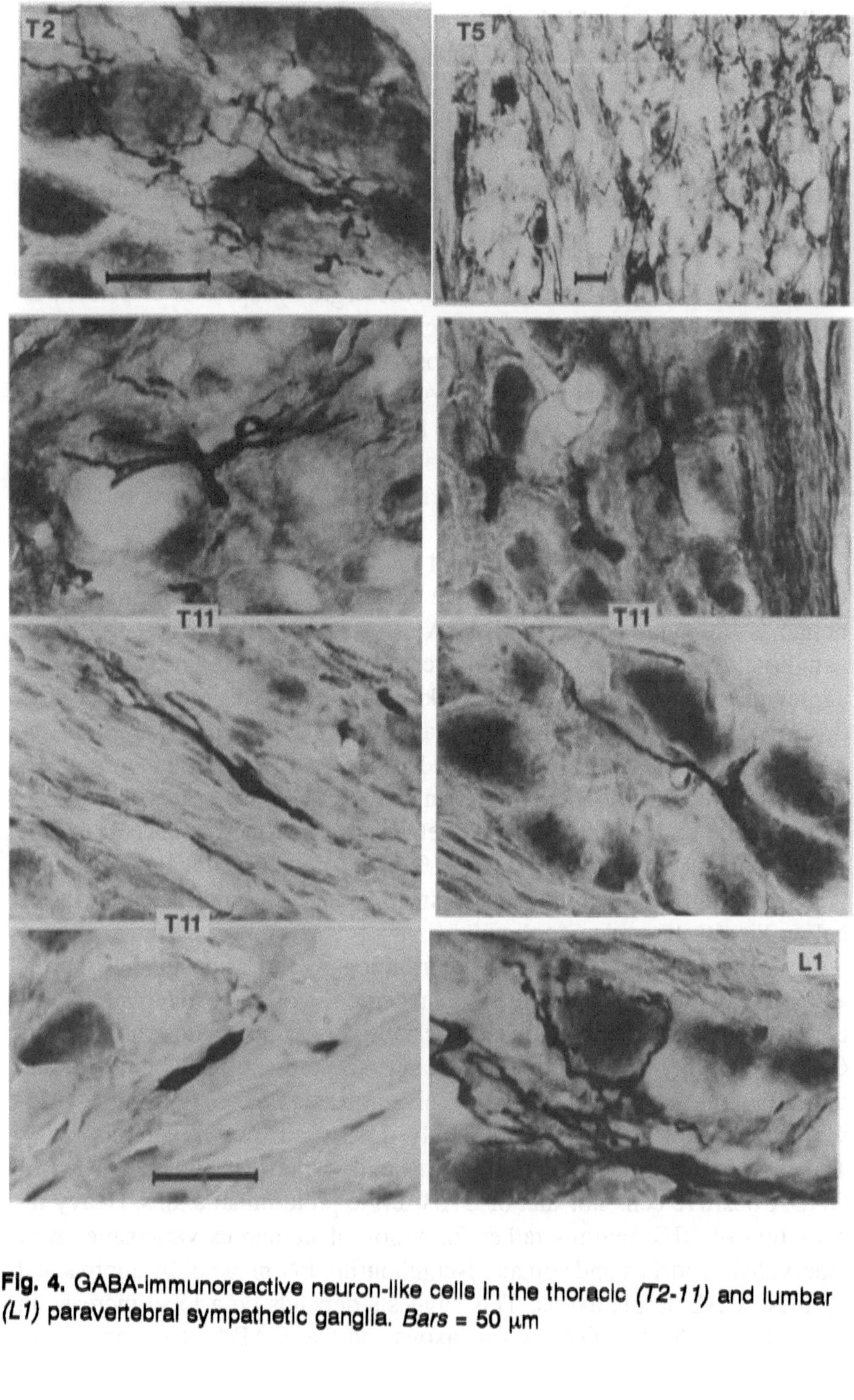

Fig. 4. GABA-immunoreactive neuron-like cells in the thoracic *(T2-11)* and lumbar *(L1)* paravertebral sympathetic ganglia. *Bars* = 50 μm

may be attracted by postsynaptic factors. A combination of GABA immunohistochemistry and retrograde transport of HRP from various postganglionic nerves showed that RIG cells project their axons to different targets [44]. Thus, the attraction of RIG cells for GABAergic innervation does not seem to be related to target-derived factors. The functional meaning of such heavy GABAergic input to small subpopulations of PN remains an enigma.

Origin and Distribution of GABA-Immunoreactive Axons

As mentioned above, GABA-immunoreactive axons may originate from small neuron-like cells. These cells were predominantly multipolar, when found within ganglia, e.g., in the SCG and cervical sympathetic complex (i.e., intermediate cervical and stellate ganglia) down to the 4th thoracic ganglion. Usually, one cell process looked like an axon and sometimes it could be followed up to distal branches and axon varicosities (Fig. 3). In other cases, the axon was obviously longer and left the section plane after some distance from the soma. Similar cells were seen in the lower thoracic and upper lumbar ganglia, but here they either did not make axons or had only short axons with few terminal branches, and varicosities (Fig. 4, L1). Consequently, this cell type (if it is uniform) seems to have axons which are highly variable in size, length, number of terminal branches and the amount of varicosities or axon terminals. For GABA- and GAD-positive SIF cells, the relation to GABAergic axons was previously obscure, although some SIF-like cells which seemed to give rise to axon-like processes were seen.

Many GABA-immunoreactive axons can be followed from cervical sympathetic ganglia into their caudal preganglionic nerve trunks where they may originate from bipolar GABA-positive cells (Fig. 3I). Other GABAergic axons possibly pass through connectives into the next ganglion. Bipolar cells were also seen within ganglia, where they were usually located in or near GABAergic axon bundles. This interpretation is supported by experiments in which the preganglionic nerve trunk of the SCG was crushed or ligated. After 6 - 11 days many GABAergic axons disappeared from the SCG, although small GABA-positive neurons and their processes survived. In addition, GABA-immunoreactive axons enlarged caudal to the ligation but vanished from the segment near the SCG [14].

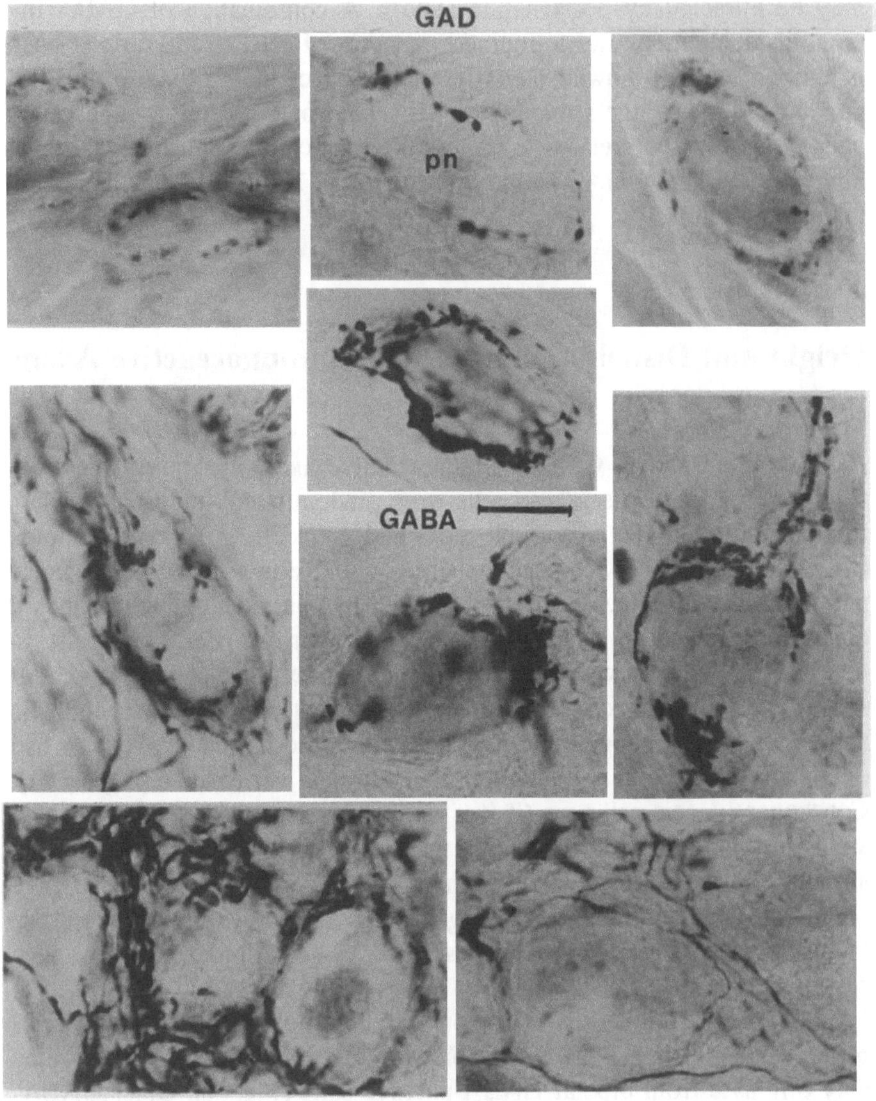

Fig. 5. Rich GABAergic basket-like innervation of special principal ganglion cells *(pn)* visualized by GAD-immunohistochemistry *(A-D)* and GABA-antibodies *(E-I)*. Note scarce to absent axon staining with GAD antibodies and absent varicosities on some perisomatic axons *(I)*. *Bar* = 20 μm

Similar results were obtained for GABA transaminase (for the ligated connective below the SCG [18]).

GABA-immunoreactive axons occur in interganglionic connectives down to the 7th thoracic ganglion, i.e., down to those thoracic ganglia which are entered by preganglionic axons from the spinal cord. Bogan et al. [3] combinated GABA immunohistochemistry with retrograde transport of HRP from injection sites in the 4th paravertebral ganglion of pigeons and the adrenal medulla and SCG of rats. These studies revealed that GABAergic axons do not originate from neurons in the intermediolateral column of the spinal cord. For the rat SCG this finding has been confirmed by our own unpublished observations. Thus, we may conclude that GABAergic axons do not originate from the spinal cord but exclusively come from small interneurons located within the paravertebral sympathetic trunk, i.e, in ganglia and in nerve trunks connecting them. Although these GABAergic neurons may have axons of considerable length (more than 1 cm), they stay within the sympathetic trunk and do not leave cervical ganglia via postganglionic nerves [14].

Quantitative analysis of the number of GABAergic axons in various levels of the sympathetic trunk (interganglionic connectives) revealed that the number of these axons decreases from about 300 below the SCG to zero below the 7th thoracic ganglion (Fig. 6). Preliminary results suggest that GABAergic neurons in upper thoracic levels and the inferior cervical ganglion a small ganglion complex, project their axons upwards and terminate in the SCG. Interestingly enough, some GABA-positive axons seem to reappear in the lowest thoracic and upper lumbar levels.

Possible Functions of GABA in Sympathetic Ganglia

The general function of GABAergic innervation in sympathetic ganglia is still obscure. It might become clearer when we know more about the characteristics of the GABA system in other autonomous ganglia, including the caudal population of paravertebral and preaortal ganglia. GABAergic synapses accumulate on RIG cells, which represent a very small subpopulation of prinicpal neurons. Therefore, GABAergic innervation may play a specific role in the function of these cells. Unfortunately, a special postganglionic function of RIG cells has not yet been identified. In the rat SCG, RIG cells apparently project to various target organs [44]. However, it cannot yet be excluded that these cells specifically innervate tissue components that occur in various target organs, e. g., those parts of blood vessels involved in the regulation of local blood flow.

GABA exerts an inhibitory effect on ganglionic transmission either by suppressing depolarization of postsynaptic membranes [11, 12, 1] or by reducing ACh release from presynaptic elements [8, 16, 17]. The characteristic augmentation of GABAergic neurons and axons from upper thoracic to cervical ganglia (Fig. 6) which are innervated by GABA-negative (preganglionic? cholinergic?) axons (Fig. 2A) prompts the speculation that these neurons represent the interneuronal effector of an ascending feedforward cascade driven by preganglionic afferents from several spinal segments. This feedforward inhibition, being maximal at a specific population of principal neurons (RIG cells, Fig. 5), might than depend on simultaneous activation of preganglionic neurons in several spinal segments. This suggests that RIG cells might be strongly inhibited during multisegmental activation of spinal neurons preganglionic to the thoracic and cervical sympathetic ganglion. However, this is not a selective criterion for RIG cells. Apart from excitatory substances, e.g., ACh [20] and substance P and VIP [23, 24], several inhibitory transmitters and/or neuromodulators have been found in sympathetic ganglia: adrenalin [9], dopamine [15], somatostatin [22], enkephalin [32]. As with GABA, some of this inhibition also seems to be mediated by presynaptic effects [32]. Thus, GABAergic mechanisms are apparently a part of a complex inhibitory system that may down-regulate sympathetic ganglion cell activity under different functional conditions. Within the ganglion chain, the GABAergic cell system may represent a specific cell population that has a specific function, which - although not recognized - may be correlated with multisegmental activation of spinal preganglionic afferents.

Apart from its role as an inhibitory transmitter, GABA may act as a trophic agent in the CNS [35] and in sympathetic ganglia [42]. This suggests yet another GABAergic function, i.e., GABA may have a role in the maintenance of excitatory afferent synapses in the sympathetic ganglion system and/or in the regulation of synaptic plasticity [38, 39, 27, 10]. Thus, further studies on the GABAergic system of the sympathetic trunk may lead to a better understanding of relationships between short-term synaptic functions (inhibition of ganglionic transmission) and long-term actions (stabilization and/or neuroplasticity of synapses) in the autonomous nervous system.

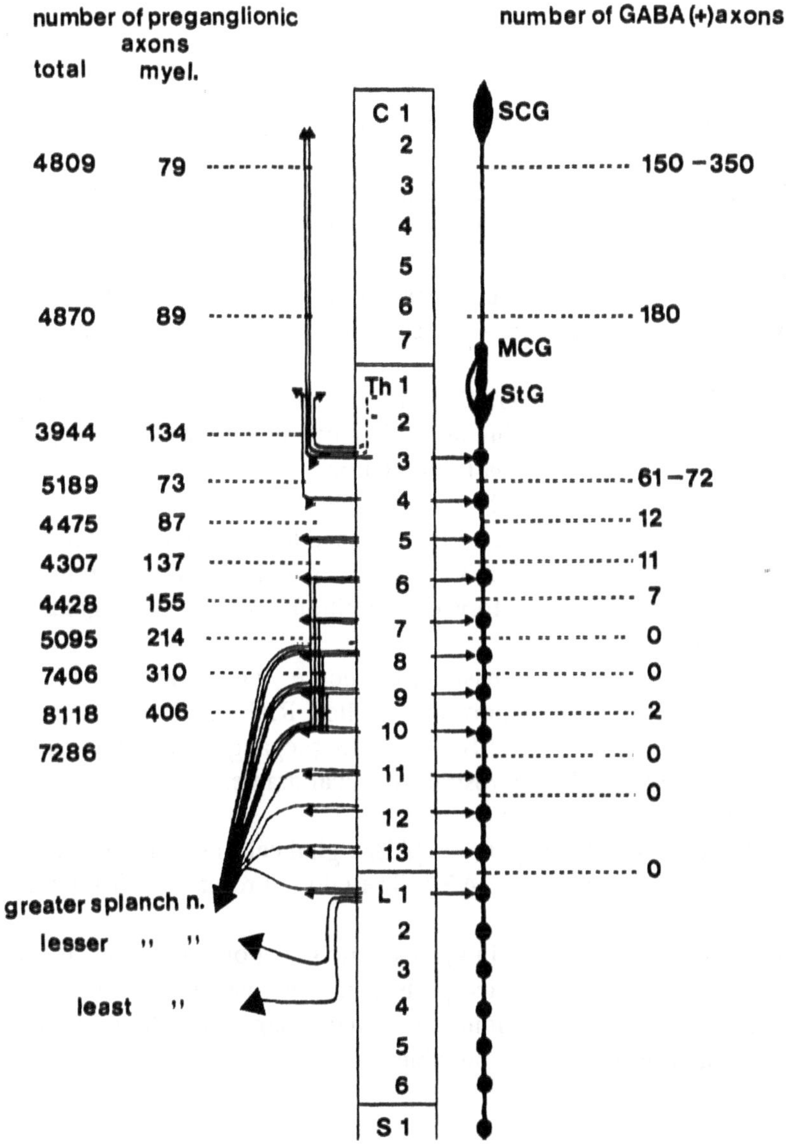

Fig. 6. Comparison between numbers of axons in interganglionic connecting nerve trunks of the paravertebral sympathetic ganglion chain. Note opposing gradients of GABA-positive *(right)* and myelinated (predominantly non-GABAergic) axons *(second column on the left)*. *Middle column* represents spinal cord segments and preganglionic fiber connnections with sympathetic paravertebral ganglia *(arrows)* *C*, cervical; *L*, lumbar; *MCG*, middle or intermediate cervical ganglion; *S*, sacral; *SCG*, superior cervical ganglion; *StG*, stellate ganglion; *T*, thoracic

Conclusions

Sympathetic ganglia apparently include a functional GABA system which consists of endogenous GABA pools, GABA synthesis through GAD acitivity, release and (re-)uptake mechanisms, both types of GABA receptors (A and B) and GABA transaminase. Immunohistochemically, GAD can be localized in small interneurons of various shape and location and in axon terminals which make synaptic junctions with principal neurons. Additionally, GAD is found in non-neuronal (mesenchymal?) cells seen a in perivascular position.

GABA antibodies stain all GAD-positive structures. In addition, they visualize axons which are continuous with axon varicosities and synapses and can be followed back to small interneurons.

GABAergic interneurons and their axons seem to form a feedforward inhibition cascade which conforms to the entrance and ascendence of spinal preganglionic afferents to cervical sympathetic ganglia, predominantly the SCG. It seems possible that this forward inhibition requires multisegmental spinal activation.

GABAergic axons form a basket-like dense innervation around a sub-population of principal neurons (so-called RIG cells). GABAergic axons are most numerous in the SCG and its preganglionic nerve trunk. Their number decreases along a rostral-to-caudal gradient and reaches zero below the 7th thoracic ganglion.

Without preloading with exogenous GABA, glial cells are seldom and incompletely immunoreactive for GABA. They do not react with GAD antibodies.

The functional role of GABA in sympathetic ganglionic functions is not yet clear. Apart from acute pre- and postsynaptic actions on synaptic transmission (inhibition), a role in (long-term) maintenance, function, and the regulation of neuroplasticity are possible.

References

1. Adams PR, Brown DA (1975) Actions of gamma-aminobutyric acid on sympathetic ganglion cells. J Physiol (Lond) 250: 85-120
2. Bertilson L, Suria A, Costa E (1976) Gamma-aminobutyric acid in rat superior cervical ganglion. Nature 260: 540-541

3. Bogan N, Mennone A, Cabot JB (1989) Light microscopic and ultrastructural localization of GABA-like immunoreactive input to retrogradely labeled sympathetic preganglionic neurons. Brain Res 505: 257-270
4. Bowery NG, Brown DA (1972) Gamma-aminobutyric acid uptake by sympathetic ganglia. Nature 238: 89-91
5. Bowery NG, Hill DR (1986) GABA mechanisms in autonomic ganglia. In: Erdö SL, Bowery NG (eds) GABAergic mechanisms in the mammalian periphery, Raven Press, New York, pp 135-152
6. Bowery NG, Brown DA, Collins GGS, Galvan M, Marsh S, Yamini (1976) Indirect effects of amino acids on sympathetic ganglion cells mediated through the release of gamma-aminobutyric acid from glial cells. Br J Pharmacol 157: 73-91
7. Briel G, Neuhoff V (1972) Microanalysis of amino acids and their determination in biological material using dansyl chloride. Hoppe Seilers Z Physiol Chem 353: 540-553
8. Brown DA, Higgins AJ (1979) Presynaptic effects of gamma-aminobutyric acid in isolated rat superior cervical ganglia. Br J Pharmacol 66: 108-109
9. Christ DD, Nishi S (1971) Effects of adrenaline on nerve terminals in the superior cervical ganglion of the rabbit. Br J Pharmacol 41: 331-339
10. Dames W, Joó F, Fehér O, Toldi J, Wolff JR (1985) Gamma-aminobutyric acid enables synaptogenesis in the intact superior cervical ganglion of the adult rat. Neurosci Lett 54: 159-164
11. De Groat WC (1970) The actions of gamma-aminobutyric acid and related amino acids on mammalian autonomic ganglia. J Pharmacol Exp Ther 172: 384-396
12. De Groat WC, Lalley PM, Block M (1971) The effects of bicuculline and GABA on the superior cervical ganglion of the cat. Brain Res 25: 665-668
13. Dobó E, Kása P, Wenthold RJ, Joó F, Wolff JR (1989) Evidence for GABAergic fibers entering the superior cervical ganglion of rat from the preganglionic nerve trunk. Histochemistry 92: 133-136
14. Dobó E, Kása P, Joó F, Wenthold RJ, Wolff JR (1990) Structures with GABA-like and GAD-like immunoreactivity in the cervical sympathetic ganglion complex of adult rats. Cell Tiss Res (in press)
15. Dun NJ, Nishi S (1974) Effects of dopamine on the superior cervical ganglion of the rabbit. J Physiol (Lond) 239: 155-164
16. Farkas Z, Balcar V, Kása P, Joó F, Wolff JR (1984) Neuropharmacalogical evidence for the involvement of GABA receptors in the inhibition of ACh release from the presynaptic axon terminals in the SCG of rat. In: Vizi ES, Magyar K (eds) Regulation of transmitter function: basic and clinical Aaspects. Akademiaia Kiado, Budapest, p 87 (Abstr)
17. Farkas Z, Kása P, Balcar VJ, Joó F, Wolff JR (1986) Type A and B GABA receptors mediate inhibition of acetylcholine release from cholinergic nerve terminals in the superior cervical ganglion of rat. Neurochem Int 8: 565-572
18. Häppölä O, Päivärinta H, Soinila S, Wu JY, Panula P (1987) Localization of L-glutamate decarboxylase and GABA transaminase immunoreactivity in the sympathetic ganglia of the rat. Neuroscience 21: 271-281
19. Häppölä O, Soinila S, Päivärinta H, Panula P (1987) MET5 enkephalin-Arg6-Phe7- and MET5 enkephalin-Arg6-Gly7-Leu8-immunoreactive nerve fibres and neurons in the superior cervical ganglion of the rat. Neuroscience 21: 283-295
20. Hebb CO, Waites GMH (1956) Choline acetylase on antero- and retrograde degeneration of cholinergic nerve. J Physiol (Lond) 132: 667-671

21. Hills JM, King BF, Mirsky R, Jessen KR (1988) Immunohistochemical localisation and electrophysiological actions of GABA in prevertebral ganglia in guinea-pig. J Auton Nerv Syst 22: 129-140
22. Hökfelt T, Elfvin L-G, Elde R, Schultzberg M, Goldstein M, Luft R (1977) Occurrence of somatostatin-like immunoreactivity in some peripheral sympathetic noradrenergic neurons. Proc Natl Acad Sci USA 74: 3587-3591
23. Hökfelt T, Elfvin L-G, Schultzberg M, Fuxe K, Said SI, Mutt V, Goldstein M (1977) Immunohistochemical evidence of vasoactive intestinal polypeptide-containing neurons and nerve fibers in sympathetic ganglia. Neuroscience 2: 885-896
24. Hökfelt T, Elfvin L-G, Schultzberg M, Goldstein M, Nilsson G (1977) On the occurrence of substance P-containing fibers in sympathetic ganglia: immunohistochemical evidence. Brain Res 132: 29-41
25. Kanazawa I, Iversen LL, Kelley JS (1976) Glutamate decarboxylase activity in the rat posterior pituitary, pineal gland, dorsal root ganglion and superior cervical ganglion. J Neurochem 27: 1267-1269
26. Karhula T, Häppölä O, Joh T, Wu JY (1988) Localization of L-glutamate decarboxylase immunoreactivity in the major pelvic ganglion and in the coeliac-superior mesentric ganglion complex of the rat. Histochemistry 90: 255-260
27. Kása P, Dames W, Rakonczay L, Gulya K, Joó F, Wolff JR (1985) Modulation of the acetylcholine system in the superior cervical ganglion of rat: effect of GABA and hypoglossal nerve implantation after in vivo GABA treatment. J Neurochem 44: 1363-1372
28. Kása P, Joó F, Dobó E, Wenthold RJ (1988) Heterogeneous distribution of GABA-immunoreactive nerve fibers and axon terminals in the superior cervical ganglion of adult rat. Neuroscience 26: 635-644
29. Kása P, Dobó E, Wolff JR (1990) Distribution of cholinergic nerve fibers and synapses in the mouse superior cervical ganglion: Light- and electronmicroscopical immunohistochemistry for choline acetyltransferase. (in prep)
30. Kato E, Morita K, Kuba K, Yamada S, Kuhara T, Shinka T, Matsumoto I (1980) Does gamma-aminobutyric acid in blood control transmitter release in bullfrog sympathetic ganglia? Brain Res 195: 208-214
31. Kenny SL, Ariano MA (1986) The immunofluorescence localization of gluatamate decarboxylase in the rat superior cervical ganglion. J Auton Nerv Syst 17: 211-215
32. Konishi S, Tsunoo A, Otsuka M (1979) Enkephalins presynaptically inhibit cholinergic transmission in sympathetic ganglia. Nature 282: 515-516
33. Lindh B, Staines W, Hökfelt T, Terenius L, Salvaterra PM (1986) Immunohistochemical demonstration of choline acetyltransferase-immunoreactive preganglionic nerve fibers in guinea pig autonomic ganglia. Proc Natl Acad Sci USA 83: 5316-5320
34. Nagata Y, Yokoi Y, Tsukada Y (1966) Studies on free amino acid metabolism in excised cervical sympathetic ganglia from the rat. J Neurochem 13: 1421-1431
35. Redburn D, Schousboe A (1987) Neurotrophic activity of GABA during development. Liss, New York
36. Stanton HC, Woodhouse FH (1960) The effect of gamma-amino-n-butyric acid and some related compounds on the cardiovascular system of anesthetized dogs. J Pharmacol Exp Ther 128: 231-243
37. Tanaka C, Taniyama K (1986) GABA transport in peripheral tissues: Uptake and efflux. In: Erdö SL, Bowery NG (eds) GABAergic mechanisms in the mammalian periphery, Raven Press, New York, pp 57-75

38. Wolff JR, Joó F, Dames W (1978) Plasticity in dendrites shown by continuous GABA administration in superior cervical ganglion of adult rat. Nature 274: 72-74
39. Wolff JR, Joó F, Dames W, Fehér O (1979) Induction and maintenance of free post-synaptic membrane thickenings in the adult superior cervical ganglion. J Neurocytol 8: 549-563
40. Wolff JR, Joó F, Kása P, Storm-Mathisen J, Toldi J, Balcar VJ (1986) Presence of neurons with GABA-like immunoreactivity in the superior cervical ganglion of the rat. Neurosci Lett 71: 157-162
41. Wolff JR, Ottersen P, Storm-Mathisen J, Joó F, Kása P (1987) Neurons and axons with GABA-like immunoreactivity in the superior cervical ganglion of adult rat. In: Elsner N, Creutzfeldt O (eds) New Frontiers in Brain Research. Thieme, Stuttgart, p 266
42. Wolff JR, Joó F, Kása P (1987) Synaptic, metabolic and morphogenetic effects of GABA in the superior cervical ganglion of rats. In: Redburn D, Schousboe A (eds) Neurotrophic activity of GABA during development. Liss, New York, pp 221-252
43. Wolff JR, Kása P, Dobó E, Joó F (1988) GABA-ergic innervation of sympathetic ganglion cells. In: Elsner N, Barth FC (eds) Sense organs: Interfaces between environment and behaviour. Thieme, Stuttgart, p 313
44. Wolff JR, Kása P, Dobó E, Wenthold RJ, Joó F (1989) Quantitative analysis of the number and distribution of neurons richly innervated by GABA-immunoreactive axons in the rat superior cervical ganglion. J Comp Neurol 282: 264-273
45. Young JAC, Brown DA, Kelly JS, Schon F (1973) Autoradiographic localization of sites of [^{3}H] gamma-aminobutyric acid accumulation in peripheral autonomic ganglia. Brain Res 63: 479-486

L-Glutamate Decarboxylase Immunoreactivity in the Sympathoadrenal System

O. Häppölä, T. Karhula, H. Päivärinta, S. Soinila, J.-Y. Wu, and M. Ahonen

Introduction

The role of GABA as an inhibitory neurotransmitter in the mammalian brain has been well established. In recent years, increasing evidence has been presented to show that GABA may act as a neurotransmitter also in parts of the peripheral nervous system [20, 21, 51]. In the myenteric plexus of the gut an autonomic neurotransmitter role for GABA has been proposed by several authors [29-32]. In sympathetic ganglia, a specific GABA uptake system [11, 62], GABA receptors [12, 13], and the depolarizing effect of GABA on sympathetic neurons [1, 12, 13] have been described in previous studies. The presence of GABA and the GABA-synthesizing enzyme L-glutamate decarboxylase (GAD) in sympathetic ganglia was also shown by biochemical [8, 33, 55] and, recently, immunohistochemical methods [14, 15, 24-28, 34, 35, 38, 44, 57, 58]. Studies have now been carried out in order to investigate the expression and distribution of GAD in the developing sympathoadrenal system and to further analyze the presence and localization of GAD in different parts of the sympathetic nervous system.

The Structure and Cell Types of the Sympathoadrenal System

The sympathoadrenal system consists of sympathetic ganglia, paraganglia, and adrenal medullae. The cells of the sympathoadrenal system are known to originate from the neural crest [41] and synthesize catecholamines. The catecholamines and/or their synthesis in these cells can be demonstrated either histochemically with the formaldehyde-induced fluorescence technique [16] or immunohistochemically with antisera raised against the catecholamine-sythesizing enzymes such as tyrosine hydroxylase (TH), the rate-limiting enzyme in catecholamine synthesis, or phenylethanolamine-N-methyltransferase (PNMT), the enzyme converting norepinephrine to epinephrine [19]. By use of these methods two types of amine-containing cells have been classified in sympathetic ganglia: the principal nerve cells (principal neurons), 20-60 μm in diameter and showing moderate TH immunofluorescence, and the small intensely fluorescent (SIF) cells, 10-20 μm in diameter, with bright TH immunofluorescence. The fine structure of SIF cells corresponds to small granule-containing (SGC) cells [23, 53] and differs in many respects from principal neurons.

Two catecholamine-containing cell types can be detected in other sympathetic tissues. Paraganglia, which usually are situated retroperitoneally, contain small, intensely TH-immunoreactive, paraganglion-type cells forming the major cell population and also large neuron-like cells with less intense TH immunofluorescence [2, 10]. Some TH-negative cells are present in paraganglia too.

In the adrenal medulla, two chromaffin cell types can be distinguished. In the rat, the majority of medullary cells synthesize epinephrine and are intensely immunoreactive for both TH and PNMT, while the minority of medullary cells synthesize norepinephrine and show TH but no PNMT immunoreactivity. A low number of larger TH-immunoreactive neurons are also situated in the adrenal medulla.

Prenatal Development of Retroperitoneal Sympathetic Tissue

The prenatal development of sympathetic tissues in different species has been described by a number of authors [2, 4, 18, 22, 45, 48, 49, 54]. The sympathetic cells are derivatives of the neural crest [7, 41]. In the rat, neural crest cells begin to migrate ventrally around embryonic day 10 (E10) [9], but the first signs of catecholamine synthesis in these migrating cells can be observed only after they have reached the region of the developing sympathetic chain.

The ontogenesis of different sympathetic tissues can be analyzed in the retroperitoneal space, where developing sympathetic ganglia, paraganglia, and adrenal glands are localized in close topographical relation. At the thoraco-lumbar level, where retroperitoneal sympathetic tissue is situated, the first TH-immunoreactive sympathetic cells appear on day E12 in chain ganglia. From the chain ganglia, a population of sympathetic cells continues migrating to the lateral and ventral aspects of the aorta and forms the perirenal ganglia and preaortic sympathetic tissue on day E13. A fraction of these cells migrates further to the developing adrenal glands on day E14, when the first TH-immunoreactive cells are found in the region of the adrenal medulla.

When catecholamine synthesis is first induced in sympathetic cells, on days E12-14, all cells exhibit TH immunoreactivity, but the intensity of staining shows great variation. About 2 days later two distinct TH-immunoreactive cell populations can be seen in these organs: one with moderate TH immunoreactivity and the other with intense immunoreactivity. In chain ganglia, the small intensely TH-immunoreactive cells are known to differentiate to ganglionic SIF cells and the larger moderately TH-immunoreactive cells to ganglionic principal neurons [48]. In the preaortic sympathetic tissue the moderately TH-immunoreactive cells are located in the cranial part of this tissue, later forming the celiac-superior mesenteric ganglion. The caudal part is mainly comprised of cells with intense TH immunoreactivity which differentiate to sympathetic paraganglia [2, 4]. In the rat adrenal medulla, two different TH-immunoreactive cell populations can be distinguished for the first time on day E16, when the cells with intense TH immunoreactivity also show PNMT immunoreactivity; cells with moderate TH immunoreactivity are not PNMT-immunoreactive [49]. Thus, the differentiation of sympathetic cells to epinephrine- and norepinephrine-synthesizing medullary cells is evident already at this developmental stage.

Expression of GAD in Developing Sympathetic Tissues

Immunohistochemical studies with specific antisera against GAD [59-61] revealed that GAD is first expressed in all retroperitoneal sympathetic tissues of the rat on day E14 (Figs. 1, 2) [3, 4]. In chain ganglia and in preaortic tissue, the first expression of GAD occurs a little later than that of TH but concomitantly with that of TH in the adrenal gland (Fig. 2). In all tissues studied, GAD was found only in a subpopulation of TH-immunoreactive cells. When GAD appears in sympathetic tissues, different TH-immunoreactive cell types cannot be distinguished. However, later, when different types of sympathetic cells can be observed, GAD is expressed only in cells with intense TH immunoreactivity. At E15-17, GAD is localized in chain ganglia in a population of intensely TH-reactive cells, which represent developing SIF cells. In preaortic sympathetic tissue, GAD is expressed by developing paraganglionic cells, localized mainly caudally. In the adrenal medulla, starting from day E16, GAD is present only in epine-

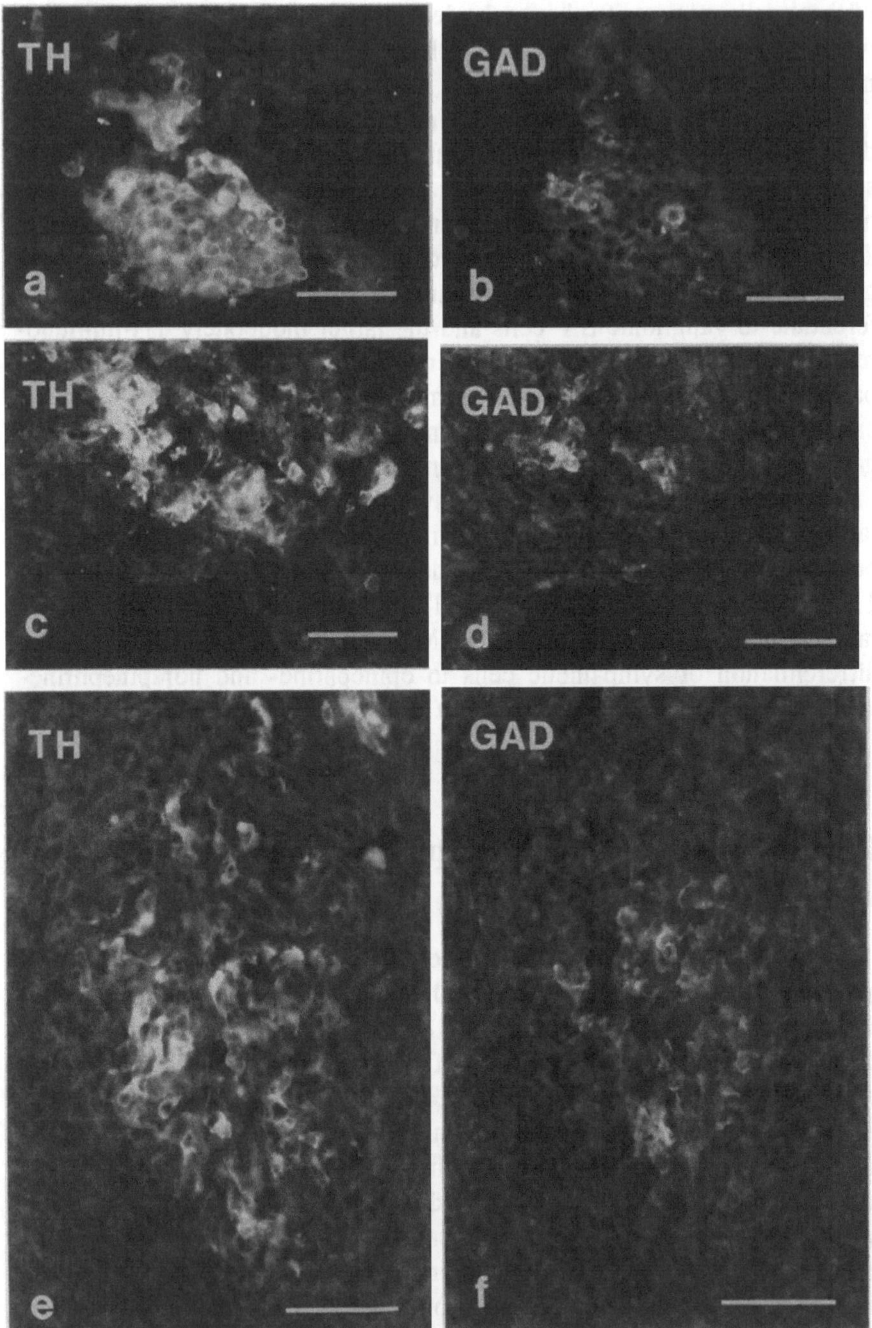
TH
a
GAD
b
TH
c
GAD
d
TH
e
GAD
f

Fig. 1. Consecutive 10 μm sections of lumbar chain ganglia (**a**, **b**), preaortic sympathetic tissue (**c**, **d**), and adrenal medulla (**e**, **f**) of E14 rats demonstrating TH and GAD immunoreactivities. Only a subpopulation of TH-immunoreactive cells express GAD. *Bars* = 50 μm

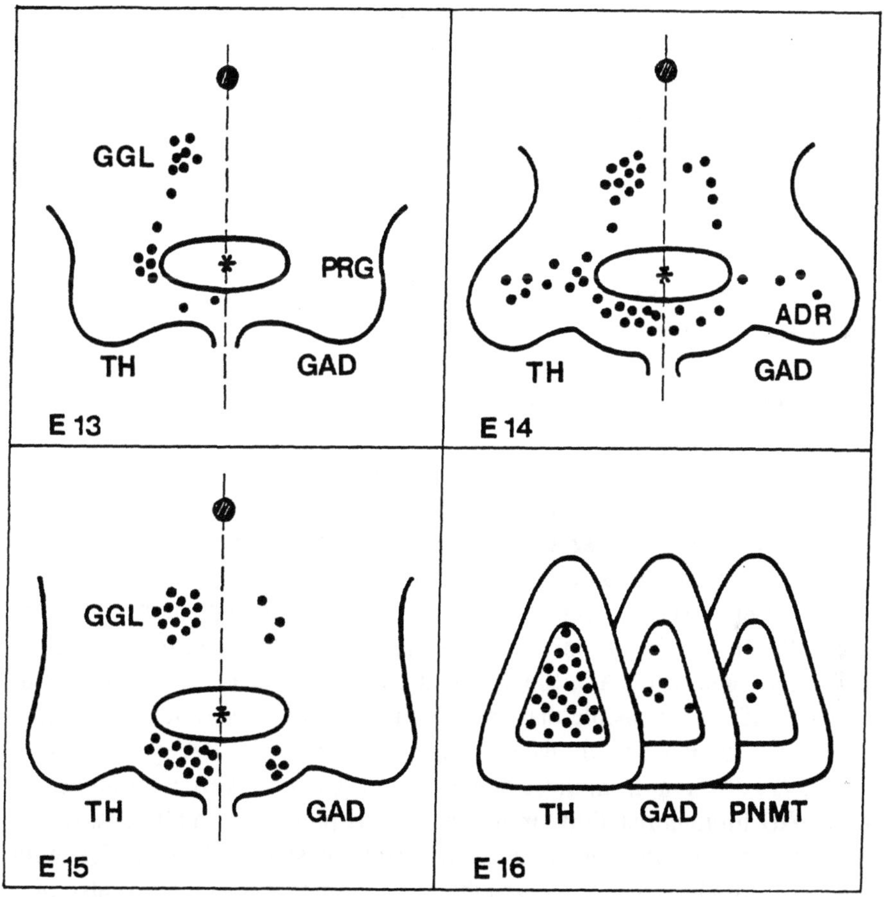

Fig. 2. TH and GAD immunoreactivities in retroperitoneal sympathetic cells of E13, E14, and E15 embryos and in E16 adrenal gland. *Black dots*, immunoreactive cells for TH (*left*) and/or GAD (*right*); *CGL*, chain ganglion; *PRG*, perirenal ganglion; *ADR*, adrenal gland; *asterisk*, aorta

phrine-synthesizing cells (Fig. 2). During prenatal development no neuron-like cells with moderate TH immunoreactivity were found to express GAD [4].

In newborn rat sympathetic chain ganglia and prevertebral ganglia, the principal nerve cells are nonreactive for GAD, but a subpopulation of SIF cells contain GAD (Fig. 3) [3]. The main retroperitoneal paraganglion of the newborn rat consists of neuron-like cells, which are nonreactive for GAD, and paraganglionic cells with intense TH immunoreactivity. A population of these paraganglionic cells are GAD-immunoreactive (Fig. 3) and some of them also express PNMT. In the newborn rat adrenal medulla, GAD is present in the PNMT-containing epinephrine cells, but not in norepinephrine cells [3, 4].

Expression of GAD in Sympathetic Tissues of Adult Animals

Superior Cervical Ganglion

In the rat superior cervical ganglion (SCG), both clustered and solitary small cells exhibited intense GAD immunoreactivity (Fig. 4a) [24-26, 38]. Studies on consecutive sections indicated that these small GAD-immunoreactive cells also display intense TH immunoreactivity (Fig. 5a, c) [26]. These observations indicate that the small GAD-immunoreactive cells are SIF cells. Interestingly, only a subpopulation of catecholamine-containing SIF cells contained GAD. Since both solitary and clustered SIF cells contained GAD, it is possible that both interneuronal [56] and neuroendocrine [47] type SIF cells contain the enzyme necessary for GABA synthesis. Our results are different from those of Kenny and Ariano (1986), who suggested that GAD immunoreactivity in the rat SCG is present in small cells, which do not contain catecholamines. Our immunohistochemical studies with GABA antibodies further showed that GABA immunoreactivity is also present in small cells of the rat SCG (Fig. 5b). These observations are in accordance with those of Wolff et al. [57] and Kása et al. [35]. Whether these GABA-immunoreactive cells represent SIF cells or principal projection neurons of small size is presently unknown. However, according to the results of Kása et al. [35] these small GABA-immunoreactive cells do not appear in cell clusters typical of neuroendocrine type SIF cells. Interestingly, an interneuronal role has been suggested for those SIF cells which appear solitarily in the ganglion [56].

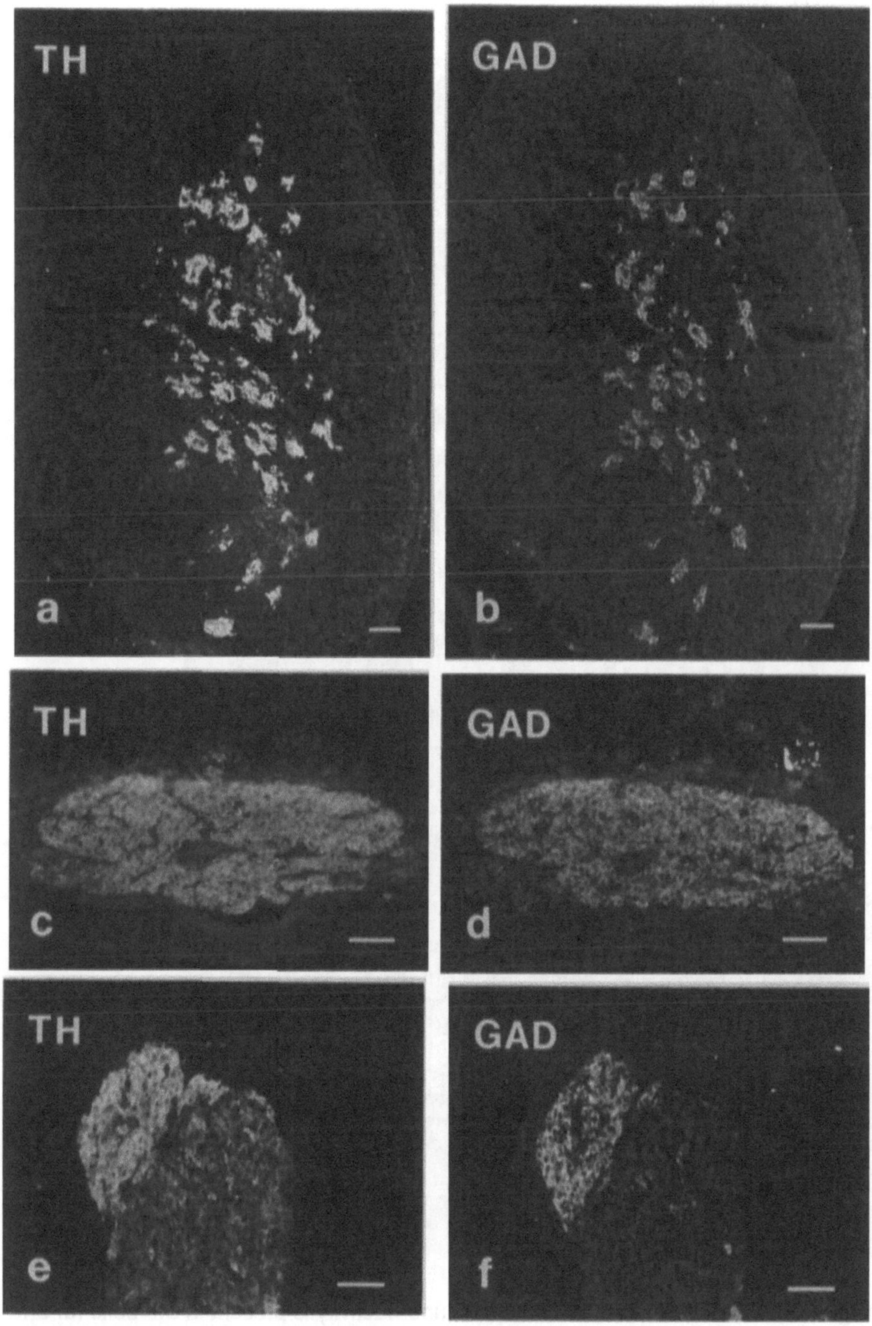

Fig. 3. Adjacent 10 µm sections of adrenal gland (**a**, **b**), the main retroperitoneal paraganglion (**c**, **d**), and the celiac-superior mesenteric ganglion (**e**, **f**) of the newborn rat stained with TH and GAD antisera. Note that in the adrenal gland, GAD is expressed only in epinephrine cells. *Bars* = 50 µm

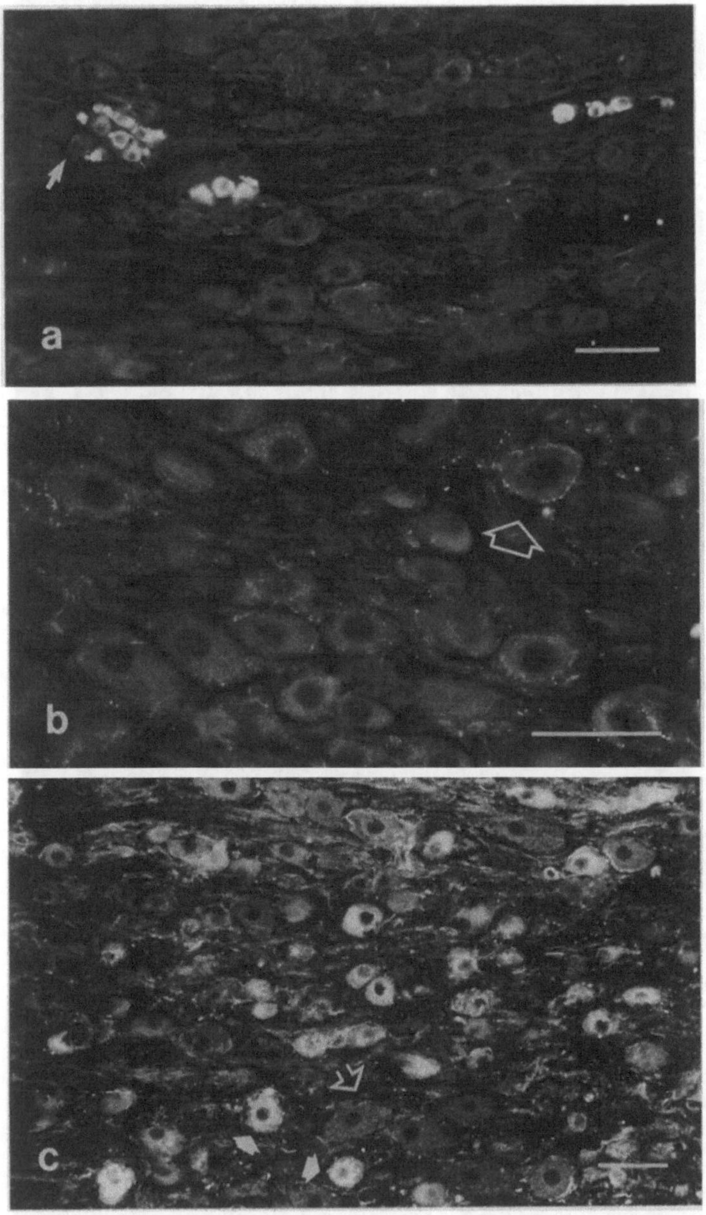

Fig. 4a-c. SCG of the adult rat. GAD immunoreactivity is present in SIF cells (**a**) and in a few nerve fibers (**b**), some of which surround principal neurons (*open arrow*). Some cells in an SIF cell cluster are nonreactive for GAD (**a**, *arrow*). Following ligation of the main postganglionic nerve trunks GAD immunoreactivity is also localized in principal neurons (**c**, *arrows*) and nerve fibers, but nonreactive cells are also seen (**c**, *open arrow*). *Bars* = 50 μm

A population of principal neurons in the rat SCG also contained GAD immunoreactivity of moderate intensity (Fig. 4c). This kind of staining pattern could, however, be observed only with some lots of GAD antisera [26, 27], while other GAD antisera tested produced only weak or no detectable staining in cell bodies of principal neurons [26, 38]. The significance of observed GAD staining in perikarya of principal neurons is presently not clear, since results obtained by antisera against GABA have revealed no GABA-immunoreactive large principal neurons in the SCG [35]. A few GAD-immunoreactive nerve fibers were found in the SCG between or surrounding principal neurons (Fig. 4b) [26], while a large number of fibers were immunoreactive to GABA [35] and GABA transaminase [26].

Celiac-Superior Mesenteric Ganglion

In the rat celiac-superior mesenteric ganglion, small GAD-immunoreactive cells were found in cell clusters or appeared as solitary cells (Fig. 6a) [34]. GAD-immunoreactive cells were also detected in paraganglia closely associated with the ganglion complex. All GAD-immunoreactive cells were also intensely labeled with TH, suggesting that they contain catecholamines. Conversely, a large proportion of TH-immunoreactive SIF cells showed no GAD immunoreactivity (Fig. 6b), indicating that only a subpopulation of SIF cells in the celiac-superior mesenteric ganglion contains the enzyme necessary for GABA synthesis. The principal neurons contained no detectable GAD immunoreactivity [26, 34].

Major Pelvic Ganglion

Previous electrophysiological studies have revealed that GABA inhibits the motility of the urinary bladder of the guinea-pig by blocking the vesical ganglia [52]. Moreover, GABA inhibits contractions of the rat urinary bladder induced by preganglionic stimulation, by inhibition of excitatory neurotransmission in the pelvic ganglia [42]. Evidence has also been presented to show that GABA affects neurotransmission in the cat [13] and dog [50] vesical ganglia.

On the basis of the previously observed profound effects of GABA on neurotransmission in pelvic ganglia as well as on the function of the urinary bladder, we investigated the presence and distribution of GABAergic systems in the rat major pelvic ganglion. GAD immunoreactivity was localized in many small-sized ganglionic cells (Fig. 6c, d, f) [34]. These GAD-

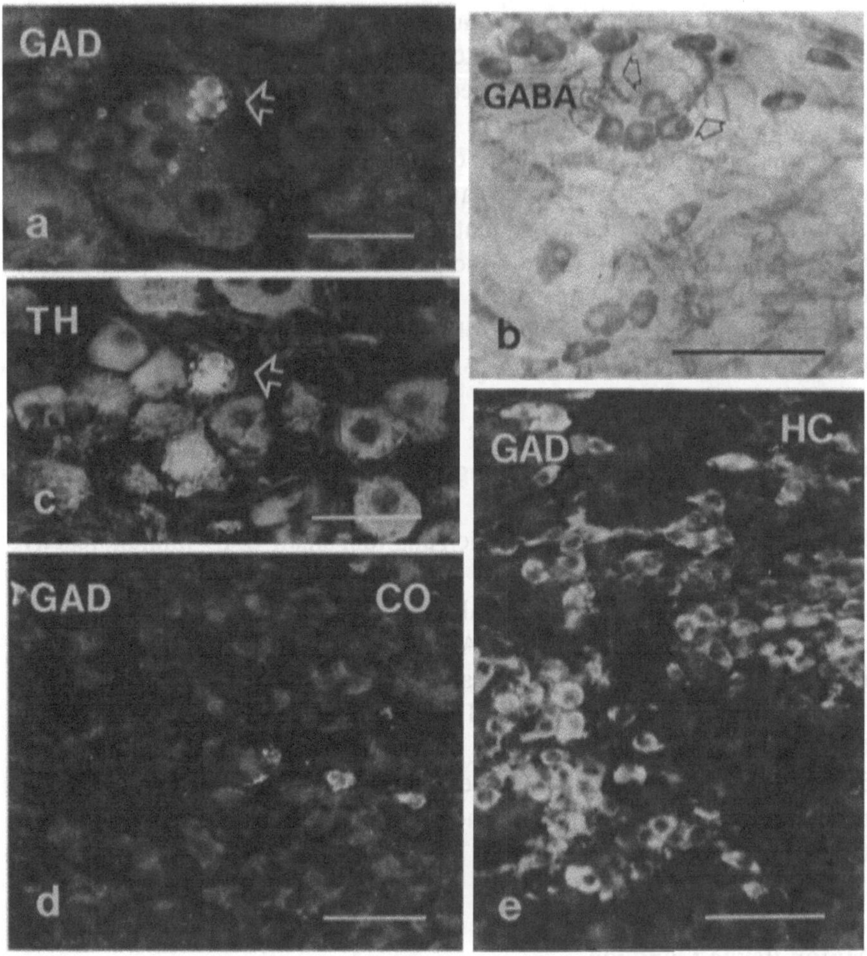

Fig. 5. Consecutive sections of the rat SCG (**a**, **c**) show GAD (**a**) and TH (**c**) immunoreactivities in the same SIF cells. A low number of GABA-immunoreactive small cells are present in the SCG, and some of them have processes (**b**, *open arrows*). A few cells in the newborn rat SCG express GAD (**d**) but, following injections with 20 μg/g hydrocortisone acetate, a dramatic increase of GAD-immunoreactive small cells is seen (**e**). *Bars* = 50 μm

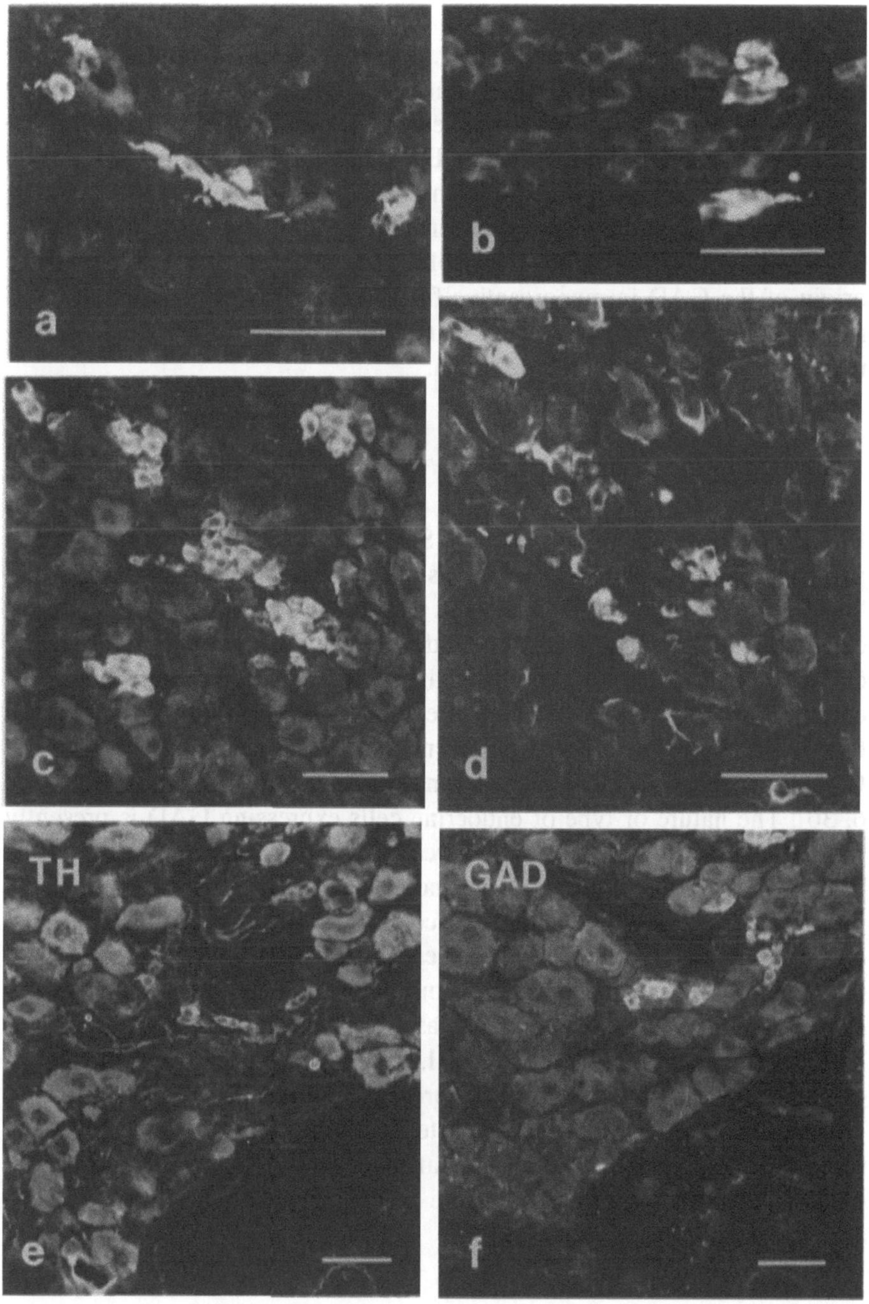

Fig. 6. GAD-immunoreactive SIF cells in the celiac-superior mesenteric ganglion (**a**, **b**). Many small cells in a cluster are nonreactive for GAD (**b**). In the rat major pelvic ganglion both clustered and solitary process-bearing small cells contain GAD (**c**, **d**, **f**). These GAD-immunoreactive cells also contain TH (**e**, **f**). *Bars* = 50 μm

containing cells were often seen in large cell clusters and they all also contained TH immunoreactivity (Fig. 6e, f), suggesting that they are catecholamine-containing small ganglionic cells, presumably SIF cells. Many GAD-immunoreactive cells also appeared as solitary cells and had processes (Fig. 6d). In addition, small GABA-immunoreactive cells were found in the major pelvic ganglion and some of them had long processes. GAD-immunoreactive nerve fibers were seen among ganglionic principal neurons, but the neurons showed no detectable GAD staining (Fig. 6d) [34]. The origin of the GAD-immunoreactive fibers among the principal neurons is not clear, but they at least in part originate from the small GAD-immunoreactive cells in the ganglion.

Adrenal Gland

In addition to principal neurons and SIF cells of sympathetic ganglia and paraganglion cells, the endocrine cells of the adrenal medulla are known to be derived from the neural crest [41]. The adrenal medullary cells and SIF cells of sympathetic ganglia are both developmentally and morphologically closely related and share similar characteristics [18]. Recently both biochemical and immunohistochemical evidence has been presented to show that the adrenal medulla of different species contains both GAD and GABA, localized in medullary chromaffin cells and/or in nerve fibers [5, 6, 36]. The nature or type of endocrine cells expressing GAD is presently unknown. Our studies, however, revealed that GAD immunoreactivity is present in the epinephrine cells but not in the norepinephrine cells of the rat adrenal medulla [3, 4]. The presence of GAD and GABA in nerve fibers of the adrenal gland in different species was reported by Alho et al. [5], but this result could not be repeated in our studies on the rat. The release of catecholamines caused by GABA was first demonstrated in the perfused bovine adrenal glands by Sangiah et al. [46]. The morphological, biochemical, and functional evidence for the presence and localization of GAD and GABA and the effect of GABA on catecholamine secretion through GABA receptors located on adrenal chromaffin cells [5, 37, 39] suggests that GABA indeed has a modulatory role of physiological significance in the adrenal gland.

The Effect of Nerve Ligations and Glucocorticoids on the Expression of GAD in Sympathetic Ganglia

Nerve Ligation Experiments on the SCG

The observed differences in staining of principal neurons of the SCG with various GAD antisera raised two possible explanations: (1) false or nonspecific binding of some GAD antisera to perikarya of principal neurons and/or (2) low levels of GAD in cell bodies of principal neurons, which could not be detected with some of the GAD antisera tested.

In order to test the latter possibility, the main postganglionic nerve trunks, the external and internal carotid nerves, were ligated to inhibit postganglionic axonal flow, which presumably should result in increased levels of GAD in cell bodies of principal neurons in the SCG (Fig. 4a, b). After a postoperative time of 3 days, the SCG of the operated animals were processed for immunocytochemical demonstration of GAD. In these experimental studies a GAD antiserum lot was used which in normal animals produced no staining in principal neurons of the SCG. Ligation of the main postganglionic nerve trunks resulted in the appearance of many GAD-immunoreactive principal neurons and nerve fibers in the ganglion (Fig. 4c). The majority of the GAD-immunoreactive principal neurons were small-sized, but larger perikarya also showed GAD staining. In addition, SIF cells immunoreactive for GAD were found in these ganglia of experimental animals. The results of the postganglionic ligation experiments show that even though some batches of GAD antisera do not show GAD staining in cell bodies and fibers of the SCG of normal animals, these structures exhibit GAD immunoreactivity after inhibition of postganglionic axonal flow. According to the present observations it may be suggested that the levels of GAD in postganglionic structures of the SCG are normally below the detection limit of some GAD antisera lots, but inhibition of postganglionic axonal flow raises GAD concentrations to such levels that GAD can be detected by the present immunohistochemical technique. Nonetheless, the possibility has to be taken into consideration that nerve ligation experiments may cause neuronal damage to axons of postganglionic neurons which may induce increased expression of GAD in these neurons.

Three days following ligation of the preganglionic cervical sympathetic trunk, an intense accumulation of GAD-immunoreactive material was observed on the spinal side of the ligature. The results of these nerve ligation experiments suggested that GAD-immunoreactive nerve fibers are present in the preganglionic sympathetic nerve trunk and these fibers project to the SCG. This is in accordance with previous studies demonstrating that GAD-immunoreactive, GABA transaminase-immunoreactive [26], and GABA-immunoreactive [14, 15, 35, 57, 58] nerve fibers are present in the rat SCG.

The Effect of Glucocorticoids on GAD Expression in Sympathetic Ganglia

Glucocorticoids and nerve growth factor have opposite effects on developing sympathetic precursor cells [40], the former favoring differentiation into SIF cells and the latter into principal neurons. Glucocorticoids, when injected into newborn rats, cause approximately a tenfold increase in the number of SIF cells [17] and induce PNMT expression in these cells [19, 43]. GAD immunoreactivity was found in a few small cells in the SCG of newborn rats (Fig. 5d). Therefore, we analyzed further GAD immunoreactivity in cells of the SCG by injecting newborn rats daily with 20 μg/g hydrocortisone acetate on postnatal days 2-6. In 7-day-old postnatal rats injected with hydrocortisone a dramatic increase of small GAD-immunoreactive cells was observed in the SCG (Fig. 5e) [44]. These results confirmed the view that the GAD-immunoreactive small cells in the SCG are SIF cells. These observations suggest that while glucocorticoids favor adrenergic expression, they also have a stimulatory effect on the expression of GAD and possibly on GABA synthesis in SIF cells of the SCG in early postnatal rats.

References

1. Adams PR, Brown DA (1975) Actions of γ-Aminobutyric acid on sympathetic ganglion cells. J Physiol 250: 85-120
2. Ahonen M, Soinila S, Joh TH (1987) Pre- and postnatal development of rat retroperitoneal paraganglia. J Auton Nerv Syst 18: 111-120
3. Ahonen M, Joh TH, Wu J-Y, Häppöla O (1989) Immunocytochemical localization of L-glutamate decarboxylase and catecholamine-synthesizing enzymes in the retroperitoneal sympathetic tissue of the newborn rat. J Auton Nerv Syst 26: 89-96
4. Ahonen M, Soinila S, Wu J-Y, Häppölä O (1989) L-glutamate decarboxylase immunoreactivity in developing sympathetic tissues of the rat. J Auton Nerv Syst 27: 155-164
5. Alho H, Fujimoto M, Guidotti A, Hanbauer I, Kataoka Y, Costa E (1986) γ-Aminobutyric acid (GABA) in the adrenal medulla: location, pharmacology and function. In: Panula P, Päivärinta H, Soinila S (eds) Neurohistochemistry: modern methods and applications. Liss, New York, pp 453-464

6. Alho H, Hanbauer I, Guidotti A, Costa E (1987) Gamma-aminobutyric acid (GABA) signalling in the adrenal medulla. In: Heym C (ed) Histochemistry and cell biology of autonomic neurons and paraganglia. Springer, Berlin, Heidelberg, New York, pp 61-66 (Experimental brain research series, vol 16)
7. Ayer-LeLievre C, Fontaine-Perus J (1982) The neural crest: its relation with APUD and paraneuron concepts. Arch Histol Jpn 45: 409-427
8. Bertilsson L, Suria A, Costa E (1976) γ-Aminobutyric acid in rat superior cervical ganglion. Nature 260: 540-541
9. Black IB (1978) Sympathetic neurone development. Trends Neurosci 1: 101-104
10. Böck P (1982) Paraganglia. Springer, Berlin, Heidelberg, New York
11. Bowery NG, Brown DA (1972) γ-Aminobutyric acid uptake by sympathetic ganglia. Nature 238: 89-91
12. Bowery NG, Brown DA (1974) Depolarizing actions of γ-aminobutyric acid and related compounds on rat superior cervical ganglia in vitro. Br J Pharmacol 50: 205-218
13. De Groat WC (1970) The actions of γ-aminobutyric acid and related amino acids on mammalian autonomic ganglia. J Pharmacol Exp Ther 172: 384-396
14. Dobó E, Kása P, Wenthold RJ, Joó F, Wolff JR (1989) Evidence for GABAergic fibers entering the superior cervical ganglion of rat from the preganglionic nerve trunk. Histochemistry 92: 133-136
15. Dobó E, Kása P, Wenthold RJ, Wolff JR (1989) Pronase treatment increases the staining intensity of GABA-immunoreactive structures in the paravertebral sympathetic ganglia. Histochemistry 93: 13-18
16. Eränkö O (1967) The practical histochemical demonstration of catecholamines by formaldehyde-induced fluorescence. J R Microsc Soc 87: 259-276
17. Eränkö L, Eränkö O (1972) Effect of hydrocortisone on histochemically demonstrable catecholamines in the sympathetic ganglia and extra-adrenal chromaffin tissue of the rat. Acta Physiol Scand 84: 125-133
18. Eränkö L, Eränkö O (1977) Developmental aspects of SIF cells. In: Costa E, Gessa GL (eds) Advances in biochemical psychopharmacology, Vol 16. Raven, New York, pp 519-524
19. Eränkö O, Pickel VM, Härkönen M, Eränkö L, Joh TH, Reis DJ (1982) Effect of hydrocortisone on catecholamines and the enzymes synthesizing them in the developing sympathetic ganglion. Histochem J 14: 461-478
20. Erdö SL (1985) Peripheral GABAergic mechanisms. TIPS 6: 205-208
21. Erdö SL, Wolff JR (1990) γ-Aminobutyric acid outside the mammalian brain. J Neurochem 54: 363-372
22. Fernholm M (1972) On the appearance of monoamines in the sympathetic systems and the chromaffin tissue in the mouse embryo. Z Anat Entwickl-Gesch 135: 350-361
23. Grillo M (1966) Electron microscopy of sympathetic tissues. Pharmacol Rev 18: 387-399
24. Häppölä O, Soinila S, Päivärinta H, Eränkö L, Wu J-Y, Panula P (1985) Glutamic acid decarboxylase (GAD) and GABA aminotransferase (GABA-T) immunoreactivity in the superior cervical ganglion of the rat. Scandinavian Neurohistochemistry Meeting, p 15 (abstr)
25. Häppölä O, Räivärinta H, Soinila S, Wu J-Y, Panula P (1986) Localization of L-glutamic acid decarboxylase (GAD) and GABA transaminase (GABA-T) immunoreactivity in the rat superior cervical ganglion. Soc Neurosci Abstr: 1535

26. Häppölä O, Päivärinta H, Soinila S, Wu J-Y, Panula P (1987) Localization of L-glutamate decarboxylase and GABA transaminase immunoreactivity in the sympathetic ganglia of the rat. Neuroscience 21: 271-281
27. Häppölä O, Päivärinta H, Soinila S, Wu J-Y, Panula P (1987) Immunocytochemical demonstration of GABA-metabolizing enzymes in the superior cervical and nodose ganglia of the rat. Exp Brain Res 16: 145-148
28. Hills JM, King BF, Mirsky R, Jessen KR (1988) Immunohistochemical localization and electrophysiological actions of GABA in prevertebral ganglia in guinea-pig. J Auton Nerv Syst 22: 129-140
29. Jessen KR, Hills JM, Saffrey MJ (1986) Immunohistochemical demonstration of GABAergic neurons in the enteric nervous system. J Neurosci 6: 1628-1634
30. Jessen KR, Mirsky R, Dennison ME, Burnstock G (1979) GABA may be a neurotransmitter in the vertebrate peripheral nervous system. Nature 281: 71-74
31. Jessen KR, Mirsky R, Hills JM (1987) GABA as an autonomic neurotransmitter: studies on intrinsic GABAergic neurons in the myenteric plexus of the gut. TINS 10: 255-262
32. Jessen KR, Hills JM, Dennison ME, Mirsky R (1983) γ-Aminobutyrate as an autonomic neurotransmitter: release and uptake of [^{3}H]-γ-aminobutyrate in guinea-pig large intestine and cultured enteric neurons using physiological methods and electron microscopic autoradiography. Neuroscience 10: 1427-1442
33. Kanazawa I, Iversen LL, Kelly JS (1976) Glutamate decarboxylase activity in the rat posterior pituitary, pineal gland, dorsal root ganglion and superior cervical ganglion. J Neurochem 27: 1267-1269
34. Karhula T, Häppölä O, Joh TH, Wu J.Y (1988) Localization of L-glutamate decarboxylase immunoreactivity in the major pelvic ganglion and in the coeliac-superior mesenteric ganglion complex of the rat. Histochemistry 90: 255-260
35. Kása P, Joó F, Dobó E, Wenthold RJ, Ottersen OP, Storm-Mathisen J, Wolff JR (1988) Heterogenous distribution of GABA-immunoreactive nerve fibers and axon terminals in the superior cervical ganglion of adult rat. Neuroscience 26: 635-644
36. Kataoka Y, Gutman Y, Guidotti A, Panula P, Wroblewski J, Cosenza-Murphy D, Wu J-Y, Costa E (1984) Intrinsic GABAergic system of adrenal chromaffin cells. Proc Natl Acad Sci USA 81: 3218-3222
37. Kataoka Y, Fujimoto M, Alho H, Guidotti A, Geffard M, Kelly GD, Hanbauer I (1986) Intrinsic gamma-aminobutyric acid receptors modulate the release of catecholamine from canine adrenal gland in situ. J Pharmacol Exp Ther 239: 584-590
38. Kenny SL, Ariano MA (1986) The immunofluorescence localization of glutamate decarboxylase in the rat cervical ganglion. J Auton Nerv Syst 17: 211-215
39. Kitayama S, Tsujimoto A (1986) Involvement of GABAergic mechanism in the catecholamine secretion from adrenal medulla. In: Erdö SL, Bowery NG (eds) GABAergic mechanisms in the mammalian periphery. Raven, New York, pp 249-259
40. Landis SC, Patterson PH (1981) Neural crest cell lineages. Trends Neurosci 4: 172-175
41. Le Douarin N (1973) A biological cell labelling technique and its use in experimental embryology. Dev Biol 30: 217-222
42. Maggi CA, Santicioli P, Meli A (1985) GABA inhibits neurotransmission in rat pelvic ganglia. J Pharm Pharmacol 37: 349-351
43. Päivärinta H, Soinila S, Eränkö O, Joh TH (1985) Phenylethanolamine-N-methyltransferase-immunoreactive cells in developing rat superior cervical ganglion and the effect of hydrocortisone on their number. Int J Dev Neurosci 3: 9-18

44. Päivärinta H, Eränkö L, Häppölä O, Soinila S, Steinbusch H, Wu J-Y, Panula P (1987) The effect of hydrocortisone on the number of 5-hydroxytryptamine- and glutamic acid decarboxylase-immunoreactive cells in the superior cervical ganglion of the rat. Exp Brain Res 16: 305-309
45. Papka RE (1972) Ultrastructural and fluorescence histochemical studies of developing sympathetic ganglia in the rabbit. Am J Anat 134: 337-363
46. Sangiah S, Borowitz JL, Yim GKW (1974) Actions of GABA, picrotoxin and bicuculline on adrenal medulla. Eur J Pharmacol 27: 130-135
47. Siegrist G, Dolivo M, Dunant Y, Foroglou-Kerameus C, de Ribaupierre F, Rouiller C (1968) Ultrastructure and fuction of the chromaffin cells in the superior cervical ganglion of the rat. J Ultrastruct Mol Struct Res 25: 381-407
48. Soinila S (1984) Pre- and postnatal development of the small intensely fluorescent cells in the rat superior cervical ganglion. Int J Dev Neurosci 2: 65-76
49. Soinila S, Ahonen M, Joh TH, Steibusch HWM (1988) 5-hydroxytryptamine and catecholamines in the developing sympathetic cells of the rat. J Auton Nerv Syst 22: 193-202
50. Stanton HC (1963) Mode of action of GABA on the cardiovascular system. Arch Int Pharmacodyn Ther 143: 195-204
51. Tanaka C (1985) γ-Aminobutyric acid in peripheral tissues. Life Sci 37: 2221-2235
52. Taniyama K, Tanaka C (1986) GABAergic mechanisms and their functional relevance in the urinary bladder. In: Erdö SL, Bowery NG (eds) GABAergic mechanisms in the mammalian periphery. Raven, New York, pp 175-183
53. Taxi J (1979) The chromaffin and chromaffin-like cells in the autonomic nervous system. Int Rev Cytol 57: 283-343
54. Verhofstad AAJ, Hökfelt T, Goldstein M, Steinbusch HWM, Joosten HWJ (1979) Appearance of tyrosine hydroxylase. Aromatic amino-acid decarboxylase, dopamine hydroxylase and phenylethanolamine-N-methyltransferase during the ontogenesis of the adrenal medulla. Cell Tissue Res 200: 1-13
55. Waniewski RA, Suria A (1977) Alterations in γ-aminobutyric acid content in the rat superior cervical ganglion and pineal gland. Life Sci 21: 1129-1142
56. Williams TH (1967) Electron microscopic evidence for an autonomic interneuron. Nature 214: 309-310
57. Wolff JR, Joó F, Kása P, Storm-Mathisen J, Toldi J, Balcar VJ (1986) Presence of neurons with GABA-like immunoreactivity in the superior cervical ganglion of the rat. Neurosci Lett 71: 157-162
58. Wolff JR, Kása P, Dobó E, Wenthold RJ, Joó F (1989) Quantitative analysis of the number and distribution of neurons richly innervated by GABA-immunoreactive axons in the rat superior cervical ganglion. J Comp Neurol 282: 264-273
59. Wu J-Y (1976) Purification and properties of L-glutamate decarboxylase (GAD) and GABA-aminotransferase (GABA-T). In: Roberts E, Chase T, Tower D (eds) GABA in nervous system function. Raven, New York, pp 7-55
60. Wu J-Y (1982) Purification and characterization of cysteic acid and cysteine sulfinic acid decarboxylase and L-glutamate decarboxylase from bovine brain. Proc Natl Acad Sci USA 79: 4270-4274
61. Wu J-Y, Lin C-T, Brandon C, Chan T-S, Möhler H, Richards JG (1982) Regulation and immunocytochemical characterization of glutamic acid decarboxylase. In: Palay S, Palay V (eds) Cytochemical methods in neuroanatomy. Liss, New York, pp 279-296

62. Young JAC, Brown DA, Kelly JS, Schon F (1973) Autoradiographic localization of sites of [^{3}H]-γ-aminobutyric acid accumulation in peripheral autonomic ganglia. Brain Res 63: 479-486

GABAergic Action on Cholinergic Axon Terminals in the Superior Cervical Ganglion[1]

P. Kása, E. Dobó, and J.R. Wolff

Introduction

Interactions between different neuronal or transmitter systems are considered to play an important role in the regulation of the function of the mammalian sympathetic ganglia. The excitatory effect of acetylcholine (ACh) in the superior cervical ganglia (SCG) has been well documented by electrophysiological techniques [18, 27, 35, 37]. The elements of the ACh system (choline acetyltransferase, ChAT [4, 7, 14, 22, 29, 39, 42]; ACh [10, 36]; acetylcholinesterase, AChE [24, 28, 29, 46]; muscarinic acetylcholine receptor, mAChR [8, 9, 26, 34, 41]; nicotinic acetylcholine receptor, nAChR [20, 21, 25, 38]) have also been demonstrated. Other neurotransmitters, such as γ-aminobutyric acid (GABA), are known to effect synaptic transmission between preganglionic axon terminals and postganglionic sympathetic neurons [1, 5, 11, 16, 23]. More recently, an immunocytochemical technique [13] was used to demonstrate GABA-positive fibers [12, 31, 45] and GABA-positive cells [44] in the SCG. What is the morphological basis of the GABAergic interaction with cholinergic nerve fibers? How does GABA affect the release of ACh from cholinergic axon terminals? What kind of GABA receptors play a role in the GABAergic-cholinergic interaction? These questions are addressed in the present paper.

[1] This work was supported by OTKA (518141), ETT (457) Ministry of Health, Hungary and the Deutsche Forschungsgemeinschaft (Wo 279/8-2), BRD.

Materials and Methods

Immunocytochemical Demonstration of Cholinergic Axons and Axon Terminals

Light Microscopic Immunohistochemistry. The SCG of adult albino mice (25-30 g) of either sex were used in these studies. After anesthesia, the animals were perfused transcardially with a fixative which contained 8% formaldehyde, 0.4% picric acid, and 0.1% glutaraldehyde in 0.12 *M* phosphate buffer (PB) pH 7.4. The SCG were dissected out and serial sections were made.

The samples were processed for ChAT immunocytochemistry [30] using a monoclonal rat anti-ChAT antibody (AB8, a gift from B.H. Wainer). Antibody-antigen complexes were visualized by the peroxidase-antiperoxidase (PAP) method [40]. The sections were mounted onto glass slides, air-dried, dehydrated, and enclosed in Entellan.

Electron Microscopic Immunocytochemistry. For ultrastructural study, the SCG sections were processed for immunocytochemistry. After completion of the PAP procedure, the samples were fixed with 1% OsO_4 in PB pH 7.4, dehydrated, and embedded in Durcupan ACM (Fluka). Thin sections were stained with uranyl acetate and lead citrate and examined with a Zeiss EM 10 electron microscope at an accelerating voltage of 80 kV.

Immunocytochemical Demonstration of GABAergic Structures

Light Microscopic Immunohistochemistry. The SCG of adult Sprague-Dawley rats of either sex were studied. After removal, the SCG were fixed by immersion in a fixative containing 2.5% glutaraldehyde in 0.1 *M* PB pH 7.4 at 4°C. After fixation, 10-20 μm samples were cut from the SCG. For immunocytochemical localization, sections were preincubated in 20% normal bovine serum. Free-floating sections were incubated with affinity-purified antibodies against GABA (a gift from R.J. Wenthold). The samples were left overnight at room temperature in a 1:150 dilution of the primary antiserum, which had been characterized previously [43]. The samples were then kept in purified rat anti-rabbit IgG (dilution 1:10) for 2 h and treated according to the PAP method [40].

Electron Microscopic Immunocytochemistry. The SCG were treated as described above for light microscopy. After application of the PAP technique, the sections were treated with 1% OsO_4 and dehydrated in a graded series of ethanol. Thin sections were examined with a Zeiss EM 10 electron microscope at an accelerating voltage of 80 kV.

Neurochemical Study of the Effect of GABA on ACh Release

SCG were dissected from lightly anesthetized rats and placed in Krebs-Ringer solution (KRS) of the following composition: 120 m*M* NaCl, 3 m*M* KCl, 2.6 m*M* $CaCl_2$, 1.2 m*M* NaH_2PO_4, 24 m*M* $NaHCO_3$, and 10 m*M* D-glucose (pH 7.3) maintained at 37°C. The experiments started with a short (2 min) preincubation with K^+ (53 m*M*). The SCG were subsequently incubated for 40 min at 37°C in KRS containing normal (3 m*M*) K^+, continually gassed with 95% O_2/5% CO_2, and including 2 n*M* [methyl-^{3}H]choline chloride. Hemicholinium (0.5 m*M*) was present in the superfusion medium to prevent reuptake of the released [^{3}H]ACh.

Release Studies. The SCG, prelabeled with [^{3}H]ACh, were stimulated with 53 m*M* K^+ and the [^{3}H]ACh released in the superfusate was collected. Radioactivity in the samples was extracted [19] and estimated by liquid scintillation counting. The ACh release was expressed as a percentage of the basal release or as pmol/g of tissue (w/w).

Pharmacological Experiments. Observations were made of the K^+-stimulated release of ACh and, in some experiments, the release of ACh in the presence of 1 m*M* GABA, 300 μ*M* (+)-bicuculline, 200 μ*M* (+)-baclofen, 300 μ*M* (+)-bicuculline plus 1 m*M* GABA plus 53 m*M* KCl, and finally 200 μ*M* (+)-baclofen and 1 m*M* GABA.

Results

Morphological Localization of ChAT-immunoreactive Fibers and GABA-Positive Structures in the SCG

At a light microscopic level, dense networks of ChAT-immunoreactive axons (Fig. 1) or GABA-positive nerve fibers (Fig. 2) could be observed around and among the principal ganglionic cells. Most of the stained fibers had a varicose appearance. Ultrastructurally, the varicose axonic areas were filled with synaptic vesicles and these areas usually made synaptic contacts

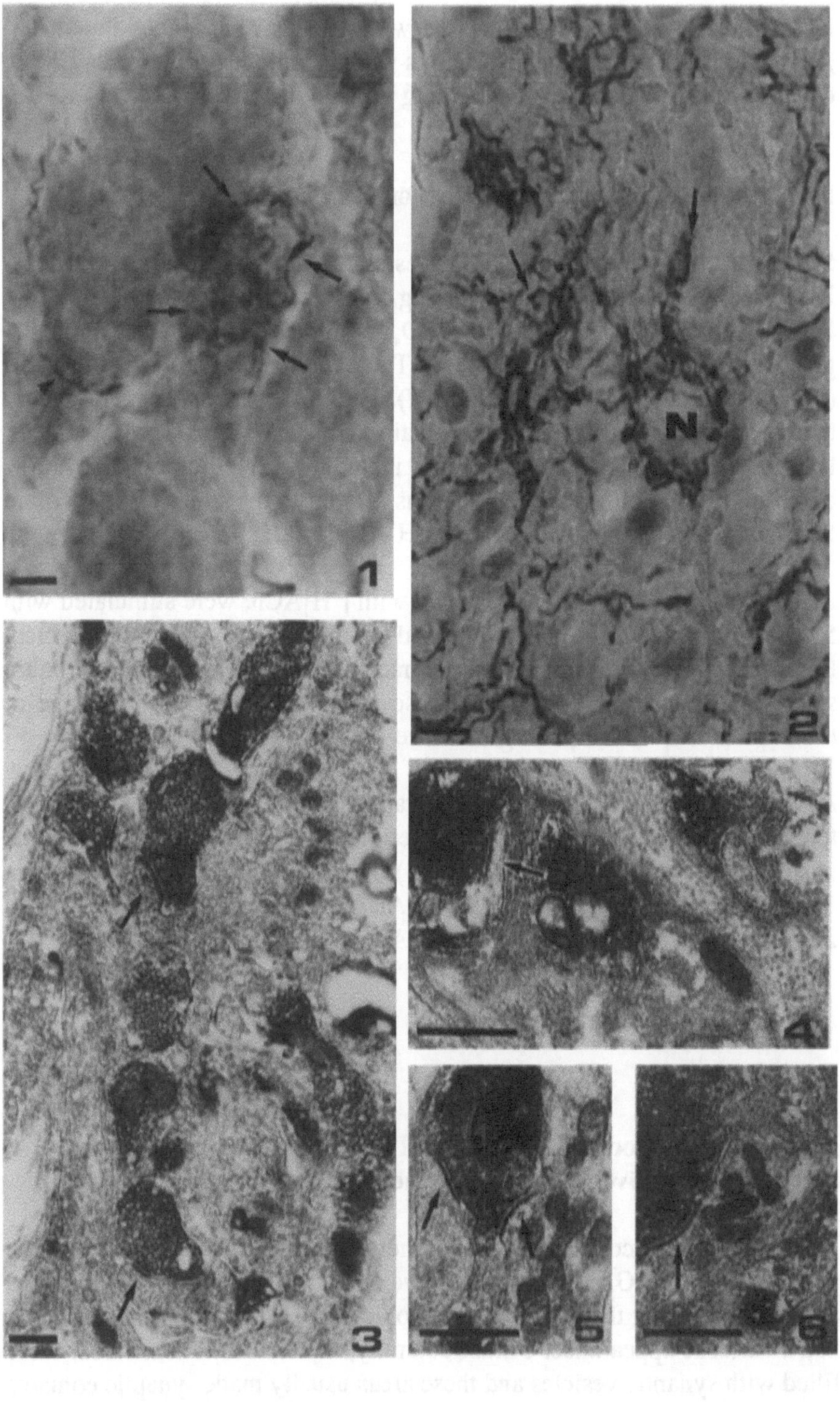
1
2
N
3
4
5
6

with different areas (dendritic spine, dendrites, and soma) of postsynaptic ganglionic cells. Most of the ChAT-positive axon terminals and/or axon varicosities appeared to make synaptic contacts with dendritic spines (Fig. 3). Similar to ChAT localization, GABA positivity was present in axon varicosities and/or axon terminals which made synaptic contacts with dendrites and dendritic spines (Figs. 4-6). Although ChAT-positive axon terminals made synaptic contacts with postsynaptic structures where many vesicles were present (Figs. 7-8), none of these postsynaptic structures fulfilled the criteria for axon terminals. Despite the great number of vesicles in the postsynaptic structures, these were regarded as dendrites or parts of the neuronal soma. No axo-axonic synapses could be seen and none of the cells in the SCG stained for ChAT.

Effect of GABA on ACh Release

In order to ascertain whether GABA has an effect on the release of [^{3}H]ACh, superfused rat SCG samples were investigated. The results of this have already been published in detail [17]. In these experiments, it was shown that the release of ACh can be reduced to GABA. In short, 53 m*M* K^+ induced a twofold increase in the rate of [^{3}H]ACh release from the SCG (Fig. 7), while 1m*M* GABA in the superfusion medium significantly reduced spontaneous release of the labeled compound (Fig. 8). (+)-Bicuculline (300 *μM*) however, reduced the inhibitory action of 1 m*M* GABA on the K^+-stimulated release (Fig. 9), but the effect was not complete. (+)-Baclofen (200 *μM*) reduced the spontaneous and the K^+-stimulated release of the labeled compound (Fig. 10). Complete inhibition was observed only in the presence of 200 *μM* (+)-baclofen and 1 m*M* GABA.

Fig. 1. Immunohistochemical localization of ChAT in the SCG of mouse. Note the presence of stained fibers between (*arrows*) and around (*arrowhead*) the principal ganglionic cells. *Bar* = 5 μm

Fig. 2. Immunohistochemical localization of GABA in the SCG of rat. GABA-positive fibers are present around the ganglionic neurons (*N*), while others are scattered among the principal nerve cells. Note the varicose appearance of the fibers (*arrows*). *Bar* = 10 μm

Fig. 3. ChAT immunocytochemistry in the SCG of mouse. The ChAT-positive axon varicosities and/or axon terminals are in synaptic contact with the dendritic spines (*arrow*). *Bar* = 0.5 μm

Figs. 4-6. GABA immunocytochemistry in the SCG of rat. Note that GABA-positive axon terminal makes synaptic junctions with dendritic spine (**4**), with dendrite and dendritic spine (**5**), and with a dendrite (**6**). *Bars* = 0.5 μm

Discussion

The results described above provide strong evidence for a specific association of GABA with cholinergic synaptic transmission in the SCG. The immunocytochemical findings, that ChAT can be found in almost all axon terminals in the SCG [30] and, similarly, that many GABA-positive fibers are present in the SCG [13, 31, 45], suggest that these substances may be colocalized. The specific association of the two transmitter systems is supported by transmitter release studies [17]. GABA appeared to reduce ACh release from the SCG [1, 6], where ACh is thought to be present in preganglionic axon terminals [15, 18, 36]. The question arises: How can GABA modulate the release of ACh?

In morphological [33] and immunocytochemical ultrastructural studies [30], no axo-axonic synapses could be seen although many axon profiles could be found side by side. Since none of the neurons (principal nerve cells and interneurons) expressed ChAT immunopositivity [30, 32], it is suggested that the most likely site of the inhibitory action of GABA on ACh release is presynaptic, i.e., on the cholinergic terminals of preganglionic fibers. From the results of the release experiment [6, 17], it

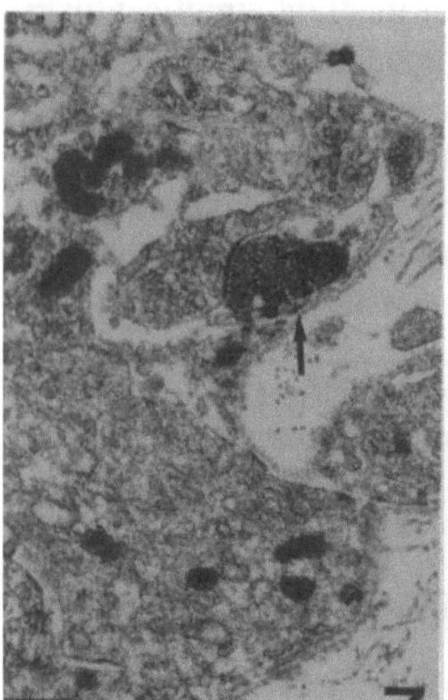
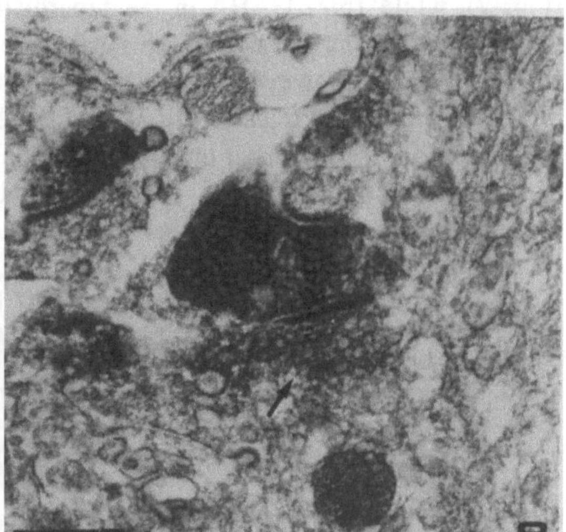

Fig. 7. ChAT immunocytochemistry in the mouse SCG. A ChAT-positive axon terminal (*arrow*) is in synaptic contact with a postsynaptic structure (possibly a dendrite), where a great number of vesicles are present. *Bar* = 0.5 μm

Fig. 8. ChAT immunocytochemistry in the SCG of mouse. Note the ChAT-positive axon terminal and the postsynaptic accumulation of vesicles (*arrow*). *Bar* = 0.5 μm

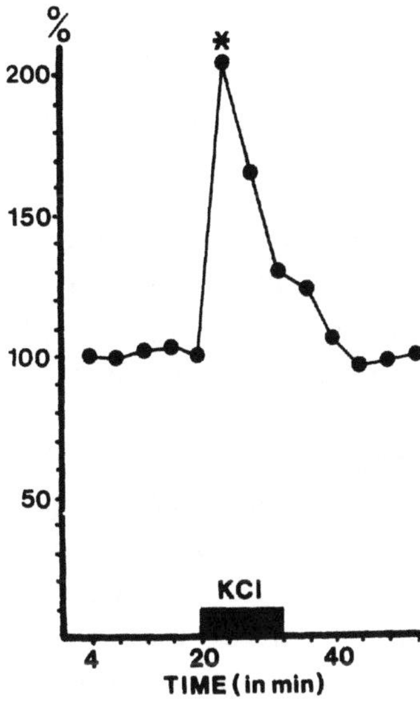

9

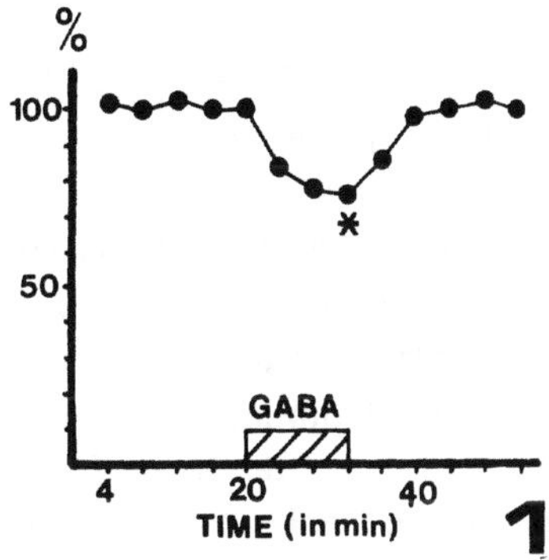

10

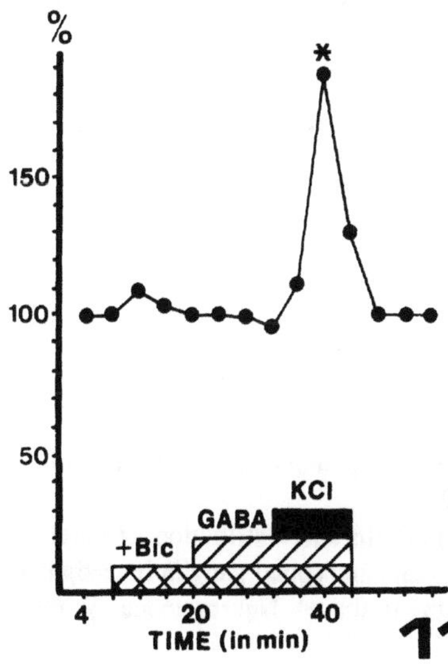

11

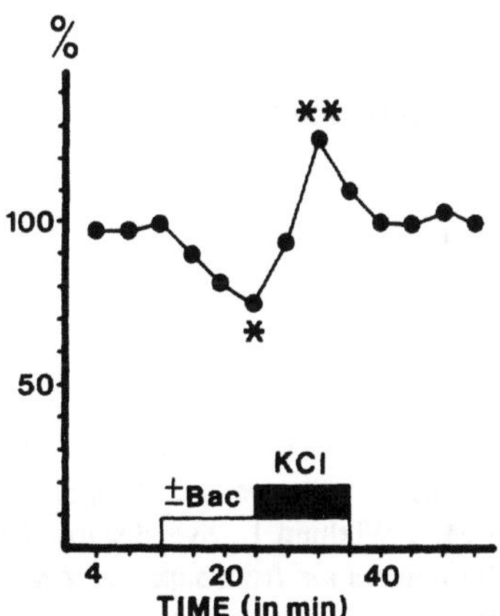

12

Fig. 9. Release of $[^3H]$ACh from the SCG in response to 53 m*M* KCl. Note the significant difference from the basal release ($p < 0.02$)

Fig. 10. Effect of 1 m*M* GABA on the basal release. The release was reduced, 23 ± 3% ($p < 0.02$)

Fig. 11. Antagonism of GABA inhibition by (+)-bicuculline

Fig. 12. Effect of (±)-baclofen on the spontaneous and on the K^+-stimulated release of ACh

can also be inferrred that preganglionic cholinergic axon terminals may have two types of GABA receptors, $GABA_A$ and $GABA_B$ [17]. These receptors may have specific roles in the regulation of ACH release from cholinergic nerve fibers in the SCG, as has already been suggested [17].

The immunocytochemical localization of ChAT and GABA in the SCG and the study of the effect of GABA on ACh release in the ganglion provide evidence for the nonsynaptic inhibitory effect of GABA on ACh release. However, since GABA-positive terminals are in synaptic contact with the principal ganglionic cells, a direct inhibitory effect of GABA on sympathetic ganglionic cells cannot be excluded.

The association of ACh with GABA in axon terminals of the preganglionic sympathetic nerve trunk may be of relevance in regulating the function of postganglionic sympathetic neurons which innervate cerebral blood vessels [2]. In view of the high endogenous GABA level [3] and its presence in preganglionic axon terminals [12, 31, 45], it is concluded that GABA may be one of the local factors involved in controlling the amount of transmitter released from cholinergic varicosities upon depolarization and as such is important in the regulation of blood circulation in the CNS. Since no axo-axonic synapses were found in the SCG [33], it is suggested that the modulatory effect of GABA on ACh release is nonsynaptic.

Acknowledgements. The expert secretarial work of Mrs Gyongyi Ugrin is gratefully acknowledged.

References

1. Adams PR, Brown DA, Saunders L (1975) Actions of gamma-aminobutyric acid on sympathetic ganglion cells. J Physiol 250: 85-120
2. Arbab MA, Wiklund L, Svendgaard NA (1986) Origin and distribution of cerebral vascular innervation from superior cervical, trigeminal and spinal ganglia investigated with retrograde and anterograde WGA-HRP tracing in the rat. Neuroscience 19: 695-708
3. Bertilsson L, Suria A, Costa E (1976) Gamma-aminobutyric acid in rat superior cervical ganglion. Nature 260: 540-541
4. Black IB, Hendry IA, Iversen LL (1971) Trans-synaptic regulation of growth and development of adrenergic neurons in a mouse sympathetic ganglion. Brain Res 34: 229-240
5. Bowery NG, Brown DA (1974) Depolarizing actions of gamma-aminobutyric acid and related compounds on rat superior cervical ganglia in vitro. Br J Pharmacol 50: 205-218

6. Brown DA, Higgins AJ (1979) Presynaptic effects of γ-aminobutyric acid in isolated rat superior cervical ganglia. Br J Pharmacol 66: 108P-109P
7. Buckley G, Consolo S, Giacobini E, Sjöqvist F (1967) Cholinacetylase in innervated and denervated sympathetic ganglia and ganglion cells of the cat. Acta Physiol Scand 71: 348-356
8. Burt DR (1978) Muscarinic receptor binding in rat sympathetic ganglia is unaffected by denervation. Brain Res 143: 573-579
9. Capuzzo A, Borasio PG, Fabbri E (1989) Presynaptic muscarinic receptors in guinea-pig superior cervical ganglion. Neurosci Lett 104: 88-92
10. Collier B (1969) The preferential release of newly synthesized transmitter by a sympathetic ganglion. J Physiol 205: 341-352
11. De Groat WC, Lalley PM, Block M (1971) The effects of bicuculline and GABA on the superior cervical ganglion of the cat. Brain Res 25: 665-668
12. Dobó E, Kása P, Wenthold RJ, Joó F, Wolff JR (1989) Evidence for GABAergic fibers entering the superior cervical ganglion of rat from the preganglionic nerve trunk. Histochemistry 92: 133-136
13. Dobó E, Kása P, Wenthold RJ, Wolff JR (1989) Pronase treatment increases the staining intensity of GABA-immunoreactive structures in the paravertebral sympathetic ganglia. Histochemistry 93: 13-18
14. Dolezalova, Giacobini E, Giacobini G, Rossi A, Toschi G (1974) Developmental variations of choline acetyl-transferase, dopamine-beta-hydroxylase and monoamineoxidase in chicken embryo and chicken sympathetic ganglia. Brain Res 73: 309-320
15. Dun NJ (1980) Ganglionic transmission: electrophysiology and pharmacology. Fed Proc 39: 2982-2989
16. Eugéne D (1987) Fast non-cholinergic depolarizing postsynaptic potentials in neurons of rat superior cervical ganglia. Neurosci Lett 78: 51-56
17. Farkas Z, Kása P, Balcar VJ, Joó F, Wolff JR (1986) Type A and B GABA receptors mediate inhibition of acetylcholine release from cholinergic nerve terminals in the superior cervical ganglion of rat. Neurochem Int 8: 565-572
18. Feldberg W, Gaddum JH (1934) The chemical transmitter at synapses in sympathetic ganglion. J Physiol 81: 305-319
19. Fonnum F (1975) A rapid radiochemical method for the determination of choline acetyltransferase. J Neurochem 24: 407-409
20. Fumagalli L, De Renzis G (1980) Alpha-bungarotoxin binding sites in the rat superior cervical ganglion are influenced by postganglionic axotomy. Neuroscience 5: 611-616
21. Fumagalli L, De Renzis G, Miani N (1976) Acetylcholine receptors: number and distribution in intact and deafferented superior cervical ganglion of the rat. J Neurochem 27: 47-52
22. Gaetani S, Mengheri E, Spadoni MA, Rossi A, Toschi G (1975) Effects of litter size on protein, choline acetyltransferase (CAT), and dopamine-beta-hydroxylase (DBH) of a mouse sympathetic ganglion. Brain Res 86: 75-84
23. Galvan M, Grafe P, Bruggencate G (1980) Presynaptic actions of 4-aminopyridine and gamma-aminobutyric acid on rat sympathetic ganglia in vitro. Naunyn Schmiedebergs Arch Pharmacol 314: 141-147
24. Gisiger V, Vigny M, Gautron J, Rieger F (1978) Acetylcholinesterase of rat sympathetic ganglion: molecular forms, localization and effects of denervation. J Neurochem 30: 501-516

25. Giotti C, Spagnoli D, Omini C, Clementi F (1985) Alpha-bungarotoxin and P15 toxin binding site in mammalian sympathetic ganglia. Neurosci Lett 57: 227-231
26. Hadjiconstantinou M, Potter PE, Neff NH (1982) Trans-synaptic modulation via muscarinic receptors of serotonin-containing small intensely fluorescent cells of superior cervical ganglion. J Neurosci 2: 1836-1839
27. Karczmar AG (1970) History of the research with anticholinesterase agents. In: Karczmar AG (ed) Anticholinesterase agents, Vol 1, Sect 13. Pergamon, New York, pp 1-44
28. Kása P, Csernovszky E (1967) Electron microscopic localization of acetylcholinesterase in the superior cervical ganglion of the rat. Acta Histochem 28: 274-285
29. Kása P, Dames W, Rakonczay Z, Gulya K, Joó F, Wolff JR (1985) Modulation of the acetylcholine system in the superior cervical ganglion of rat: effects of GABA and hypoglossal nerve implantation after in vivo GABA treatment. J Neurochem 44: 1363-1372
30. Kása P, Dobó E, Wolff JR (1990) Distribution of cholinergic fibers and synapses in the mouse superior cervical ganglion: light and electron microscopical immunohistochemistry for choline acetyltransferase. Neuroscience (to be published)
31. Kása P, Joó F, Dobó E, Wenthold RJ, Ottersen OP, Storm-Mathisen J, Wolff JR (1988) Heterogenous distribution of GABA-immunoreactive nerve fibers and axon terminals in the superior cervical ganglion of adult rat. Neuroscience 26: 635-644
32. Lindh B, Staines W, Hökfelt T, Terenius L, Salvaterra PM (1986) Immunohistochemical demonstration of choline acetyltransferase-immunoreactive preganglionic nerve fibers in guinea-pig autonomic ganglia. Proc Natl Acad Sci USA 83: 5316-5320
33. Matthews MR (1983) The ultrastructure of junctions in sympathetic ganglia of mammals. In: Elfvin LG (ed) Autonomic ganglia. Wiley, Chichester, pp 27-66
34. Newberry NR, Priestley T, Woodruff GN (1985) Pharmacological distinction between two muscarinic responses on the isolated superior cervical ganglion of the rat. Eur J Pharmacol 116: 191-192
35. Paton WDM, Perry WLM (1953) The relationship between depolarization and block in the cat's superior cervical ganglion. J Physiol 119: 43-57
36. Perry WLM (1953) Acetylcholine release in the cat's superior cervical ganglion. J Physiol 119: 439-454
37. Prigioni I, Casella C (1984) Acetylcholine-induced back-firing in the preganglionic trunk of the rat superior cervical ganglion. J Auton Nerv Syst 10: 19-25
38. Smolen AJ (1983) Specific binding of alpha-bungarotoxin synaptic membranes in rat sympathetic ganglion: computer best-fit analysis of electron microscope autoradiographs. Brain Res 289: 177-188
39. Sorimachi M, Kataoka K (1974) Developmental change of choline acetyltransferase in the ciliary and the superior cervical ganglion of the chick. Brain Res 70: 123-130
40. Sternberger LA (1979) Immunocytochemistry, 2nd edn. Wiley, New York
41. Taniguchi T, Kurahashi K, Fujiwara M (1983) Alterations in muscarinic cholinergic receptors after preganglionic denervation of the superior cervical ganglion in cats. J Pharmacol Exp Ther 224: 674-678
42. Viana GB, Kauffman FC (1984) Cholinesterase activity in the rat superior cervical ganglion: effect of denervation and axotomy. Brain Res 304: 37-45
43. Wenthold RJ, Zempel JM, Parakkal MH, Reeks KA, Altschuler RA (1986) Immunocytochemical localization of GABA in the cochlear nucleus of the guinea pig. Brain Res 380: 7-18

44. Wolff JR, Joó F, Kása P, Storm-Mathisen J, Toldi J, Balcar V (1986) Presence of neurons with GABA-like immunoreactivity on the superior cervical ganglion of the rat. Neurosci Lett 71: 157-162
45. Wolff JR, Kása P, Dobó E, Wenthold RJ, Joó F (1989) Quantitative analysis of the number and distributions of neurons richly innervated by GABA-immunoreactive axons in the rat superior cervical ganglion. J Comp Neurol 282: 264-273
46. Yamauchi A, Lever JD (1971) Correlations between formol fluorescence and acetylcholinesterase (AChE) staining in the superior cervical ganglion of normal rat, pig and sheep. J Anat 110: 435-443

GABA in Non-Neural Tissues

Non-Neuronal GABA Systems: An Overview

S.L. Erdö

Introduction

Free amino acids are substrates for numerous metabolic pathways in mammals and, as such, they were once considered to be of primarily metabolic relevance. Therefore, the discovery of γ-aminobutyric acid (GABA) and its specific inhibitory action in the central nervous system (CNS) 40 years ago [6, 65, 73] evoked significant attention. Comprehensive studies during the following two decades revealed that GABA is a major transmitter of synaptic inhibition in the vertebrate CNS [52]. In 1979, Jessen et al. provided partial evidence to show that this role is not restricted to the CNS, but that GABA may also be a neurotransmitter in the peripheral nervous system (PNS), namely, in the myenteric plexus of the intestine [45]. However the general recognition of GABA's central and peripheral neurotransmitter function, led, until recently, to the neglect of its possible extraneuronal functions. This chapter is, therefore, an attempt to summarize the recent knowledge of non-neuronal GABA systems in mammals and to emphasize that GABA may play a specific mediator role in a number of non-neuronal tissues.

This overview is written in an encyclopedic style, in order to minimize overlapping with other chapters (Gilon and Remacle; Fogel and Erdö; Minuk; Dobó et al.; Oset-Gasque et al.; László et al.; Boldizsár et al. this volume).

γ-Aminobutyric Acid

The presence of GABA outside the CNS was first demonstrated by Tallan and coworkers in several tissues of the cat in 1954 [70]. More than a decade later, Zachman et al. reported the presence of detectable GABA levels in several human organs [78]. This later report initiated intensive research during the subsequent years, which led to the demonstration of GABA in a series of organs of different mammals. GABA concentrations in the periphery, as determined by various assay techniques, have been summarized in a recent review [21]. These results show that cerebral GABA levels are usually higher than those in the periphery, with few exceptions. For instance, GABA is more abundant in the rat oviduct and pancreatic islets than in the whole brain.

Most of the quantitative results, however, failed to provide information about the distribution of GABA within a particular organ. Thus, apart from the information obtained in studies with dispersed and cultured cells (blood cells, hepatocytes, adrenal chromaffin cells etc.), the cellular and intracellular distribution of GABA in the periphery remained chiefly unknown until recently [17, 21].

During the last 5 years, immunocytochemical studies provided evidence that GABA-like immunoreactivity (GABA-LI) may be located not only in nervous tissue (CNS, PNS), but also in certain non-neuronal elements of various organs. For instance, endocrine cells of the antral gastric mucosa and epithelial cells of the renal collecting tubules exhibit heavy, specific GABA immunostaining (Fig. 1). As summarized in Table 1, the GABA-containing cell types are of different developmental origin and include neuroendocrine, endocrine, exocrine, and other epithelial cells or even gametocytes (for a review see also [25]).

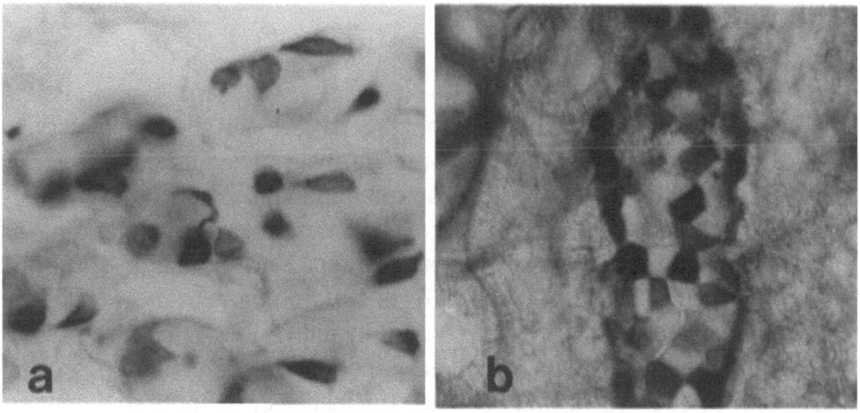

Fig. 1 a, b. Examples of non-neuronal cell types exhibiting strong specific GABA-like immunoreactivity (GABA-LI). **a** GABA-positive endocrine cells in the rat antral gastric mucosa; **b** GABA-positive epithelial cells of a collecting tube in the rat kidney. (E. Dobó and S. L. Erdö, original illustrations)

Table 1. GABA in non-neuronal cells

Organ/tissue	Cell type	Reference
Adrenals	Chromaffin cells	[47]
Blood	Platelets, erythrocytes	[60]
Gut	Endocrine	[12, 46]
Kidney	Tubular epithelium	[32] ED
Liver	Hepatocytes	[54]
Oviduct	Ciliated epithelium	[28]
Pancreas		
Endocrine	Beta cells	[38, 58, 71]
Exocrine	Acinar cells	[36]
Pineal gland	Pinealocytes	[16]
Semen	Spermatozon	HKB
Skin	Keratinocytes	[8]
Stomach, antrum	Endocrine cells	[41, 46]

ED, E. Dobó et al., this volume; HKB, H. K. Boldizsár et al., this volume

Glutamate Decarboxylase

The origin of GABA in non-neuronal cells has not been extensively studied. It is likely, however, that both intrinsic biosynthesis and active uptake systems may contribute. The predominant pathway of GABA formation in the CNS and PNS, i.e., the glutamate decarboxylase (GAD)-catalyzed one, does not necessarily represent the major metabolic route in peripheral tissues [3, 4, 9, 33]. In certain non-neuronal cell types, such as the pancreatic beta cell (see P. Gilon and C. Remacle, this volume) or the epithelial cells of the oviductal mucosa [18, 28, 30], GABA and GAD seem to coexist. In other tissues, for instance, in endocrine cells of the gastrointestinal mucosa, such a coexistence, may not necessarily be found (see P.Gilon and C.Remacle, this volume). Interestingly, in the gastrointestinal mucosa, the distribution of GABA showed a better correlation with the distribution of diamino oxidase (DAO) than with that of GAD (see W.A.Fogel and S.L.Erdö, this volume). This suggests that a major portion of GABA located in the endocrine cells of the gastrointestinal mucosa may be formed from putrescine rather than from glutamate (see W. A. Fogel and S. L. Erdö, this volume).

In mammalian tissues, several biochemically and immunochemically different forms of GAD have been distinguished [75, 77]. The heterogeneity of the enzyme has not helped to establish the relative contribution of GAD and its different forms to the synthesis of non-neuronal GABA pools. In fact, the GAD obtained from brain tissue (neuronal form) showed only weak immunochemical cross-reaction with renal GAD (non-neuronal form), thus making the precise immunocytochemical localization of the enzyme in the kidney, using antisera against neuronal GAD, impossible [77] (E. Dobó, unpublished observation). Surprisingly, in several other peripheral organs, e.g., the female reproductive tract, the antisera raised against brain GAD proved to be suitable for the immunocytochemical visualization of the enzyme even in non-neuronal cell types (Fig. 2). The cell types which have been shown to contain GAD in biochemical and immunocytochemcial studies are summarized in Table 2.

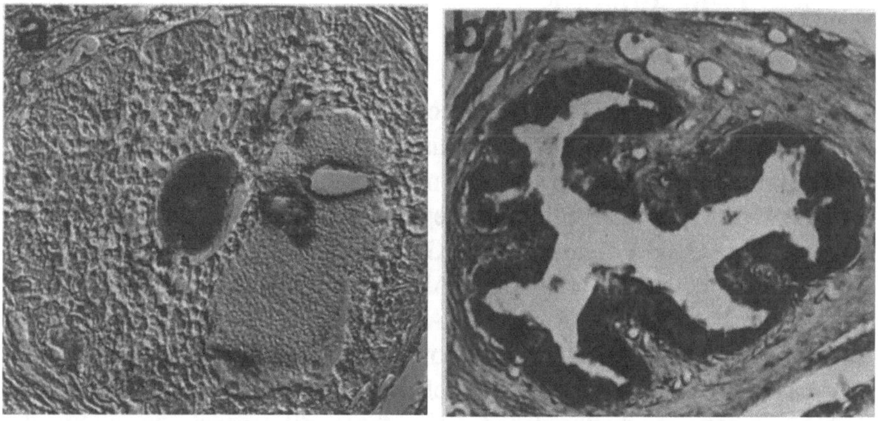

Fig. 2 a, b. Examples for non-neuronal localization of GAD. **a** An ovarian follicle with heavy GAD-positive immunostaining in the oocyte; **b** GAD-positive epithelial cells in the mucosa of rat oviductal ampulla. Immunocytochemistry was performed using an antiserum that had been raised against neuronal GAD (for methodological details see [30]). (Modified from [30])

Table 2. Glutamate decarboxylase in non-neuronal cells

Organ/tissue	Cell type	Reference
Adrenals	Chromaffin cells	[47, 77]
Blood	Erythrocytes	[60]
Blood vessels	Endothelial cells	[20, 51]
Gut	Endocrine cells	PG, ED
Liver	Hepatocytes	[54]
Ovary	Oocyte, granulosa cells	[30]
Oviduct	Epithelial cells	[18, 30]
Pancreas		
Endocrine	Beta cells	[58, 74] PG
Exocrine	Acinar cells	[36]
Stomach, antrum	Endocrine cells	PG, ED

PG, see P. Gilon and C. Remacle, this volume;
ED, E. Dobó, unpublished observation

GABA Transaminase

The information available with respect to the pathways of GABA's further transformation (degradation, catabolism) in non-neuronal tissues is even more limited than that for its biosynthesis. In fact, there is no clear-cut evidence to suggest that significant degradative pathways other than the one catalyzed by GABA transaminase (GABA-T), which is known to be the predominant catabolic route for GABA in nervous tissue [77], exist in non-neuronal tissues. In fact, GABA-T seems to be the most important catabolic enzyme of GABA also in non-neuronal tissues [33]. Although, on the basis of mainly biochemical characteristics, different subtypes of GABA-T can be distinguished [77], these differences do allow the immunocytochemical visualization of the enzyme in both neuronal and non-neuronal tissues [76, 77]. In addition, enzyme histochemical techniques could also be used for the visualization of GABA-T in peripheral organs [20, 28]. The non-neuronal cell types of mammals which were found to contain GABA-T are summarized in Table 3. The utilization of GABA for the synthesis of GABA-containing macromolecules in keratinocytes [8] and other minor pathways in several non-neuronal cell types [33], however, may also play some role in the further metabolism of GABA in the mammalian periphery.

Table 3. GABA transaminase in non-neuronal cells

Organ/tissue	Cell type	Reference
Adrenals	Chromaffin cells	[47, 77]
Blood	Platelets, lymphocytes	[60]
Blood vessels	Endothelial cells	[20, 51]
Hair follicle	Follicle cells	[5]
Kidney	Tubular epithelial cells	[77]
Liver	Hepatocytes	[51, 77]
Oviduct	Epithelial cells	[18, 28]
Pancreas, endocrine	Beta cells	[34, 74, 77]

Uptake

Biochemical and autoradiographic studies using radiolabeled substrates have revealed that several non-neuronal cell types possess the capacity to accumulate GABA from the external media, at least in vitro (Table 4). Furthermore, evidence has been presented to show that in several cell types multiple GABA uptake mechanisms may operate: one with high affinity and moderate capacity and one with low affinity but high capacity (e.g., in adrenal chromaffin cells [61]; see also [72]). The substrate specificity, pharmacological sensitivity, and ionic requirements of non-neuronal uptake systems for GABA have not been extensively studied, except for a few cell types (e.g., hepatocytes, renal tubular cells, adrenal chromaffin cells, oviductal mucosal gland cells). In most of these cell types, the features of uptake not only differed from those of neuronal GABA uptake, but also failed to resemble the characteristics of glial GABA uptake [19, 37, 54].

Table 4. GABA uptake by non-neuronal cells

Organ	Cell type	Reference
Adrenals	chromaffin cells	[61]
Blood	Platelets	[60]
Gut	Endocrine cells	[39, 50]
Kidney	Tubular epithelial cells	[42]
Liver	Hepatocytes	[54-56]
Oviduct	Epithelial cells	[19]
Pancreas, endocrine	Somatostatin cells	[38, 40, 58, 63]
Pineal gland	Pinealocytes	[66]
Semen	Spermatozoon	SLE
Stomach, antrum	Endocrine cells	[41, 46]
Thyroid gland	Entodermal follicle cells	[37]

SLE, S.L. Erdö, unpublished data

Release

The stimulus-evoked release from non-neuronal cell types is a condition of crucial importance for the putative function of GABA as a mediator between various cell types. Therefore, the efflux of GABA from peripheral tissues that was evoked by depolarizing stimuli (high K^+ concentration, ouabain, etc.) was extensively studied in isolated tissue preparations in vitro (reviewed in [72]). Although the cellular origin of GABA could not be exactly determined in most of these studies, the biochemical and/or pharmacological characteristics of the release processes suggested that GABA was released from nerve terminals in the majority of organs. Non-neuronal GABA-releasing cells could be identified in very few cases (Table 5).

Table 5. GABA release from non-neuronal cells

Organ/tissue	Cell type	Reference
Adrenals	Chromaffin cells	[47]
Kidney	Tubular epithelial cells	[32]
Oviduct	Epithelial cells	[27]
Pineal gland	Pinealocytes	[66]
Stomach, antrum	Endocrine cells	[24]

Receptors

Both GABA-gated chloride channels ($GABA_A$ receptors) and bicuculline-insensitive, baclofen-sensitive $GABA_B$ receptors have been demonstrated in a number of peripheral organs. The $GABA_A$ and $GABA_B$ receptors in non-neuronal tissues were found to possess biochemical and pharmacological features similar to those of the respective GABA receptor types in the CNS and PNS [7, 17, 21, 43, 59] (see also N.G.Bowery, this volume). The pharmacological responses mediated by non-neuronal GABA receptors and their possible functional relevance are discussed in other chapters of this book.

As demonstrated on the basis of pharmacological effects, such as release, contractile, and electrophysiological responses [43, 59], or as revealed by in vitro ligand binding techniques [1, 2, 57], GABA receptors may be located on a number of mammalian cell types other than neurons (Table 6). It should be noted, however, that the lines of evidence suggesting the presence of non-neuronal target cells in certain organs (e.g., gut, oviduct, blood vessels) are mainly indirect. To obtain more direct information on the localization of peripheral GABA receptors, immunocytochemical techniques using specific antisera against GABA receptors and/or their subunits (e.g., in pancreas [67]) appear to be a promising new tool for future studies.

Table 6. GABA receptors on non-neuronal cells

Organ	Cell type	Reference
Adrenals	Chromaffin cells	[10, 11, 15, 47-49]
Blood vessels	Smooth muscle cells	[51, 57, 69]
	Endothelial cells	[51]
Gut (ileum)	Smooth muscle cells	[62]
Liver	Hepatocytes	[54, 56]
Ovary	Oocytes, granulosa cells	[1, 14]
Oviduct	Smooth muscle cells	[18, 26]
Pancreas		
Endocrine	Non-beta cells	[58, 63, 67]
Exocrine	Epithelial cells	[63]
Pituitary		
Anterior	Lactotrophs	[44]
Intermediate lobes	Melanothrops	[13]
Pineal gland	Pinealocytes	[16, 66]
Semen	Spermatozoon	[23]
Stomach	Epithelial cells	[29]
Urinary bladder	Smooth muscle cells	[31]
Uterus	Smooth muscle cells	[2, 53, 64]

Conclusions

From the studies summarized above, it is apparent that GABA is present in a number of cell types other than neurons. The GABA within such cells is stored, at least in part, in releasable compartments. The depolarization-induced efflux into the extracellular space provides the possibility for GABA to evoke functional responses in various target cells bearing GABA receptors. The target cells may be identical to the GABA-releasing cells (e.g., adrenal chromaffin cells) or in the proximity of the GABA-releasing cells (e.g., alpha and beta cells in pancreatic islets). GABA may also reach more distant target cells via tissue fluids. Thus, as a hormone, GABA may be involved in endocrine functions of organs such as the stomach and the duodenum. Moreover, non-neuronal GABA, as a humoral factor, may participate in exocrine regulation. For instance, such a role was recently hypothesized for the GABA of renal tubular origin which, transported with urine, might affect the contractility of the urinary bladder [32].

Furthermore, GABA of non-neuronal origin may interact with neuronal GABA receptors, and the GABA released from peripheral nerve terminals may exert direct effects on non-neuronal cells, suggesting the possibility of a fairly complex modulatory role for GABA.

The above-mentioned putative roles of GABA are proposals based on pharmacological responses evoked in isolated organs and tissue preparations or, in certain cases, on the basis of receptor binding experiments with specific radioligands [17, 25, 43, 59]. Consequently, the physiological relevance of these mechanisms is still a matter of some debate. Nevertheless, the available information is compatible with a multiple role for GABA in both neuronal and non-neuronal tissues of mammals. Moreover, irrespective of their physiological significance, they represent new targets for drug effects. Thus, the research into non-neuronal GABA systems deserves to be continued.

References

1. Amenta F, Erdő SL, Mione MC, Napoleone P (1986) 3H-muscimol binding sites within guinea pig ovary: a histoautoradiographic study. Pharmacology 32: 202-207
2. Amenta F, Cavallotti C, Ferrante F, Erdő SL (1988) Autoradiographic visualization of the $GABA_A$ receptor agonist 3H-muscimol in the rat uterus. Pharmacol Res Commun 20: 863-868
3. Andersson AC, Henningson S (1980) In vitro metabolism of putrescine by diamine oxidase in tissues of the pregnant rat. Agents Actions 10: 104-106
4. Andersson AC, Henningson S, Järhult J (1980) Diamine oxidase activity and gamma-aminobutyric acid formation in medullary carcinoma of the thyroid. Agents Actions 10: 299-301

5. Armijo (1989) 4-Aminobutyrate-2-ketoglutarate aminotrasferase (GABA-T) in human hair follicle. Rev Esp Fisiol 45: 337-342
6. Awapura J, Launda AJ, Fuerst R, Seale B (1950) Free gamma-aminobutyric acid in brain. J Biol Chem 187: 35-39
7. Bowery NG (1989) $GABA_B$ receptors and their significance in mammalian pharmacology. Trends Pharmacol Sci 10: 401-406
8. Canellakis ZN, Milstone LM, Marsh LL, Young RP, Bondy PK (1983) GABA from putrescine is bound in macromolecular form in keratinocytes. Life Sci 33: 599-603
9. Caron PC, Kremzner LT, Cote LJ (1987) GABA and its relationship to putrescine metabolism in the rat brain and pancreas. Neurochem Int 10: 219-229
10. Castro E, Oset-Gasque MJ, Canadas S, Gimenez G, Gonzalez MP (1988) $GABA_A$ and $GABA_B$ sites in bovine adrenal medulla membranes. J Neurosci Res 20: 241-245
11. Castro E, Oset-Gasque MJ, Gonzalez MP (1989) $GABA_A$ and $GABA_B$ receptors are functionally active in the regulation of catecholamine secretion by bovine chromaffin cells. J Neurosci Res 23: 290-296
12. Davanger S, Ottersen OP, Storm-Mathisen JP (1989) GABA-immuno-reactive cells in the rat gastrointestinal epithelium. Anat Embryol 179: 221-226
13. Demeneix BA, Felz P, Loeffler JP (1986) GABAergic mechanisms and their functional relevance in the pituitary. In: Erdö SL, Bowery NG (eds) GABAergic mechanisms in the mammalian periphery. Raven, New York, pp 261-289
14. Dolci S, Eusebi F, Siracusa G (1985) Gamma-aminobutyric acid sensitivity of mouse and human oocytes. Dev Biol 109: 242-246
15. Doroshenko PA (1989) Gamma-aminobutyric acid elevates cytosolic Ca in bovine chromaffin cells. Neurosci Lett 10: 83-87
16. Ebadi M, Hama Y, Govitrapong P, Awad A (1986) Biochemical and functional characterization of the GABA-benzodiazepine receptor complex in mammalian pineal glands. In: Erdö SL, Bowery NG (eds) GABAergic mechanisms in the mammalian periphery. Raven, New York, pp 291-303
17. Erdö SL (1985) Peripheral GABAergic mechanisms. Trend Pharmacol Sci 6: 205-208
18. Erdö SL (1986) GABAergic mechanisms and their possible role in oviduct and uterus. In: Erdö SL, Bowery NG (eds) GABAergic mechanisms in the mammalian periphery. Raven, New York, pp 291-303
19. Erdö SL, Amenta F (1986) Characterization and localization of high-affinity GABA uptake in the rabbit oviduct. Eur J Pharmacol 130: 287-294
20. Erdö SL, Amenta F (1986) Specific vascular localisation of GABA-transaminase in the guinea pig lung. Neurosci Lett 68: 202-206
21. Erdö SL, Bowery NG (eds) (1986) GABAergic mechanisms in the mammalian periphery. Raven, New York
22. Erdö SL, Kiss B (1986) Presence of GABA, glutamate decarboxylase and GABA-transaminase in peripheral organs: a collection of quantitative data. In: Erdö SL, Bowery NG (eds) GABAergic mechanisms in the mammalian periphery. Raven, New York, pp 17-34
23. Erdö SL, Wekerle L (1990) $GABA_A$ type binding sites on membranes of spermatozoa. Life Sci 47: 1147-1151
24. Erdö SL, Wolff JR (1988) Releasable non-neuronal GABA pool in rat stomach. Eur J Pharmacol 156: 165-168
25. Erdö SL, Wolff JR (1990) Gamma-aminobutyric acid outside the mammalian brain. J Neurochem 54: 363-372

26. Erdö SL, Riesz M, Karpati E, Szporny L (1984) GABA-$_B$ receptor- mediated stimulation of the contractility of isolated rabbit oviduct. Eur J Pharmacol 99: 333-336
27. Erdö SL, Kiss B, Riesz M, Szporny L (1986) Stimulus-evoked release of GABA from preloaded slices of the rabbit oviduct. Eur J Pharmacol 130: 295-303
28. Erdö SL, Somogyi J, Hámori J, Amenta F (1986) Light and electron microscopic localization of gamma-aminobutyric acid and GABA-transaminase within the rat oviduct: predominant occurrence in epithelum. Cell Tissue Res 244: 621-626
29. Erdö SL, De Vincentis G, Amenta F (1989) Autoradiographic localization of ^{3}H-muscimol binding sites in rat stomach: evidence for mucosal GABA$_A$ receptors. Eur J Pharmacol 175: 351-354
30. Erdö SL, Jóo F, Wolff JR (1989) Immunocytochemical localization of glutamate decarboxylase in the rat oviduct and ovary: further evidence for non-neural GABA systems. Cell Tissue Res 255: 431-434
31. Erdö SL, Mione MC, Amenta F, Wolff JR (1989) Binding of ^{3}H-muscimol to GABA$_A$ sites in the guinea-pig urinary bladder: biochemical assay and autoradiography. Br J Pharmacol 96: 313-318
32. Erdö SL, Dobó E, Párducz Á, Wolff JR (1991) Releasable GABA in tubular epithelium of rat kidney. Experientia 47: 227-229
33. Fogel WA (1986) GABA and polyamine metabolism in peripheral tissues. In: Erdö SL, Bowery NG (eds) GABAergic mechanisms in the mammalian periphery. Raven, New York, pp 35-56
34. Garry DJ, Coulter HD, McIntee TJ, Wu JY, Sorenson RL (1987) Immunoreactive GABA transaminase within the pancreatic islet is localized in mitochondria of the B-cell. J Histochem Cytochem 35: 831-836
35. Garry DJ, Sorensen RL, Coulter HD (1987) Ultrastructural localization of gamma-aminobutyric acid immunoreactivity in B cells of the rat pancreas. Diabetologia 30: 115-119
36. Garry DJ, Garry MG, Sorensen RL (1988) Ultrastructural immunocytochemical localization of L-glutamate decarboxylase and GABA in rat pancreatic zymogen granules. Cell Tissue Res 252: 191-197
37. Gebauer H, Pabst MA (1981) Autoradiographic localization of ^{3}H-GABA uptake in the thyroid gland of the rat. Cell Tissue Res 220: 873-879
38. Gilon P, Remacle C, Janssens de Varebeke Ph, Rauwels G, Hoet JJ (1987) GABA content and localisation of high-affinity GABA uptake during the development of the rat pancreas. Cell Mol Biol 33: 573-585
39. Gilon P, Reusens-Billen B, Remacle C, Janssens de Varebeke Ph, Pauwels G, Hoet JJ (1987) Localization of high-affinity GABA uptake and GABA content in the rat duodenum during development. Cell Tissue Res 249: 593-600
40. Gilon P, Remacle C (1989) High-affinity GABA uptake in a subpopulation of somatostatin cells in rat pancreas. J Histochem Cytochem 37: 1133-1139
41. Gilon P, Mallefet J, DeVriendt C, Pauwels S, Geffard M, Campistron G, Remacle C (1990) Immunocytochemical and autoradiographic studies of the endocrine cells interacting with GABA in the rat stomach. Histochemistry 93: 645-654
42. Goodyer PR, Rosen R, Scriver CR (1985) A gamma-aminobutyric acid-specific transport mechanism in mammalian kidney. Biochim Biophys Acta 818: 45-54
43. Hill DR, Bowery NG (1986) GABA-receptor-mediated functional responses in peripheral tissues. In: Erdö SL, Bowery NG (eds) GABAergic mechanisms in the mammalian periphery. Raven, New York, pp 87-97

44. Inenaga K, Mason WT (1987) Gamma-aminobutyric acid modulates chloride channel activity in cultured primary bovine lactotrophs. Neuroscience 23: 649-660
45. Jessen KR, Mirsky R, Dennison ME, Burnstock G (1979) GABA may be a neurotransmitter in the vertebrate peripheral nervous system. Nature 281: 71-74
46. Jessen KR, Hills JM, Limbrick AR (1988) GABA immunoreactivity and ^{3}H-GABA uptake in mucosal epithelial cells of the rat stomach. Gut 29: 1549-1556
47. Kataoka Y, Gutman Y, Guidotti A, Panula P, Wroblewski J, Cosenza-Murphy D, Wu JY, Costa E (1984) Intrinsic GABAergic system of adrenal chromaffin cells. Proc Natl Acad Sci USA 81: 3218-3222
48. Kitayama S, Tsujimoto A (1986) Involvement of GABAergic mechanisms in the catecholamine secretion from adrenal medulla. In: Erdö SL, Bowery NG (eds) GABAergic mechanisms in the mammalian periphery. Raven, New York, pp 249-259
49. Kitayama S, Morita K, Dohi T, Tsujimoto A (1990) Enhancement by GABA of the stimulation-evoked catecholamine release from cultured bovine adrenal chromaffin cells. Naunyn Schmiedebergs Arch Pharmacol 341: 414-418
50. Krantis A, Webb T (1989) Autoradiographic localization of ^{3}H-gamma-aminobutyric acid in neuronal elements of the rat gastric antrum and intestine J Auton Nerv Syst 29: 41-48
51. Krause DN (1986) Involvement of local GABA mechanisms in vascular regulation. In: Erdö SL, Bowery NG (eds) GABAergic mechanisms in the mammalian periphery. Raven, New York, pp 193-203
52. Krnjevic K (1974) Chemical nature of synaptic transmission in vertebrates. Physiol Rev 54: 418-540
53. Majewska MD, Falkay G, Baulieu EE (1989) Modulation of uterine $GABA_A$ receptors during gestation and by tetrahydroprogesterone. Eur J Pharmacol 174: 43-47
54. Minuk GY (1986) GABAergic mechanisms ant their functional importance in the liver. Erdö SL, Bowery NG (eds) GABAergic mechanisms in the mammalian periphery. Raven, New York, pp 325-337
55. Minuk GY, Vergalla J, Ferenci P, Jones EA (1984) Identification of an acceptor system for gamma-aminobutyric acid on isolated rat hepatocytes. Hepatology 4: 180-185
56. Minuk GY, Bear CE, Sarjeant EJ (1987) Sodium-independent, bicuculline-sensitive ^{3}H-GABA binding to isolated rat hepatocytes. Am J Physiol 252: G642-G647
57. Napoleone P, Erdö SL, Amenta F (1987) Autoradiographic localization of the $GABA_A$ receptor agonist ^{3}H-muscimol in rat cerebral vessels. Brain Res 423:109-115
58. Okada Y (1986) Localization and function of GABA in the pancreatic islets. In: Erdö SL, Bowery NG (eds) GABAergic mechanisms in the mammalian periphery. Raven, New York, pp 223-240
59. Ong J, Kerr DIB (1990) GABA receptors in peripheral tissues. Life Sci 46: 1489-1501
60. Oset-Gasque MJ, Launay JM, Gonzalez MP (1986) GABAergic mechanisms in blood cells: their possible role. In: Erdö SL, Bowery NG (eds) GABAergic mechanisms in the mammalian periphery. Raven, New York, pp 305-324
61. Oset-Gasque MJ, Aunis D (1989) Gamma-aminobutyric acid uptake and localization in bovine chromaffin cells in primary culture. Biochem Pharmacol 38: 2227-2232
62. Pencheva N, Venkova K, Radomirov R (1990) $GABA_B$ receptor-mediated contractile effects resistant to tetrodotoxin in isolated cat ileum. Eur J Pharmacol 182: 199-202
63. Reusens-Billen B, Pirlot X, Remacle C, Hoet JJ, de Gasparo M (1984) Localization of GABA high-affinity binding sites in the pancreas of neonatal rat. Cell Tissue Res 235: 503-508

64. Riesz M, Erdö SL (1985) GABA-$_B$ receptors in the rabbit uterus may mediate contractile responses. Eur J Pharmacol 119: 199-204
65. Roberts E, Farnkel S (1950) Gamma-aminobutyric acid in brain: its formation from glutamic acid. J Biol Chem 187: 55-63
66. Rosenstein RE, Sanjurjo C, Cardinali DO (1989) Gamma aminobutyric acid uptake, release and effect on $^{36}Cl^-$ influx in bovine pineal gland. J Neural Transm 77: 141-152
67. Rorsman P, Berggren PO, Bokvist K, Ericson H, Mohler H, Östenson CG, Smith PA (1989) Glucose inhibition of glucagon secretion involves activation of GABA$_A$ receptor chloride channels. Nature 341: 233-236
68. Sakaue M, Saito N, Tanaka C (1987) Immunohistochemical localization of gamma-aminobutyric acid (GABA) in the rat pancreas. Histochemistry 86: 365-369
69. Shirakawa J, Hosoda K, Taniyama K, Matsumoto S, Tanaka C (1989) Gamma-aminobutyric acid-A receptor-mediated supression of 5-hydroxytryptamine-induced guinea-pig basilar artery smooth muscle contractility. Blood Vessels 26: 129-136
70. Tallan HH, Moore S, Stein WH (1954) Studies on the free amino acids and related compounds in the tissues of the cat. J Biol Chem 211: 927-939
71. Taniguchi H, Okada Y, Seguchi H, Shimida C, Seki M, Tsutou A, Baba S (1979) High concentration of gamma-aminobutyric acid in pancreatic beta cells. Diabetes 28: 629-633
72. Tanaka Ch, Taniyama K (1986) GABA transport in peripheral tissues: uptake and efflux. In: Erdö SL, Bowery NG (eds) GABAergic mechanisms in the mammalian periphery. Raven, New York, pp 57-75
73. Udenfriend S (1950) Identification of gamma-aminobutyric acid in human brain by the isotope derivative method. J Biol Chem 187: 65-69
74. Vincent SR, Hökfelt T, Wu JY, Elde RP, Morgan LM, Kimmel JR (1983) Immunohistochemical studies of the GABA system in the pancreas. Neuroendocrinology 36: 197-204
75. Wu JY, Chude O, Wein J, Roberts E (1978) Glutamate decarboxylase from neural and non-neural tissues. J Neurochem 30: 849-857
76. Wu JY, Moss LG, Chude O (1978) Distribution and tissue specificity of GABA-2-ketoglutarate transaminase. A new assay method. Neurochem Res 3: 207-219
77. Wu JY, Lin CT, Lin H, Xu Y, Liu JW, Hwang BH, Wei SC (1986) Immunochemical characterization and immunohistochemical localization of glutamate decarboxylase and GABA transaminase in peripheral tissues. In: Erdö SL, Bowery NG (eds) GABAergic mechanisms in the mammalian periphery. Raven, New York, pp 19-34
78. Zachmann M, Tocci P, Nyhan WL (1966) The occurrence of gamma-aminobutyric acid in human tissues other than brain. J Biol Chem 241: 1355-1358

GABA and Endocrine Cells of the Gastro-entero-pancreatic System: A Review

P. Gilon and C. Remacle

Introduction

γ-Aminobutyric acid (GABA) is not only a major inhibitory neurotransmitter in the central nervous system of vertebrates, but it is also a peripheral neuromodulator, as in the enteric nervous system [49]. A few years ago, the presence of GABA was thought to be restricted to nerve cells only. However, recent studies have revealed its presence in various non-neural organs such as the ovary, adenohypophysis, adrenal medulla, liver, digestive tract, and pancreas. In these organs, several categories of (neuro)endocrine cells display both morphological and physiological indications of direct interaction with GABA, i.e., prolactin-secreting cells [4, 5, 60, 69] and adrenal chromaffin cells [11, 16, 51, 52]. GABA has also been involved in the regulation of hormonal secretion in the gastro-entero-pancreatic (GEP) system, but its precise localization and mechanisms of action remain unclear.

This chapter is a review of the sites of synthesis, storage, and uptake of GABA in the GEP system. Besides its role as neuromodulator, special attention is paid to the interactions between GABA and endocrine cells and to possible related functional roles.

Pancreas

GABA Content

The GABA concentrations expressed in nmol/g of wet tissue are listed in Table 1. The GABA content is very low in the whole pancreas in comparison with the central nervous system, which contains 1000 - 10 000 nmol GABA/g wet tissue. Briel et al. [10] and Panten et al. [79] were the first authors to investigate the GABA content of the endocrine pancreas of obese hyperglycemic mice (*ob/ob*), but the content they recorded was relatively low. Later, Okada et al. [77] and Taniguchi et al. [99, 100] reported values in rat islets similar to those of the brain.

Localization of GABA, GAD, and GABA-T

Endocrine Pancreas. The precise localization of GABA within the endocrine pancreas has been studied by biochemical and immunocytochemical techniques.

Biochemical Approaches. Okada et al. [77] suggested that B cells (insulin secreting cells) constitute the main source of GABA since an insulinoma region of human pancreas contained 10 times more GABA and 70 times more GAD (glutamate decarboxylase, enzyme of GABA synthesis) activity than the non-tumor region. Conversely, treatment of rats with streptozotocin, which specifically destroys B cells, led to a fall of 90% of the insular GABA concentration. These results were confirmed by other authors using the same drug [31, 32, 78, 99, 100, 101] or alloxan, another specific B cell cytotoxic agent [25].

Recently, we measured pancreatic GABA levels at different developmental stages of the rat by an ion-exchange column chromatography coupled to HPLC analysis [33]. Modifications of GABA concentrations during development followed an evolution similar to that of the volume density of B cells in the whole pancreas [61]. This correlation also suggests that B cells are the main source of GABA in the pancreas.

The presence of a high GAD activity inside microdissected rat islets suggests that GABA is at least partly synthesized from glutamate [99]. The rat islet glutamate concentration averages 7040 nmol/g wet tissue, a four times higher concentration than that of GABA [76]. The GAD of the islets is of neuronal type (GAD I) since GAD of a human insulinoma requires the presence of pyridoxal-5'-phosphate as a cofactor and its activity is inhi-

Table 1. Pancreatic GABA levels (nmol/g wet tissue) reported in the literature

Type of tissue	Species	GABA levels	References
Whole pancreas	Cat	68	[97]
	Human	281[a]	[77]
	Rabbit	3 (body)	[32]
		14 (tail)	[32]
	Mouse	4	[17]
	Rat	40	[66]
		41.2	[31]
		87 (tail)	[99]
		94	[33]
		125	[76]
		153 (tail)	[101]
		251[a]	[77]
Exocrine pancreas	Rat	197[a]	[101]
		204[a]	[100]
Islets of Langerhans	Human	830[a]	[78]
	Rat	191	[31,32]
		1460[a]	[100]
		1550-1820[a]	[76]
		1630[a]	[78]
		1894[a]	[101]
	Mouse	300[a]	[78]
	ob/ob mouse	140-210[a]	[10,40]
		551[a]	[79]
Brockmann body	Catfish *(Ictalurus nebulosis)*	121	[31,32]
	Seriora quinqueradiata	670	[78]

[a] Value based on the assumption that 1 g wet tissue ~ 100 mg dry tissue ~ 100 mg protein

bited by increasing Cl^- concentration [75]. Moreover, the enzyme is recognized by specific antibodies against brain GAD which do not recognize non-neuronal GAD [113].

Immunocytochemical Approaches. GABA immunodetections reveal that insulin cells, located in the center of the islets of Langerhans, are the only cells containing GABA [26, 87]. Pretreatment of rats with streptozotocin leads to a marked decrease in the number of immunoreactive cells [87]. By using immunogold staining, it was found that GABA is highly concentrated in the nucleus, cytosol, and mitochondria of B cells. However, the lack of significant GABA immunoreactivity in the insulin secretory granules suggests that, if GABA is released from B cells, it is due to a nongranular secretory process [27, 35].

By means of antibodies directed against purified GAD of rat brain [73], we observed an intense cytoplasmic labeling of insulin cells located in the central core of mouse and rat pancreatic islets (unpublished results). Double immunodetections showed that glucagon-, somatostatin-, and pancreatic polypeptide-immunoreactive cells are not immunopositive for GAD. These results agree with another study using anti-mouse GAD and anti-insular peptide antibodies on serial sections [111] and with a triple labeling experiment, i.e., anti-rat GAD, anti-γ-MSH (which cross-reacts with pancreatic polypeptide) and anti-somatostatin [96].

Finally, the use of antibodies directed against purified GABA-T (GABA transaminase, enzyme of GABA degradation into succinic semialdehyde) of mouse brain reveals an intense labeling of B cells which is exclusively located in the mitochondrial compartment [28, 111].

Exocrine Pancreas and Innervation. Our immunodetections of GABA and GAD revealed no labeling of the exocrine pancreas and pancreatic innervation even in colchicine pretreated rats (unpublished results). Contrary to the gastroenteric system, the pancreas is thus devoid of GABAergic innervation. These results agree with those of Vincent et al. [111] and Okada [76], but disagree with the weak GAD and GABA immunoreactivities of zymogen granules of the rat pancreas reported by Garry et al. [29].

GABA Uptake

Two transport systems have been described in the central nervous system: a high-affinity but low capacity system which is dependent on temperature, energy, and Na^+ ($K_m < 5.10^{-5}$ *M*, low V_{max}) and a low-affinity but high capacity system ($K_m > 10^{-4}$ *M*, higher V_{max}) which is not very specific for GABA [65, 98].

High-Affinity Uptake. The high-affinity uptake of [^{3}H]GABA (0.5 μM) has been studied by autoradiography at different development stages of the rat (from 15.5 days of fetal life up to 105 days post natum) [33]. At early stages, the uptake was localized in intralobular ducts, acini, centroacinar cells, immature endocrine cells, and mature somatostatin cells. In the innervation, intense uptake was restricted to Schwann cells; even if scarce perikarya and nerve fibers were seen labeled, they did not seem to reflect the presence of GABAergic innervation since neither GABA nor GAD immunoreactivities were detected in pancreatic innervation. At adult stages, the Schwann cells and somatostatin cells are the only cells which take up GABA [33, 36]. These results agree with analogous studies performed on neonatal [83] and adult rats [102, 109]. The case of the B cell, which contains and synthesizes GABA without possessing high-affinity carriers for this amino acid, is unusual but not without precedent. The neurohypophysis possesses fibers synthesizing GABA [74, 110] but lacking a high-affinity uptake system [82]. Contrary to the nervous system [114], the autoradiographic study of the high-affinity uptake of GABA in the pancreas does not allow one to presume the GABAergic nature of the cells, i.e., that they synthesize and release GABA to act on neighbouring cells. Nevertheless, it indicates probable interaction with GABA.

In the presence of ß-alanine (1 m*M*), which inhibits the glial high-affinity uptake of GABA [65, 90, 98] or the low-affinity uptake of GABA [30], [^{3}H]GABA (0.5 μM) is only taken up by immature endocrine cells and by a subpopulation of mature D cells (somatostatin cells). The percentage of D cells which take up GABA is higher at early postnatal stages than at the adult stage. In order to better define the nature of the transport system of GABA in somatostatin cells, we added different pharmacological substances (at final concentration of 1 m*M*) to incubation medium containing [^{3}H]GABA (0.5 μM). The persisting autoradiographic labeling of D cells in the presence of ß-alanine suggests that the carriers are of a nonglial nature. They even seem to be neuronal since labeling is strongly diminished in the presence of DABA (L-diaminobutyric acid) and completely disappears in the presence of nipecotic acid. The difference between these two neuronal uptake inhibitors could result from the fact that nipecotic acid is a more powerful inhibitor than DABA [57], probably owing to its rigid molecular structure specific for the neuronal carrier [2]. A similar effect of nipecotic acid has been observed on the [^{3}H]GABA uptake by rat adenohypophysical cells [5]. Finally, the lack of inhibition of [^{3}H]GABA uptake by muscimol, baclofen, and bicuculline methiodide suggests that, as in the central nervous system, these substances do not basically interfere with the high-affinity transport of GABA in D cells.

As far as GABA degradation in somatostatin cells is concerned, it should follow an uncommon metabolic pathway as GABA-T is absent from the D cell [28, 111]. Even if GABA is taken up in vivo by a high-affinity system, since its plasmatic circulating concentration averages 0.6-0.9 μM [18], it does not seem to be stored in the somatostatin cells as seen by GABA immunodetections [26, 35, 87]. The role of this uptake remains unclear. It could not serve to prevent any variation of blood GABA level from affecting A (glucagon) or B cell function, if the blood stream follows the direction B cells → A cells → D cells [88, 95]. However, it is not excluded that it could prevent or alleviate any paracrine effect of GABA released by B cells on cells close to D cells or on the D cells themselves. Finally, it is possible that GABA transport in the somatostatin cells is regulated by insulin. Insulin enhances synaptosomal uptake of GABA [84], whereas it inhibits astroglial uptake [8].

Low-Affinity Uptake. A low-affinity uptake in the pancreas has been suggested by Gerber and Hare [32] since rabbit pancreatic pieces incubated for 30 min in the presence of 10^{-3} *M* GABA showed a significant rise in their GABA content, whereas those in the presence of 10^{-5} or 10^{-7} *M* did not. Malaisse et al. [64] have also reported that rat islets can oxidize exogenous GABA but the concentration used was not mentioned.

If a low-affinity uptake exists, its physiological importance seems to be minor owing to the relatively low circulating GABA levels and the relatively high blood concentration of taurine (about 200 μM) [62], a powerful inhibitor of low-affinity uptake of GABA [30, 98].

GABA Receptors

By using the binding techniques of radioligands ([^{3}H]muscimol, 5-20 n*M*; [^{3}H]GABA, 10 n*M*), we have investigated the presence of $GABA_A$ and $GABA_B$ receptors in homogenates of rat brain, pancreas (7.5 days post natum and adult) and neonatal cultured islets. Contrary to the brain, no specific binding could be detected in the pancreas and in the islets. Similarly the autoradiographic techniques with apposition of Ultrofilm [70] or emulsion-coated coverslips [115] did not yield any specific labeling of [^{3}H]muscimol (25 n*M*) in pancreatic tissue. These negative results agree with those of Vincent and Brown [109] who used [^{3}H]GABA (50 n*M*) as the radioligand with the Ultrofilm technique.

Recently, Rorsman et al. [86] performed immunodetections with a mouse monoclonal antibody (bd-17) specific for the ß-subunit of the $GABA_A$/benzodiazepine receptor complex [89]. They detected the presence of $GABA_A$ receptors on certain somatostatin and glucagon cells obtained in culture from guinea-pig islets, but not on insulin cells. Owing to the difficulty in realizing specificity controls with such antibodies, in situ hybridization experiments with probes specific for the mRNAs of the α-, ß-, or γ-sub units of the $GABA_A$ receptors [72, 112] and immunodetections with anti-idiotypic antibodies for the GABA receptors would be useful.

Function of GABA in the Pancreatic Islets

It has been suggested that GABA could play several roles in the endocrine pancreas:

1. Regulator of protein biosynthesis [76, 101, 104] or of cellular replication [20, 23]
2. Alternative energy source for the B cell by entering the Krebs'cycle in the form of succinate after degradation by mitochondrial GABA-T [28, 111]
3. Endocrine, paracrine, or autocrine modulator after release by the insulin cell to regulate B cell activity or to act on a neighboring cell [76, 85, 86]

GABA as a Regulator of Protein Biosynthesis or Cellular Replication. Two types of "trophic" effects of GABA have been described from in vitro experiments performed on brain tissue: an effect of intracellular GABA on global protein synthesis, maybe mediated by an action on aminoacyl-tRNA synthetase or on tRNA [6, 38, 94], and an effect of extracellular GABA on the synthesis of certain proteins, mediated by an action on GABA receptors [63, 67]. Taniguchi et al. [101, 103, 104] have tested the effect of intracellular GABA on the protein synthesis of isolated pancreatic islets of adult rats. Incubation of islets for 2 h or 3 days in the presence of 10^{-4} *M* γ-vinyl GABA, a specific GABA-T inhibitor, raised the intraislet GABA level to 177% and 206%, respectively, but did not affect the insulin release or content of islets cultivated for 3 days in the presence of the drug [104]. A 4 h incubation in the presence of γ-vinyl GABA (10^{-4} *M*) enhanced (by 69%) the incorporation of [^{3}H]leucine into the insulin-immunoreactive fraction without affecting the total islet protein synthesis. By using gabaculine, another specific GABA-T inhibitor [81, 107] which, at 10^{-4} *M*, raises the intracellular GABA content of cultured neonatal islets to 180%, we also observed that this drug did not affect the insulin release or content of these islets after a 4 h incubation. However, we could not detect any significant effect of gabaculine on the incorporation of [^{3}H]leucine into the insulin immunotrapped by an anti-insulin immunoaffinity gel and into the proteins precipitated with TCA. Our results do not support the stimulatory effect of GABA on insulin biosynthesis as was suggested by Taniguchi et al. [104].

The effect of extracellular GABA on the somatostatin content was also examined [103, 105]. A 3 day incubation of rat islets in the presence of GABA (0.1 and 1 m*M*) produced an increase (30% and 38%, respectively) of their somatostatin content. This seems to result from an activation of $GABA_A$ receptors rather than from an intracellular effect of GABA taken up by the D cells, as a similar elevation was reproduced by muscimol (1 μM) and the elevation induced by 0.1 m*M* GABA was antagonized by 1 μM bicuculline.

Recently, we investigated the effect of extra- or intracellular GABA on the capacity of proliferation of cultivated rat neonatal islets by means of autoradiography after [^{3}H]thymidine incorporation. A 2 day incubation of the islets in the presence of GABA (10^{-7}-5.10^{-3} *M*) or gabaculine (10^{-7}-10^{-3} *M*) did not affect their labeling index in highly (RPMI 1640 containing 11.1 m*M* glucose) or weakly (MEM containing 2.8 m*M* glucose) proliferative media.

GABA as an Energy Source. Malaisse et al. [64] have observed that, when adult rat islets are incubated for 120 min in the presence of [^{14}C]glutamine (10 m*M*), the specific radioactivity of GABA in the islets represents ± 52% of that of glutamine. It is, however, lower than that of glutamate which averages 88%. Vincent et al. [111] have suggested that the GABA shunt (a deviation of a portion of the Krebs'cycle, namely the oxidation of α-ketoglutarate to succinate through GABA metabolism) could play a metabolic role in the B cell and that GABA formed from glutamate would constitute an energy source alternative to glycolysis. In terms of ATP formed, this hypothesis seems unlikely owing to the energy balance generated by the possible pathways of conversion from glutamate to succinate and to the very low intrinsic ability of GABA to supply energy as compared to glucose and other amino acids, which are the main fuels for the B cell. In the B cell, glutamate can be converted to α-ketoglutarate by transamination or oxidative deamination [64]. Transamination does not allow the formation of ATP, whereas glutamate oxidation by glutamate dehydrogenase forms one NAD(P)H generating three ATP. Thereafter, the conversion of α-ketoglutarate into succinate via the Krebs'cycle produces four ATP. The energy supply of these pathways is thus four ATP if glutamate is transaminated and seven ATP if it is oxidized. On the other hand, the transformation of glutamate into succinate produces only three ATP through the GABA shunt and the energy balance of this pathway is unfavorable. Malaisse et al. [64] have suggested that the turnover rate of GABA is very low in the B cell in the presence of glutamine.

GABA can also be synthesized from glucose and leucine. When obese-hyperglycemic mouse (*ob/ob*) islets are incubated for 120 min in the presence of [^{14}C]glucose or [^{14}C]leucine, a fraction of the radioactivity is found in GABA, and its specific radioactivity (from [^{14}C]glucose) is enhanced by a high glucose concentration (20 m*M* vs 5 m*M*) [41, 42]. Surprisingly, however, the specific radioactivity of GABA is higher than that of glutamate, suggesting a compartmentation of glutamate inside the B cell.

Finally, it is possible that GABA can be formed from putrescine [93] but this has never been proved in B cells.

GABA as an Endocrine, Paracrine, or Autocrine Modulator. This hypothesis seems the most attractive one. However, until now, no experiment has irrefutably clarified the mechanisms by which GABA could be released from the B cell.

Gerber and Hare [32] observed that incubation of rabbit pancreatic pieces in a medium containing a high glucose concentration (27.8 m*M*) induces the appearance of GABA in the medium. Isolated rat islets incubated in the presence of glutamine (10 m*M*) release GABA, whereas insulin secretion is not enhanced [64]. The stimulation of islets of obese-hyperglycemic mice (*ob/ob*) by glucose produces a drop in the intraislet GABA content. However, stimulation of these islets by L-leucine, α-ketoisocaproate (deamination product of L-leucine), or glibenclamide (powerful insulinotropic agent) does not significantly affect the GABA content [41, 42, 79]. From these experiments, it can be concluded that the output of GABA from B cells is not strictly coupled to insulin release. This conforms with GABA immunodetections which did not detect any GABA immunoreactivity in the insulin secretory granules, so dismissing the hypothesis of GABA cosecretion with insulin.

Several physiological studies in vivo and in vitro suggest that GABA affects the function of insular cells but they provide controversial results.

In Vivo Studies. In dogs, oral (5 g) or intravenous administration of GABA affects neither insulinemia nor glucagonemia [71]; intravenous infusion of progabide (a mixed $GABA_A$ and $GABA_B$ receptor agonist) diminishes the pancreatic polypeptide response during insulin-induced hypoglycemia [106]. In humans, oral administration of GABA (5 or 10 g) increases plasma levels of insulin, C-peptide, and glucagon, but these effects are not reproduced by muscimol (5 mg) [12]. Also in humans, a high oral dose of baclofen (20 mg) increases the insulin response to glucose and raises basal glucagon level [80]; valproate, a GABA-T inhibitor which raises the plasma

GABA concentration, decreases somatostatinemia but does not affect the plasma C-peptide concentration [58].

In Vitro Studies Performed on Perfused Pancreas or Incubated Pancreatic Pieces. In the perfused dog pancreas, GABA produces a dose-dependent (1-100 μM) inhibition of insulin secretion and a transitory rise (lasting 1 min) of that of somatostatin (occurring at 10-100 μM GABA), but it does not affect glucagon secretion. Surprisingly however, bicuculline also inhibits insulin secretion and does not prevent the effect of GABA [53]. On the other hand, in the perfused rat pancreas, muscimol (1 μM) inhibits (by 38%) glucose-stimulated (16.7 mM) somatostatin release without affecting that of insulin [85]. Finally, in a study performed on rabbit pancreatic pieces, GABA tended to facilitate glucose-stimulated insulin secretion and to decrease glucopenic-stimulated glucagon release, but these effects were not significant [32].

The discrepancies between these data could proceed from interspecific animal differences or from the diversity of the experimental models that are used. These studies never allow one to assert that the observed effects are solely due to a direct action of GABA on insular cells. An action at the level of the central nervous system can occur in in vivo studies, especially with substances which easily cross the blood brain barrier such as baclofen, progabide, and bicuculline. Bereiter et al. [7] reported the existence of a GABAergic control of the endocrine pancreas activity at the level of the nucleus ambiguus (NA) in the central nervous system. The NA is one important source of vagal motoneurons, the stimulation of which increases insulin release. These neurons are under tonic GABAergic inhibition since, in the presence of phentolamine, bilateral bicuculline infusion into the NA region of anesthetized rats led to a prompt increase of insulinemia. A GABAergic action on organs other than the pancreas can also account for effects observed on insular cells. GABA can induce catecholamine release by the adrenal medulla [3, 24], and it is well known that adrenalin inhibits insulin and somatostatin secretion whereas it enhances that of glucagon [1, 108]. Within the endocrine pancreas, an action of GABA on the secretion of a cell type can indirectly affect the secretion of another cell type. Insulin inhibits glucagon secretion; somatostatin inhibits insulin and glucagon secretion; and glucagon stimulates insulin and somatostatin secretion [88, 91]. Finally, it must be kept in mind that bicuculline is an inhibitor of acetylcholinesterase and can exert cholinomimetic effects [9, 68].

In Vitro Studies Performed on Isolated Islets or Cells. In perifusion studies, neither GABA (1-100 μM) nor bicuculline (10-100 μM) affects the insulin release of rat islets. Also, their insulin secretion is not influenced by GABA (0.1-10 m*M*) in short-term (1 h) incubations and by GABA (0.1-10 m*M*) or muscimol (0.1-10 m*M*) in long-term (3 days) incubations [104]. These results are in keeping with our experiments performed on mouse islets (unpublished results), but they are at variance with those of Sehlin and Sandström [92] who reported a potentiating effect of GABA (50 μM) on glucose-induced (10 m*M*) insulin secretion of obese-hyperglycemic mouse (*ob*/*ob*) islets. The effect of GABA on these islets is accompanied by an increase of Cl^- efflux and, transiently, of Ca^{2+} uptake.

In perifusion studies performed on rat islets, GABA (100 μM) induces a 30% suppression of glucose-stimulated (16.7 m*M*) somatostatin release which is antagonized by bicuculline (10 and 100 μM) [105]. In guinea-pig islets, GABA (100 μM) does not affect arginine-induced (10 μM) somatostatin secretion but bicuculline (100 μM) increases (69%) it [86].

As far as glucagon secretion is concerned, Rorsman et al. [86] have clearly shown, by means of patch-clamp techniques, that GABA induces a current of Cl^- ions which hyperpolarizes the A cell of guinea-pig islets and inhibits its spiking activity. As a result, it decreases (by 30%) arginine stimulated (10 m*M*) glucagon secretion of these islets. The GABA recptor involved is of the $GABA_A$ type since it is sensitive to bicuculline, picrotoxin, diazepam, and muscimol. These results agree with our experiments performed on mouse islets (unpublished results). If GABA is released from the B cell, these data predict that any impairment of B cell function, as encountered in diabetes, could enhance glucagon secretion.

Stomach and Intestine

GABA in the Mucosal Epithelium

By means of GABA immunodetections, epithelial cells were seen labeled in the oxyntic and pyloric regions of the stomach and in the duodenum, jejunum, ileum, and colon of the rat [14, 35, 37, 50]. Electron microscopic analysis revealed that they correspond to a subpopulation of non-enterochromaffin (non-EC) endocrine cells [35, 37] but their hormonal nature could not be identified.

The presence of GABA in endocrine cells is in keeping with a previous report of Hakanson et al. [43] describing a GAD immunoreactivity in endocrine cells of the gastric antrum of the rat, rabbit, cat, dog, and monkey. This also agrees with our experiments performed on the rat stomach with an antibody directed against brain GAD (unpublished results).

Autoradiographic studies have also shown an intense high-affinity uptake of GABA in epithelial cells of the oxyntic and pyloric regions of the stomach and of the upper small intestine of the rat [34, 37, 50, 56]. In the duodenum, this uptake is very frequent and intense at the perinatal stages [34]. The carriers involved are of nonglial nature since the addition of ß-alanine (1 m*M*) to the incubation medium does not block the uptake of [^{3}H]GABA (0.5 μM) in these cells [34, 37]. These latter correspond to a subpopulation of non-EC endocrine cells and are most frequent in the oxyntic region of the stomach [37]. In order to identify their hormonal nature, autoradiographic detection of [^{3}H]GABA uptake was coupled to immunodetection of hormonal peptides in the stomach [37] and duodenum (unpublished results). This revealed that the cells which take up GABA belong to a subpopulation of somatostatin cells and other unidentified cells which, in the stomach, are not immunoreactive for gastrin and which, in the duodenum, are not immunoreactive to anti-cholecystokinin 39, anti-gastrin G-34, anti-pancreatic polypeptide, anti-secretin, and anti-GIP antibodies.

Until now it was not known if the cells which take up GABA synthesize it. However, on the basis of morphologic analysis of the secretory granules [37] and coupling of GABA immunodetection with autoradiography [50], it was suggested that three populations of endocrine cells exist: cells which synthesize GABA but do not take it up, cells which do not synthesize it but take it up, and cells which display both abilities.

GABAergic Innervation

The presence of a GABAergic innervation along the gastrointestinal tract is well characterized [47, 48, 49] and will not be discussed. Briefly, let us say that this innervation is at least intrinsic to the gastrointestinal tract since GABAergic perikarya are located in the myenteric plexus. GABAergic fibers are also present in the inner circular muscle wall and the submucosal and mucosal layers.

Function of GABA in the Stomach and Intestine

Besides its modulatory effects in muscular contraction [49, 54] and ulceration [21, 59], GABA can affect the secretion of endocrine cells.

In the stomach, several studies suggest that GABA can modulate the secretion of endocrine cells by an indirect action, but the results depend on the experimental models that are used. Experiments performed on rat antral mucosal fragments reveal that GABA inhibits somatostatin secretion and stimulates gastrin secretion via an action on $GABA_A$ receptors since these effects are antagonized by bicuculline [44]. This action of GABA on G (gastrin) and D cells could be indirect and involve activation of post-ganglionic cholinergic neurons [45]. Atropine and pirenzepine (muscarinic antagonists) prevent the stimulation of gastrin secretion and the inhibition of somatostatin secretion induced by a very low GABA concentration (10^{-8} *M*). Conversely, physostigmine (eserine, an acetylcholinesterase inhibitor) increases the sensitivity of the antral preparation to GABA stimulation tenfold. Hexamethonium (ganglionic nicotinic receptor antagonist) has no effect on GABA-stimulated (10^{-8} *M*) gastrin release. Finally, the fact that tetrodotoxin, which inhibits neuronal depolarization by blocking Na^+-dependent action potentials, prevents the stimulation of GABA-induced (10^{-8} *M*) gastrin secretion suggests that the effect of GABA is secondary to a neuronal activation.

Studies performed on the model of the isolated perfused rat stomach yield controversial results. Koop and Arnold [55] report that GABA (10^{-6}-10^{-4} *M*) and muscimol (10^{-5} *M*) decrease basal somatostatin secretion but have no effect on gastrin secretion. Guo et al. [39] find that low doses of GABA (10^{-9}-10^{-6} *M*) modify neither basal somatostatin and gastrin secretions nor bombesin-stimulated (10 n*M*) gastrin secretion but dose-dependently inhibit bombesin-induced (10 n*M*) somatostatin secretion. Del Tacca et al. [15] reveal a stimulatory effect of baclofen on gastrin secretion which is fully prevented by bicuculline, atropine, and tetrodotoxin. These last results question the nature of the peripheral $GABA_A$ and $GABA_B$ receptors.

In an experiment performed on canine duodenal mucosa culture, GABA (10^{-9}-10^{-6} *M*) dose-dependently increases secretin secretion by an action via $GABA_A$ receptors since this effect is antagonized by bicuculline [46]. Carbachol alone is unable to stimulate secretin secretion. However, the stimulation of GABA-induced (10^{-8} *M*) secretin release is significantly potentiated by carbachol and is prevented by atropine (but not hexamethonium). Harty and Murthy [46] suggest that the S (secretin) cells possess both GABA and acetylcholine receptors which mutually potentiate their actions. However, they propose that an indirect effect could also occur since GABA-induced secretin release is prevented by tetrodotoxin.

If the action of GABA on the endocrine cells is exerted via the innervation, it could take place at least at two different levels: the myenteric plexus, which is the only one containing GABAergic perikarya [13, 47], and the submucosal and mucosal innervations. This latter assumption ensues from the experiments of Harty and Franklin [44, 45]. Since the antral mucosal fragments they employed are devoid of myenteric plexus, GABA would essentially act inside the nerve fiber bundles running through the submucosal layer and the lamina propria by modulating the activity of neighboring cholinergic fibers.

If the action of GABA on cholinergic fibers physiologically occurs, the sensitivity of the receptors involved would be different from that of the receptors in the central nervous system. Most of the experiments described above were performed with very low GABA concentrations (Harty and Franklin [45] and Harty and Murthy [46], 10^{-8} *M*; Guo et al. [39], 10^{-9}-10^{-6} *M*). Although in the central nervous system most of the physiological effects are observed at GABA concentrations between 10^{-6} and 10^{-3} *M* and the plasmatic GABA concentration averages 6-9.10^{-7} *M*, it is quite conceivable that GABA concentrations from 10^{-9} to 10^{-6} *M* can physiologically occur. Glial cells of the gastrointestinal nervous system possess a high-affinity uptake system of GABA. It could contribute to keeping very low GABA concentrations around the nerve fibers allowing any variation of its concentration to exert physiological effects. This is consistent with the existence of a single class of high-affinity GABA receptors (K_D of 23 n*M* and 15.1 n*M* for GABA and muscimol, respectively) in the antral portion of the rat stomach [21].

Despite these results, a direct effect of GABA on endocrine cells cannot be excluded for several reasons: (1) Koop and Arnold [55] used higher GABA concentrations (10^{-6}-10^{-9} *M*) and observed an inhibitory effect of GABA on somatostatin secretion which was partly atropine-resistant. (2) The high-affinity uptake of [^{3}H]GABA, observed in a subpopulation of somatostatin cells and other unidentified endocrine cells, could reflect the existence of a direct interaction between GABA and these cells. (3) Finally, the presence of specific $GABA_A$ receptors on gastric glands (about 10% of the total specific binding of the stomach) also strengthens this hypothesis [21, 22].

As far as the origin of GABA is concerned, it could be released by GABAergic nerve fibers or GABA containing endocrine cells. Recently, Erdö and Wolff [19] revealed the existence of a releasable non-neuronal GABA pool (L-DABA-resistant) in the gastric antrum, but they could not identify the nature of the cells involved in this process.

Summary and Conclusions

The pancreas is devoid of GABAergic innervation. The insulin cells are the only cells which contain, synthesize (by GAD), and degrade (by GABA-T) GABA. This amino acid seems to be absent from their secretory granules and the mechanisms of its release remain unclear. In the islets of Langerhans, the high-affinity uptake of GABA is restricted to a subpopulation of somatostatin cells. Contrary to the insulin cells, the glucagon and somatostatin cells seem to possess $GABA_A$ receptors. Physiologically, the role of GABA is much debated. The most direct experiments (performed on isolated islets or insular cells) indicate that GABA does not affect the secretion of insulin but inhibits that of somatostatin and glucagon (Fig. 1).

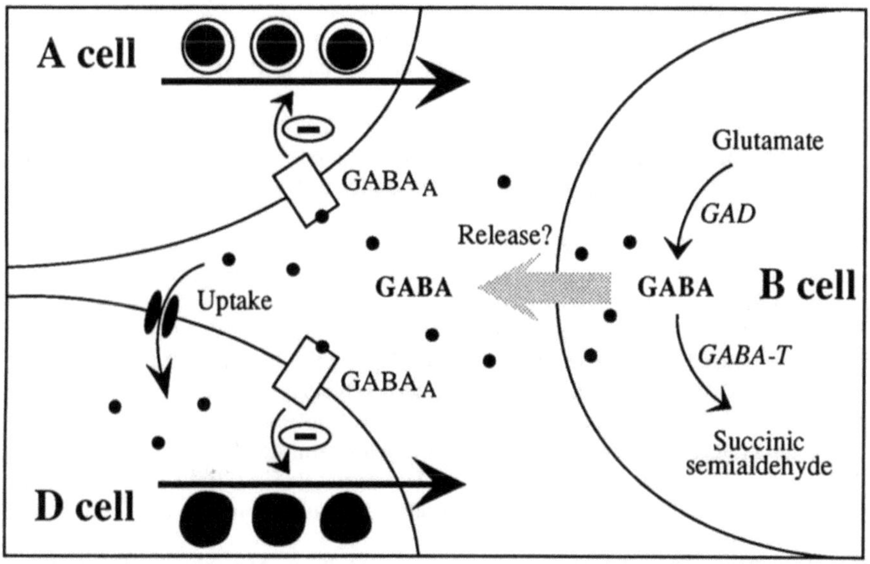

Fig. 1. GABA in the endocrine pancreas. See text for details

In the gastrointestinal tract, an important GABAergic innervation is at least partly intrinsic to the myenteric plexus. In the epithelium, different subpopulations of endocrine cells contain and synthesize GABA. A subpopulation of somatostatin cells and other unidentified endocrine cells (non-EC cells) take it up. Physiologically, besides its modulatory roles in ulceration and, indirectly, in muscle contraction by $GABA_A$ or $GABA_B$ receptors activation, GABA affects the secretion of certain endocrine cells (gastrin, somatostatin, secretin cells) via the innervation. A direct effect on these cells is still unclear.

These data support the hypothesis that GABA is not only a neurotransmitter, but also an endocrine or paracrine factor in the GEP system. This duality of action is not without precedent since many substances known to be neurotransmitters in the central nervous system can play a double role in the intestine, acting as neurotransmitters and as enteric hormones.

References

1. Ahrén B, Taborsky GJ, Porte D (1986) Neuropeptidergic versus cholinergic and adrenergic regulation of islet hormone secretion. Diabetologia 29: 827-836
2. Althaus JS, Martin DL (1989) Entropy as a factor in the binding of γ-aminobutyric acid and nipecotic acid to the γ-aminobutyric acid transport system. Neurochem Res 14: 311-316
3. Amenta F, Collier WL, Erdö SL, Giuliani S, Maggi CA, Meli A (1988) $GABA_A$ receptor sites modulating catecholamine secretion in the rat adrenal gland: evidence from ^{3}H-muscimol autoradiography and in vivo functional studies. Pharmacology 37: 394-402
4. Anderson R, Mitchell R (1986) Biphasic effect of $GABA_A$ receptor antagonists on prolactin secretion: evidence for two types of $GABA_A$ receptor complex on lactotrophes. Eur J Pharmacol 124: 1-9
5. Apud JA, Cocchi D, Locatelli V, Masotto C, Müller EE, Racagni G (1989) Biochemical and functional aspects on the control of prolactin release by the hypothalamo-pituitary GABAergic system. Psychoneuroendocrinology 14: 3-17
6. Baxter CF, Tewari S, Raeburn S (1972) The possible role of gamma-aminobutyric acid in the synthesis of protein. Adv Biochem Psychopharmacol 4: 195-216
7. Bereiter DA, Berthoud HR, Becker MJA, Jeanrenaud B (1982) Brain stem infusion of the γ-aminobutyric acid antagonist bicuculline increases plasma insulin levels in the rat. Endocrinology 111: 324-328
8. Bouhaddi K, Thomopoulos P, Fages C, Khelil M, Tardy M (1988) Insulin effect on GABA uptake in astroglial primary cultures. Neurochem Res 13: 1119-1124
9. Breuker E, Johnston GAR (1975) Inhibition of acetylcholinesterase by bicuculline and related alkaloids. J Neurochem 25: 903-904
10. Briel G, Gylfe E, Hellman B, Neuhoff V (1972) Microdetermination of free amino acids in pancreatic islets isolated from obese-hyperglycemic mice. Acta Physiol Scand 84: 247-253
11. Castro E, Oset-Gasque MJ, Cañadas S, Gimenez G, González MP (1988) $GABA_A$ and $GABA_B$ sites in bovine adrenal medulla membranes. J Neurosci Res 20: 241-245
12. Cavagnini F, Pinto M, Dubini A, Invitti C, Cappelletti G, Polli EE (1982) Effects of gamma-aminobutyric acid (GABA) and muscimol on endocrine pancreatic function in man. Metabolism 31: 73-77
13. Davanger S, Ottersen OP, Storm-Mathisen J (1987) Immunocytochemical localization of GABA in cat myenteric plexus. Neurosci Lett 73: 27-32

14. Davanger S, Ottersen OP, Storm-Mathisen J (1989) GABA-immunoreactive cells in the rat gastrointestinal epithelium. Anat Embryol 179: 221-226
15. Del Tacca M, Blandizzi C, Bernardini MC, Martelli F, Della Longa A (1989) GABA-A-mediated gastrin release induced by baclofen in the isolated vasculary perfused rat stomach. Experientia 45: 734-736
16. Doroshenko PA (1989) γ-Aminobutyric acid elevates cytosolic Ca in bovine chromaffin cells. Neurosci Lett 104: 83-87
17. Drummond RJ, Phillips AT (1974) L-Glutamic acid decarboxylase in non-neural tissues of the mouse. J Neurochem 23: 1207-1213
18. Erdö SL, Kiss B (1986) Presence of GABA, glutamate decarboxylase, and GABA transaminase in peripheral tissues: a collection of quantitative data. In: Erdö SL, Bowery NG (eds) GABAergic mechanisms in the mammalian periphery. Raven, New York, pp 5-17
19. Erdö SL, Wolff JR (1988) Releasable, non-neural GABA pool in rat stomach. Eur J Pharmacol 156: 165-168
20. Erdö SL, Wolff JR (1990) γ-Aminobutyric acid outside the mammalian brain. J Neurochem 54: 363-372
21. Erdö SL, Ezer E, Matuz J, Wolff JR, Amenta F (1989) $GABA_A$ receptors in the rat stomach may mediate mucoprotective effects. Eur J Pharmacol 165: 79-86
22. Erdö SL, De Vincentis G, Amenta F (1990) Autoradiographic localization of [^{3}H]muscimol binding sites in the rat stomach: evidence for mucosal $GABA_A$ receptors. Eur J Pharmacol 175: 351-354
23. Fogel WA (1986) GABA and polyamine metabolism in peripheral tissues. In: Erdö SL, Bowery NG (eds) GABAergic mechanisms in the mammalian periphery. Raven, New York, pp 35-56
24. Fujimoto M, Kataoka Y, Guidotti A, Hanbauer I (1987) Effect of γ-aminobutyric $acid_A$ receptor agonists and antagonists on the release of enkephalin-containing peptides from dog adrenal gland. J Pharmacol Exp Ther 243: 195-199
25. Fung KP, Lee KW, Choy YM (1985) Pancreatic and brain gamma-amino-butyric acid in alloxan-induced diabetic rats. Nutr Reports Int 32: 1267-1271
26. Garry DJ, Sorenson RL, Elde RP, Maley BE, Madsen A (1986) Immunohistochemical colocalization of GABA and insulin in ß-cells of rat islet. Diabetes 35: 1090-1095
27. Garry DJ, Sorenson RL, Coulter HD (1987) Ultrastructural localization of gamma amino butyric acid immunoreactivity in B cells of the rat pancreas. Diabetologia 30: 115-119
28. Garry DJ, Coulter HD, MacIntee TJ, Wu JY, Sorenson RL (1987) Immunoreactive GABA transaminase within the pancreatic islet is localized in mitochondria of the B-cell. J Histochem Cytochem 35: 831-836
29. Garry DJ, Garry MG, Sorenson RL (1988) Ultrastructural immunocytochemical localization of L-glutamate decarboxylase and GABA in rat pancreatic zymogen granules. Cell Tissue Res 252: 191-197
30. Gebauer H (1981) GABA transport in the rat thyroid. Naunyn Schmiedebergs Arch Pharmacol 317: 61-66
31. Gerber JC, Hare TA (1979) Gamma-aminobutyric acid in peripheral tissue, with emphasis on the endocrine pancreas. Presence in two species and reduction by streptozotocin. Diabetes 28: 1073-1076
32. Gerber JC, Hare TA (1980) GABA in peripheral tissues: presence and actions in endocrine pancreatic function. Brain Res Bull 5: 341-346

33. Gilon P, Remacle C, Janssens de Varebeke P, Pauwels G, Hoet JJ (1987) GABA content and localization of high-affinity GABA uptake during the development of the rat pancreas Cell Mol Biol 33: 573-585
34. Gilon P, Reusens-Billen B, Remacle C, Janssens de Varebeke P, Pauwels G, Hoet JJ (1987) Localization of high-affinity GABA uptake and GABA content in the rat duodenum during development. Cell Tissue Res 249: 593-600
35. Gilon P, Campistron G, Geffard M, Remacle C (1988) Immunocytochemical localization of GABA in endocrine cells of the rat entero-pancreatic system. Biol Cell 62: 265-273
36. Gilon P, Remacle C (1989) High-affinity GABA uptake in a subpopulation of somatostatin cells in rat pancreas. J Histochem Cytochem 37: 1133-1139
37. Gilon P, Mallefet J, De Vriendt C, Pauwels S, Geffard M, Campistron G, Remacle C (1990) Immunocytochemical and autoradiographic studies of the endocrine cells interacting with GABA in the rat stomach. Histochemistry 93: 645-654
38. Goertz B (1979) Effect of γ-aminobutyric acid on cell-free protein synthesizing systems from mouse brain. Exp Brain Res 34: 365-372
39. Guo YS, Thompson JC, Singh P (1989) Effect of γ-aminobutyric acid on bombesin-evoked release of somatostatin and gastrin from isolated rat stomach. Regul Pept 24: 179-186
40. Gylfe E (1974) Changes of free amino acids in pancreatic ß-cells after starvation and substrate deprivation. Acta Endocrinol 75: 105-118
41. Gylfe E, Hellman B (1974) Role of glucose as a regulator and precursor of amino acids in the pancreatic ß-cells. Endocrinology 94: 1150-1156
42. Gylfe E, Sehlin J (1976) Interactions between the metabolism of L-leucine and D-glucose in the pancreatic ß-cells. Horm Metab Res 8: 7-11
43. Hakanson R, Alumets J, Stewart JA, Sundler F (1978) Does glutamate decarboxylase occur in endocrine cells of gut and pancreas? Scand J Gastroenterol 13: 74
44. Harty RF, Franklin PA (1983) GABA affects the release of gastrin and somatostatin from rat antral mucosa. Nature 303: 623-624
45. Harty RF, Franklin PA (1986) Cholinergic mediation of γ-aminobutyric acid-induced gastrin and somatostatin release from rat antrum. Gastroenterology 91: 1221-1226
46. Harty RF, Murthy SNS (1986) GABAergic mechanisms in the endocrine stomach and duodenum: their possible functional significance. In: Erdö SL, Bowery NG (eds) GABAergic mechanisms in the mammalian periphery. Raven, New York, pp 339-352
47. Hills JM, Jessen KR, Mirsky R (1987) An immunohistochemical study of the distribution of enteric GABA-containing neurons in the rat and guinea-pig intestine. Neuroscience 22: 301-312
48. Jessen KR, Mirsky R, Dennison ME, Burnstock G (1979) GABA may be a neurotransmitter in the vertebrate peripheral nervous system. Nature 281: 71-74
49. Jessen KR, Mirsky R, Hills JM (1987) GABA as an autonomic neurotransmitter: studies on intrinsic GABAergic neurons in the myenteric plexus of the gut. TINS 10: 255-262
50. Jessen KR, Hills JM, Limbrick AR (1988) GABA immunoreactivity and ^{3}H-GABA uptake in mucosal epithelial cells of the rat stomach. Gut 29: 1549-1556
51. Kataoka Y, Fujimoto M, Alho H, Guidotti A, Geffard M, Kelly GD, Hanbauer I (1986) Intrinsic gamma aminobutyric acid receptors modulate the release of catecholamine from canine adrenal gland in situ. J Pharmacol Exp Ther 239: 584-590

52. Kataoka Y, Ohara-Imaizumi M, Ueki S, Kumakura K (1988) Stimulatory action of γ-aminobutyric acid on catecholamine secretion from bovine adrenal chromaffin cells measured by a real-time monitoring system. J Neurochem 50: 1765-1768
53. Kawai K, Unger RH (1983) Effects of γ-aminobutyric acid on insulin, glucagon, and somatostatin release from isolated perfused dog pancreas. Endocrinology 113: 111-113
54. Kerr DIB, Ong J (1986) GABAergic mechanisms in the gut: their role in the regulation of gut motility. In: Erdö SL, Bowery NG (eds) GABAergic mechanisms in the mammalian periphery. Raven, New York, pp 153-174
55. Koop H, Arnold R (1986) Control of rat gastric somatostatin release by γ-aminobutyric acid (GABA). Horm Metab Res 18: 94-97
56. Krantis A, Webb T (1989) Autoradiographic localization of[^{3}H]γ-aminobutyric acid in neuronal elements of the rat gastric antrum and intestine. J Auton Nerv Syst 29: 41-48
57. Krogsgaard-Larsen P, Falch E, Larsson OM, Schousboe A (1987) GABA uptake inhibitors: relevance to antiepileptic drug research. Epilepsy Res 1: 77-93
58. Kusunoki M, Yamamura T, Ichii S, Fujita S, Nakai T, Utsunomiya J (1988) The effects of sodium valproate on plasma somatostatin and insulin in humans. J Clin Endocrinol Metab 67: 1060-1063
59. Lloyd KG, Prouteau M, Delahaye M, Magnier G, Voltz C (1986) GABA, gastric ulceration, and gastric acid secretion. In: Erdö SL, Bowery NG (eds) GABAergic mechanisms in the mammalian periphery. Raven, New York, pp 353-364
60. MacCann SM, Rettori V (1986) Gamma aminobutyric acid (GABA) controls anterior pituitary hormone secretion. In: Racagni G, Donoso AO (eds) GABA and endocrine function. Raven, New York, pp 173-189
61. MacEvoy RC, Madson KL (1980) Pancreatic insulin-, glucagon-, and somatostatin-positive islet cell populations during the perinatal development of the rat. Biol Neonate 38: 248-254
62. MacEvoy-Bowe E, Hislop J, Wiggins D, Lund P (1987) Amino acid profiles during development of the fetal rat. Biol Neonate 52: 135-140
63. Madtes P, Redburn DA (1983) GABA as a trophic factor during development. Life Sci 33: 979-984
64. Malaisse WJ, Sener A, Carpinelli AR, Anjaneyulu K, Lebrun P, Herchuelz A, Christophe J (1980) The stimulus-secretion coupling of glucose-induced insulin release. XLVI. Physiological role of L-glutamine as a fuel for pancreatic islets. Mol Cell Endocrinol 20: 171-189
65. Martin DL (1976) Carrier-mediated transport and removal of GABA from synaptic regions. In: Roberts E, Chase TN, Tower DB (eds) GABA in nervous system function. Raven, New York, pp 347-386
66. Martin del Rio R, Caballero AL (1980) Presence of γ-aminobutyric acid in rat ovary. J Neurochem 34: 1584-1586
67. Meier E, Jφrgensen OS (1986) γ-Aminobutyric acid affects the developmental expression of neuron-associated proteins in cerebellar granule cell cultures. J Neurochem 46: 1256-1262
68. Meldrum BS (1975) Epilepsy and γ-aminobutyric acid-mediated inhibition. Int Rev Neurobiol 17: 1-36
69. Melis GB, Mais V, Paoletti AM, Antinori D, de Ruggiero A, Fioretti P (1986) GABAergic control of anterior pituitary function in humans. In: Racagni G, Donoso AO (eds) GABA and endocrine function. Raven, New York, pp 219-242

70. Miachon S, Staunton D, Chat M (1985) Anatomical distribution and characteristics of [^{3}H]muscimol high affinity binding sites in the dorsolateral pontine tegmentum of the rat brain. Neurochem Int 7: 765-775
71. Minuk GY, Sarjeant EJ (1985) Insulin and glucagon secretion by the dog pancreas during intravenous and oral administration of gamma aminobutyric acid (GABA). Horm Metab Res 17: 313-314
72. Mohler H, Malherbe P, Draguhn A, Richards JG (1990) $GABA_A$-receptors: structural requirements and sites of gene expression in mammalian brain. Neurochem Res 15: 199-207
73. Oertel WH, Schmechel DE, Tappaz ML, Kopin IJ (1981) Production of a specific antiserum to rat brain glutamatic acid decarboxylase (GAD) by injection of an antigen-antibody complex. Neuroscience 6: 2689-2700
74. Oertel WH, Mugnaini E, Tappaz ML, Weise VK, Dahl AL, Schmechel DE, Kopin IJ (1982) Central GABAergic innervation of neurointermediate pituitary lobe: biochemical and immunocytochemical study in the rat. Proc Natl Acad Sci USA 79: 675-679
75. Okada Y (1979) Properties of glutamate decarboxylase (GAD) in the human insulinoma, a non-neural tissue. Proc Jpn Acad 55: 514-517
76. Okada Y (1986) Localization and function of GABA in the pancreatic islets. In: Erdö SL, Bowery NG (eds) GABAergic mechanisms in the mammalian periphery. Raven, New York, pp 223-240
77. Okada Y, Taniguchi H, Shimada C (1976) High concentration of GABA and high glutamate decarboxylase activity in rat pancreatic islets and human insulinoma. Science 194: 620-622
78. Okada Y, Taniguchi H, Baba S (1982) High concentration of GABA in the pancreatic islets with special emphasis on B cells. In: Okada Y, Roberts E (eds) Problems in GABA research from brain to bacteria. Excerpta Medica, Amsterdam, pp 379-386
79. Panten U, Kriegstein EV, Poser W, Schönborn J, Hasselblatt A (1972) Effects of L-leucine and α-ketoisocaproic acid upon insulin secretion and metabolism of isolated pancreatic islets. FEBS Lett 20: 225-228
80. Passariello N, Giugliano D, Torella R, Sgambato S, Coppola L, Frascolla N (1982) A possible role for γ-aminobutyric acid in the control of the endocrine pancreas. J Clin Endocrinol Metab 54: 1145-1149
81. Rando RR, Bangerter FW (1977) The in vivo inhibition of GABA-transaminase by gabaculine. Biochem Biophys Res Commun 76: 1276-1281
82. Reisert I, Wöhrle M, Pilgrim C (1985) Quantitative assessment of GABA-uptake sites in the neural lobe by electron-microscopic autoradiography. Cell Tissue Res 241: 581-584
83. Reusens-Billen B, Pirlot X, Remacle C, Hoet JJ, de Gasparo M (1984) Localization of GABA high-affinity binding sites in the pancreas of neonatal rat. Cell Tissue Res 235: 503-508
84. Rhoads DE, Di Rocco RI, Osburn LD, Peterson NA, Ragupathy E (1984) Stimulation of synaptosomal uptake of neurotransmitter aminoacids by insulin: possible role of insulin as a neuromodulator. Biochem Biophys Res Commun 119: 1198-1204
85. Rorsman P, Grouse LH, Sorenson RL, Elde RP (1981) Effect of muscimol on glucose-stimulated somatostatin and insulin release from the isolated, perfused rat pancreas. Diabetes 30: 168-171

86. Rorsman P, Berggren PO, Bokvist K, Ericson H, Möhler H, Östenson CG, Smith PA (1989) Glucose-inhibition of glucagon secretion involves activation of $GABA_A$-receptor chloride channels. Nature 341: 233-236
87. Sakaue M, Saito N, Tanaka C (1987) Immunohistochemical localization of gamma-aminobutyric acid (GABA) in the rat pancreas. Histochemistry 86: 365-369
88. Samols E, Stagner JI (1988) Intra-islet regulation. Am J Med 85: 31-35
89. Schoch P, Richards JG, Häring P, Takacs B, Stähli C, Staehelin T, Haefeley W, Möhler H (1985) Co-localization of $GABA_A$ receptors and benzodiazepine receptors in the brain shown by monoclonal antibodies. Nature 314: 168-171
90. Schon F, Kelly JS (1975) Selective uptake of [^{3}H]ß-alanine by glia: association with the glial uptake system for GABA. Brain Res 86: 243-257
91. Schuit FC, Derde MP, Pipeleers DG (1989) Sensitivity of rat pancreatic A and B cells to somatostatin. Diabetologia 32: 207-212
92. Sehlin J, Sandström PE (1988) GABA-stimulated chloride channels in the pancreatic B cells. Diabetologia 31: 542A
93. Seiler N, Al-Therib MJ (1974) Putrescine catabolism in mammalian brain. Biochem J 144: 29-35
94. Snodgrass SR (1973) Studies on GABA and protein synthesis. Brain Res 59: 339-348
95. Stagner JI, Samols E, Marks V (1989) The anterograde and retrograde infusion of glucagon antibodies suggests that A cells are vascularly perfused before D cells within the rat islet. Diabetologia 32: 203-206
96. Staines WA, Meister B, Melander T, Nagy JI, Hökfelt T (1988) Three-color immunofluorescence histochemistry allowing triple labeling within a single section. J Histochem Cytochem 36: 145-151
97. Tallan HH, Moore S, Stein WH (1954) Studies on the free amino acids and related compounds in the tissues of the cat. J Biol Chem 211: 927-939
98. Tanaka C, Taniyama K (1986) GABA transport in peripheral tissues: uptake and efflux. In: Erdö SL, Bowery NG (eds) GABAergic mechanisms in the mammalian periphery. Raven, New York, pp 57-75
99. Taniguchi H, Okada Y, Shimada C, Baba S (1977) GABA in pancreatic islets. Arch Histol Jpn 40: 87-97
100. Taniguchi H, Okada Y, Seguchi H, Shimada C, Seki M, Tsutou A, Baba S (1979) High concentration of gamma-aminobutyric acid in pancreatic beta cells. Diabetes 28: 629-633
101. Taniguchi H, Okada Y, Kobayashi T, Murakami K, Baba S (1979) High concentration of γ-aminobutyric acid and its role on B-cells of pancreatic islets. In: Baba S, Kaneko T, Yanaihara N (eds) Proinsulin, insulin, C-peptide. Excerpta Medica, Amsterdam, pp 335-347
102. Taniguchi H, Okada Y, Hosoya Y, Baba S (1980) Comparison of uptake of γ-aminobutyric acid by pancreatic islets and by substantia nigra. Biomed Res 1: 175-179
103. Taniguchi H, Yoshioka M, Tsutou A, Ejiri K, Tamagawa M, Murakami K, Utsumi M, Baba S, Okada Y (1980) Effect of aminobutyric acid on somatostatin and insulin content of rat cultured islets. Biomed Res 1: 180-182
104. Taniguchi H, Murakami K, Yoshioka M, Ejiri K, Ishihara K, Baba S, Okada Y (1982) GABA and insulin in pancreatic islets. In: Okada Y, Roberts E (eds) Problems in GABA research from brain to bacteria. Excerpta Medica, Amsterdam, pp 387-405

105. Taniguchi H, Yoshioka M, Ejiri K, Ishihara K, Tamagawa M, Hirose Y, Ishihara K, Utsumi M, Baba S, Okada Y (1982) Suppression of somatostatin release and increase of somatostatin content in pancreatic islets by GABA. In: Okada Y, Roberts E (eds) Problems in GABA research from brain to bacteria. Excerpta Medica, Amsterdam, pp 406-412
106. Thirbly RC, Stevens MH, Blair AJ, Petty F, Crawford IL, Taylor IL, Walsh JH, Feldman M (1988) Effect of GABA on basal and vagally mediated gastric acid secretion and hormone release in dogs. Am J Physiol 254: G723-G731
107. Tunnicliff G (1989) Inhibitors of brain GABA aminotransferase. Comp Biochem Physiol 93A: 247-254
108. Unger RH, Orci L (1989) Glucagon secretion and metabolism in man. In: De Groot LJ (ed) Endicrinology, vol.2. Saunders, Philadelphia, pp 1318-1332
109. Vincent SR, Brown JC (1988) Autoradiographic studies of the γ-aminobutyric acid (GABA) system in the rat pancreas. Histochemistry 88: 171-173
110. Vincent SR, Hökfelt T, Wu JY (1982) GABA neuron systems in hypothalamus and the pituitary gland. Neuroendocrinology 34: 117-125
111. Vincent SR, Hökfelt T, Wu JY, Elde RP, Morgan LM, Kimmel JR (1983) Immunohistochemical studies of the GABA system in the pancreas. Neuroendocrinology 36: 197-204
112. Wisden W, Morris BJ, Darlison MG, Hunt SP, Barnard EA (1988) Distinct $GABA_A$ receptor α subunit mRNAs show differential patterns of expression in bovine brain. Neuron 1: 937-947
113. Wu JY, Lin CT, Lin H, Xu Y, Liu JW, Hwang BH, Wei SC (1986) Immunochemical characterization and immunohistochemical localization of glutamate decarboxylase and GABA transaminase in peripheral tissues. In: Erdö SL, Bowery NG (eds) GABAergic mechanisms in the mammalian periphery. Raven, New York, pp 19-34
114. Yazulla S, Studholme KM, Vitorica J, de Blas AL (1989) Immunocytochemical localization of $GABA_A$ receptors in goldfish and chicken retinas. J Comp Neurol 280: 15-26
115. Young W, Kuhar M (1979) A new method for receptor autoradiography: ^3H-opioid receptors in rat brain. Brain Res 179: 255-270

GABA and Diamine Oxidase Activity in the Gastrointestinal Mucosa: Possible Links

W.A. Fogel and S.L. Erdö

Introduction

Gamma-aminobutyric acid (GABA) has a transmitter function in the mammalian enteric nervous system. The intestine contains endogenous GABA [1, 24, 32] and its synthesizing and degrading enzymes, i.e., glutamic acid decarboxylase (GAD, EC 4.1.1.15) [24, 32] and GABA-transaminase (GABA-T, EC 2.6.1.19) [32], respectively. Endogenous and labeled GABA may be released from the intestine by electrical transmural stimulation [44] or by the application of substance P [43]. Finally, the presence of GABA-ergic neurons in the intestine has been demonstrated by immunohistochemical methods [3, 10, 23]. GABA may act on both inhibitory and excitatory neurons of the enteric plexus [26] and these effects appear to be mediated by both subtypes of GABA receptors, i.e., $GABA_A$ and $GABA_B$ [25].

In the mammalian small and large bowel, the enteric nervous system forms an extensive neural network within the muscular layers of the intestine. Studies by Miki et al. [32] have shown that endogenous GABA levels and GAD activity are at their highest in the Auerbach's plexus and significantly lower in the muscular and mucosal layers (Table 1). These data suggest that, in addition to the neuronal GABA pool, the gastrointestinal tract may also possess minor, non-neuronal pools for GABA, e.g., in the mucosal layer. This is in line with an early report on the presence of GAD activity in the gastrointestinal mucosa [47].

GABA as an Intermediate in Polyamine Metabolism

The intestine is also one of the richest sources of diamine oxidase (DAO, EC 1.4.3.6) in most of the mammalian species examined, including mice, rats, pigs, rabbits, dogs, and humans [6]. DAO is localized mainly in the mucosal layer, with less than 5% of the total activity being confined to muscular layers [7].

Zeller [51] reported, in 1938, that hog kidney preparations and various animal tissue extracts are capable of destroying by oxidative deamination not only histamine but also diamines, such as putrescine, cadaverine and agmatine. The term "histaminase" was substituted with "diamine oxidase".

Table 1. GABA content and GAD activity in individual layers of cat and human colon (modified from [32])

Layer	GABA nmol/mg protein		GAD nmol/mg protein/h	
CAT				
Auerbach's plexus	4.60	(100)	45.68	(100)
Musosa	0.14	(3.0)	3.08	(6.7)
Circular muscle	0.19	(4.19)	3.44	(7.5)
Longitudinal muscle	0.24	(5.2)	2.93	(6.4)
HUMAN				
Auerbach's plexus	2.52	(100)	10.35	(100)
Musosa	0.23	(9.1)	1.92	(18.6)
Circular muscle	0.15	(6.0)	1.61	(15.6)
Longitudinal muscle	0.34	(13.5)	1.62	(15.7)

Numbers in parentheses represent percentage values, with the respective values for the Auerbach's plexus being considered 100%.

Nonetheless, conclusive studies on the possible role of putrescine as a GABA precursor in the brain were not carried out until three decades later.

In fact, DAO catalyzes the first step in the conversion of putrescine to GABA. However, putrescine (1,4-diaminobutane) is also a precursor of two polyamines, spermidine and spermine. These compounds are important in cell growth and differentiation and are susceptible to rapid modulation by various hormones in a number of target tissues [19, 21].

Seiler and coworkers [41] originally described the incorporation of putrescine carbon in GABA in the rat liver and brain in vivo. Since the end product of [^{14}C]putrescine in vivo is $^{14}CO_2$ [21], GABA can be regarded as an intermediate in the catabolism of putrescine [20, 39], which is rapidly degraded via the Krebs' cycle to carbon dioxide [48]. Inhibitors of DAO and to a smaller extent inhibitors of monoamine oxidase are capable of blocking $^{14}CO_2$ formation after the administration of radioactive putrescine [39]. These findings suggest that two different pathways of putrescine catabolism may operate in vertebrates.

Alternative Pathways of GABA Formation from Putrescine

The first pathway is the direct oxidative deamination by DAO to gamma-aminobutyraldehyde which may then be further oxidized to GABA, presumably by an aldehyde dehydrogenase. This metabolic route has been demonstrated in rat liver [39] and ovary [1], in other peripheral tissues of rats (45) and mice [17], in the guinea-pig liver [2, 16], and in the human placenta [2]. The second pathway involves acetylation of putrescine by N-acetyltransferase. Subsequent steps comprise oxidative deamination of monoacetylputrescine by monoamine oxydase, (MAO, EC 1.4.3.4), the oxidation of the aldehyde product to N-acety-GABA, and deacetylation. This pathway has been reported to occur in mouse brain and liver and in the intestine and kidney of the rat [37-39].

Putrescine and Glutamate as GABA Precursors: Their Relative Contribution

The contribution of putrescine-derived GABA to the total GABA pool in the CNS is considered to be insignificant, i.e., not exceeding 0.5% [37]. However, this does not seem to be true for the periphery. There is convincing evidence to show that in rat adrenal gland GABA is almost exclusively derived from putrescine. Aminoguanidine, a potent inhibitor of DAO [34], not only eliminated the formation of radioactive GABA from labeled putrescine, but also depleted endogenous GABA from the adrenals [9]. In additon, the turnover time of GABA in adrenals (5.6 h) remarkably differs from that in the brain (1 h), suggesting that different metabolic pathways may operate in these organs [9].

Studies on the distribution of radioactive GABA following putrescine injection into rats [45] and mice [17] indicated that the intestine is a main source of circulating GABA. Indeed, if the radioactivity that was confined to newly formed GABA in the blood was defined to be of 1 arbitrary unit, the values for the rat and mouse intestine were 64 and 125, respectively. Much lower activities were determined in all other organs. For comparison, the second highest values obtained were 7.9 in the rat spleen and 13.3 in the mouse liver.

The Origin of GABA in the Gastrointestinal Mucosa

The intestinal mucosa fulfills all criteria for being a potential source of putrescine-derived GABA. Under normal conditions not only a high DAO activity, but also a high activity of the putrescine synthesizing enzyme ornithine decarboxylase (ODC, EC 4.1.1.17) is present in this tissue (Table 2). Moreover, DAO and ODC show parallel localization both along [6, 4] (see also Tables 2 and 4) and across [5, 14, 42] the intestinal mucosa. These two enzymes are found in highest quantities in the small intestine and are present almost exclusively in non dividing mature cells of the villus tip and midvillus.

The activity of ODC is unusually high in the small intestine; it approximates that of fetal tissue or regenerating rat liver [4]. However, this already high activity can still be further increased to a remarkable extent by exposure to various trophic stimuli. Thus, enzyme activity is increased after partial resection of the small bowel [30], during lactation [50], at the time of weaning [29], in response to intestinal obstruction [36], and in other instances associated with intensive mucosal growth [14, 35]. In all these cases, the polyamine content of the mucosa is also increased.

Putrescine concentrations in the gastrointestinal mucosa may be regulated not only by intracellular biosynthesis via the ODC pathway, but also by the uptake of putrescine ingested with food or produced by the gastrointestinal flora. It has recently been demonstrated that isolated villous enterocytes accumulate putrescine against a concentration gradient [27]. This uptake is dependent on energy and temperature and exhibits saturation kinetics.

Table 2. Ornithine decarboxylase activity and the level of polyamines in the mucosa of rat gastrointestinal tract (modified from [4])

Tissue	ODC activity pmol/50 min/mg protein	Putrescine	Spermidine μmol/g dry tissue	Spermine
Stomach	3	0.58	5.45	19.50
Duodenum	541	0.57	4.41	3.58
Jejunum	334	0.50	2.65	1.54
Ileum	1063	0.52	3.94	3.96
Colon	23	0.05	1.84	2.25

Moreover, it is not connected with the transporter of amino acids or Na^+ and is independent of the Na^+ gradient in the cell membrane. Interestingly enough, the capacity of putrescine uptake and the activity of DAO appear to show a similar pattern of distribution along the gut, the highest values being found for the distal segment of the small intestine [6, 27] (see also Table 4).

According to Canellakis et al. [8] the daily dietary intake of either of the three polyamines, putrescine, spermidine, and spermine is roughly 100 μmole. In this context it should be mentioned that higher polyamines can be reutilized for putrescine formation under physiological conditions [40]. The conversion of spermine into spermidine and further to putrescine is achieved by acetylation followed by oxidative deamination via polyamine oxidase.

The distribution patterns of polyamine oxidase activity (Table 3), polyamine concentration, and ODC activity (Table 2) along the gastrointestinal tract suggest that the above interconversion of polyamines is especially active in the proximal duodenum and in the large intestine. Similarly, it appears that the transformation of putrescine to GABA via the acetylation pathway, which involves the participation of MAO, would also be more active in these parts of the alimentary canal (Table 3).

Recently published data of Berlo et al. [46] confirms that the intestinal (both small and large) mucosa contributes to the production of endogenous GABA. Using normal and germ-free Wistar rats the authors have shown that the venous effluent obtained from bowels contains GABA in an appreciable amount. They have also demonstrated that normal rats exhibit approximately the same GABA concentrations in blood samples of both arterial and portal origin as the GABA levels in the respective blood samples of germ-free rats. These findings clearly indicate that the majority

Table 3. Monoamine oxidase and polyamine oxidase activities in the mucosa of the rat alimentary tract

Tissue	Monoamine oxidase	Polyamine oxidase
	pmol/mg protein/min	
Pars prepylorica	381 ± 32	92 ± 24
Duodenum		
Proximal	821 ± 111	206 ± 11
Distal	564 ± 50	133 ± 5
Small intestine		
Proximal	432 ± 27	142 ± 11
Middle	426 ± 33	109 ± 12
Distal	531 ± 44	81 ± 10
Large intestine		
Proximal	628 ± 40	131 ± 15
Distal	706 ± 51	186 ± 35

The values are mean + SEM of 6-8 determinations. Original data. MAO activity was measured according to the radioassay method of Wurtman and Axelrod [49] with [14C] tryptamine (0.1 m*M*) as substrate. PAO activity was assayed by the fluorometric technique of Matsumoto et al. [31] using 0.1 m*M* *N*-l-acetylspermine as substrate.

of GABA in the blood is not of bacterial origin, in contrast to an earlier suggestion (33).

Moreover, it has been shown that the presence of GAD [32, 47] and DAO [6, 17, 45] in the intestinal mucosa may provide efficient sources of GABA, irrespective of a minor bacterial contribution. These studies, however, did not elucidate which of the two enzymic pathways is more important, or whether one of them may even predominate. Thus, the question of whether glutamate or putrescine is the major GABA precursor in the gastrointestinal mucosa remained open. To elucidate this point, the segmental distribution of endogenous GABA and the activities of GAD and DAO were examined comparatively in the mucosa of the rat alimentary canal (Table 4).

The distribution profile of DAO obtained in our experiments is in agreement with earlier observations [6]. As shown in Table 4, the activity of DAO progressively rose from the pars prepylorica through the duodenum towards the middle and distal parts of the intestine and showed a steep decline in the colon.

Table 4. Segmental distribution of GABA content and the activities of GAD and DAO in the gastrointestinal mucosa of the rat.

Segment	GABA pmol/mg protein	GAD pmol/mg protein/min	DAO pmol/mg protein/min
Pars prepylorica	278 ± 37	50 ± 11	40 ± 14
Duodenum			
Proximal	627 ± 39	72 ± 25	450 ± 40
Distal	649 ± 70	79 ± 29	1040 ± 170
Small intestine			
Proximal	628 ± 23	61 ± 14	1150 ± 120
Middle	748 ± 54	88 ± 21	1920 ± 360
Distal	849 ± 90	109 ± 29	1940 ± 360
Large intestine			
Proximal	587 ± 49	72 ± 16	670 ± 140
Distal	491 ± 35	92 ± 24	300 ± 130

Original data. Each value represents the mean ± SEM obtained for 7 rats. The GABA content in mucosal scrapings was determined by HPLC/fluorometry [12]; the activities of DAO and GAD were determined by radioassay procedures [13, 18]. Estimates of protein content were made according to Lowry et al. [28].

The segmental distribution of GABA and GAD in the mucosa was found to be similar to that of DAO, with the changes being somewhat more attenuated. The statistical evaluation of the positive correlations obtained for DAO and GABA (r = 0.9057, r < 0.002) and for GAD and GABA (r = 0.7302, p < 0.04) suggests that both pathways may well contribute to the synthesis of the mucosal GABA pool. However, taking into account that the mucosal activity of DAO is five to ten times higher than the activity of GAD in the respective tissue, the predominance of the polyamine pathway seems very likely. Further studies are in progress to substantiate this observation by using appropriate inhibitors of polyamine and glutamate catabolism.

Summary

The GABA content of the gastrointestinal mucosa may be derived from both glutamate and putrescine. Considering the distribution patterns and the activities of the respective enzymes, the putrescine pathway appears to prevail. Although the physiological significance of the mucosal GABA pool has not yet been elucidated, it seems reasonable to assume that GABA has some role in the differentiation and maturation of mucosal cells [15] and/or in gastric/enteric mucoprotection [12a].

References

1. Andersson AC, Henningson S, von Scheele C (1980) Formation of GABA via oxidative deamination of putrescine in ovaries of immature rats after gonadotropin stimulation. Agents Actions 10: 106-107
2. Andersson AC, Henningson S, Persson L, Rosengren E (1978) Aspects of diamine oxidase activity and its determination. Acta Physiol Scand 102: 159-166
3. Beatge G, Gershon MD (1986) GABA in the PNS: demonstration in enteric neurons. Brain Res Bull 16: 421-424
4. Ball WJ, Balis ME (1976) Changes in ornithine decarboxylase activity in rat intestines during aging. Cancer Res 36: 3312-3316
5. Baylin SB, Stevens SA, Shakir KMM (1978) Association of diamine oxidase and ornithine decarboxylase with maturing cells in rapidly proliferating epithelium. Biochim Biophys Acta 541: 415-419
6. Biegdanski T (1983) Biochemical, physiological and pathophysiological aspects of intestinal diamine oxidase. Acta Physiol Pol 34: 139-154
7. Biegdanski T, Kuche J, Lorenz W, Hersterberg R, Stahlknecht C-D, Feussner KD (1983) Distribution and properties of human intestinal diamine oxidase and its relevance for histamine catabolism. Biochim Biophys Acta 756: 196-203
8. Canellakis ES, Viceps-Madore D, Kyriakidis DA, Hellers JS (1979) The regulation and functions of ornithine decarboxylase and of the polyamines. Curr Top Cell Regul 15: 155-202
9. Caron PCV, Cote LJ, Kremzner LT (1988) Putrescine, a source of gamma-aminobutyric acid in the adrenal gland of the rat. Biochem J 251: 559-562
10. Davanger S, Ottersen OP, Storm-Mathisen (1987) Immunocytochemical localisation of GABA in cat myenteric plexus. Neurosci Lett 73: 27-32
11. Del Rio RM, Caballero AL (1980) Presence of gamma-aminobutyric acid in rat ovary. J Neurochem 34: 1584-1586
12. Erdö SL, Wolff JR (1989) A comparison of the postnatal changes in aspartate and glutamate levels in cerebral cortex of the rat. Neurosci Res Commun 4: 51-56.

12a. Erdö SL, Ezer E, Matuz J, Wolff JR, Amenta F (1989) $GABA_A$ receptors in the rat stomach may mediate mucoprotective effects. Eur J Pharmacol 165: 79-86
13. Ferenci P, Jacobs R, Pappas SC, Schafer DF, Jones EA (1984) Enzymes of cerebral GABA metabolism and synaptosomal GABA uptake in acute liver failure in the rabbit: evidence for decreased cerebral GABA-transaminase activity. J Neurochem 42: 1479-1490
14. Fitzpatrick LR, Wang P, Eikenburg BE, Haddox MK, Johnson LR (1986) Effect of refeeding on polyamine biosynthesis in isolated enterocytes. Am J Physiol 250: G709-G713
15. Fogel WA (1986) GABA and polyamine metabolisms in peripheral tissues. In: Erdö SL, Bowery NG (eds) GABAergic Mechanisms in the mammalian periphery. Raven, New York, pp 35-56
16. Fogel WA, Biegdanski T, Maslinski C (1979) Effects of inhibitors of aldehyde metabolizing enzymes on putrescine metabolism in guinea pig liver homogenates. Agents Actions 9: 42-44
17. Fogel WA, Biegdanski T, Schayer RW, Maslinski C (1981) Involvement of diamine oxidase in catabolism of ^{14}C-putrescine in mice in vivo with special reference to the formation of gamma-aminobutyric acid. Agents Actions 16: 679-684
18. Fogel WA, Ulatowska M, Adach K, Osinska C (1985) A sum of ^{14}C-putrescine metabolites as a measure of DAO activity. Column chromatography assay. Agents Actions 16: 99-101
19. Heby O (1981) Role of polyamines in control of cell proliferation and differentiation. Differentiation 19: 1-20
20. Henningsson S, Rosengren E (1976) The effect of nandrolone, an anabolic steroid on putrescine metabolism in the mouse. Br J Pharmacol 58: 401-406
21. Jänne J (1967) Studies on the biosynthetioc pathway of polyamines in rat liver. Acta Physiol Scand [Suppl] 300: 1-71
22. Jänne J, Pösö H, Raina A (1978) Polyamines in rapid growth and cancer. Biochim Biophys Acta 473: 241-293
23. Jessen KR, Hills JM, Saffrey MJ (1986) Immunohistochemical demonstration of GABAergic neurons in the enteric nervous system. J Neurosci 6: 1628-1634
24. Jessen KR, Mirsky R, Dennison ME, Burnstock G (1979) GABA may be a neurotransmitter in the vertebrate peripheral nervous system. Nature 281: 71-74
25. Kerr DIB, Ong J (1986) GABAergic mechanisms in the gut: their role in the regulation of gut motility. In: Erdö SL, Bowery NG (eds) GABAergic mechanisms in the mammalian periphery. Raven, New York, pp 153-174
26. Krantis A, Costa M, Furness JB, Orbach J (1980) Gamma-aminobutyric acid stimulates intrinsic inhibitory and excitatory nerves in the guinea pig intestine. Eur J Pharmacol 67: 461-468
27. Kumai J, Johnson LR (1988) Characteristics of putrescine uptake in isolated rat enterocytes. Am J Physiol 254: G81-G86
28. Lowry OH, Rosebrough NJ, Farr AL, Randall RJ (1951) Protein measurement with the Folin phenol reagent. J Biol Chem 193: 265-275
29. Luk GD, Baylin SB (1982) Ornithine decarboxylase in intestinal maturation, recovery and adaptation. In: Robinson JWL, Dowling RH, Rieken EO (eds) Mechanisms of intestinal maturation. MTP Press, Lancaster pp 65-78
30. Luk GD, Baylin SB (1983) Polyamines and intestinal growth: increased polyamine biosynthesis after jejunectomy. Am J Physiol 245: G656-G660

31. Matsumoto T, Furuta T, Niumra Y, Suzuki O (1982) Increased sensitivity of the fluorometric method of Snyder and Hendley for oxidase assays. Biochem Pharmacol 31: 2207-2209
32. Miki Y, Taniyama K, Tanaka C (1983) GABA, glutamic acid decarboxylase and GABA-transaminase levels in the myenteric plexus in the intestine of humans and other mammals. J Neurochem 40: 861-865
33. Schäfer DF, Fowler JM, Jones EA (1981) Colonic bacteria. A source of gamma-aminobutyric acid in blood. Proc Soc Exp Biol Med 167: 301-303
34. Schuler W (1952) Zur Hemmung der Diaminooxidase (Histaminase). Experientia 8: 230-233
35. Seidel ER (1986) Hormonal regulation of postprandial induction of gastrointestinal ornithine decarboxylase activity. Am J Physiol 251: G460-G466
36. Seidel ER, Haddox MK, Johnson LR (1984) Polyamines in response to intestinal obstruction. Am J Physiol 246: G649-G653
37. Seiler N (1980) On the role of GABA in vertebrate polyamine metabolism. Physiol Chem Phys Med NMR 12: 411-429
38. Seiler N, Al-Therib B (1975) Putrescine catabolism in mammalian brain. Biochem J 144: 29-35
39. Seiler N, Eichentopf B (1975) 4-Aminobutyrate in mammalian putrescine catabolism. Biochem J 152: 201-210.
40. Seiler N, Bolkenius FN, Rennert OM (1981) Interconversion, catabolism and elimination of the polyamines. Med Biol 59: 334-346
41. Seiler N, Weichmann M, Fischer HA, Werner G (1971) The incorporation of putrescine carbon into gamma-aminobutyric acid in rat liver and brain in vivo. Brain Res 28: 317-325
42. Shakir KMM, Margolis S, Baylin SB (1977) Localization of histaminase (diamine oxidase) in rat small intestine mucosa: site of release by heparin. Biochem Pharmacol 25: 317-325
43. Tanaka C, Taniyama K (1985) Substance P provoked gamma-aminobutyric acid release from the myenteric plexus of the guinea pig small intestine. J Physiol 362: 319-329
44. Taniyama K, Miki Y, Kusunoki M, Saito N, Tanaka C (1983) Release of endogenous and labeled GABA from isolated guinea pig ileum. Am J Physiol 245: G717-G721
45. Tsuji M, Nakajima T (1978) Studies on the formation of gamma-aminobutyric acid from putrescine in rat organs and purification of its synthetic enzyme from rat intestine. J Biochem 83: 1407-1412
46. Van Berlo CLH, De Jonge HR, Van den Bogaard AEJM (1987) Gamma-aminobutyric acid production in small and large intestine of normal and germ-free Wistar rats. Gastroenterology 93: 472-499
47. Whaler BC (1955) The metabolism of amino acids by the small intestine. J Physiol 130: 278-290
48. Wilson WE, Hill RJ, Koeppe RE (1959) The metabolism of gamma-amino-butyric acid-4-C^{14} by intact rats. J Biol Chem 234: 278-290
49. Wurtman RJ, Axelrod J (1963) A sensitive and specific assay for the estimation of monoamine oxidase activity. Biochem Pharmacol 12: 1439-1441
50. Yang P, Baylin SB, Luk GD (1984) Polyamines and intestinal growth: absolute requirement for ODC activity in adaptation and lactation. Am J Physiol 247: G553-G557
51. Zeller EA (198) Über den enzymatischen Abbau von Histamin und Diaminen. Helv Chir Acta 23: 1502-1508

GABA and the Liver: The First 40 Years[1]

G.Y. Minuk

Introduction

Despite the fact that it has been approximately 40 years since high concentrations of GABA were first described in the liver, relatively little is known about the role GABA plays in hepatobiliary function and disease. Indeed, much of the effort over the past 40 years has consisted of documenting hepatic GABA concentrations in various animal species and quantitating the amount of enzymes responsible for GABA synthesis and metabolism. Only in the past 10 years has a concerted effort been made to determine the functional importance of GABA in the liver. The preliminary results of this more recent work suggest that GABAergic innervation of the liver not only exists, but may play an important role in regulating hepatic development and function. The purpose of this review will be to discuss the results of the past decade of research regarding GABA in the liver and, in the process, identify specific directions that future research might take.

GABA Levels in the Liver

Detectable amounts of GABA have been reported in the livers of all animal species reported to date including the cat, mouse, rabbit, and rat [4, 5, 11, 30, 31]. Although the concentrations vary, in general, the values tend to range between 15 and 100 nmol/g liver tissue (Table 1). The exception

[1] The work described herein was made possible by grants from the Medical Research Council of Canada and the Manitoba Medical Services Foundation

appears to be the human in whom liver GABA concentrations were reported to approach 300 nmol/g tissue [35]. The reason for the higher levels in humans is unclear. Differences in the time required to obtain and prepare specimens for analysis between experimental animals and humans might in part explain these findings as might differences in the techniques employed to carry out the analyses [32, 7].

Table 1. GABA, GAD and GABA-T levels

Organ and Tissue	Species	GABA (nmol/g)	GAD (nmol/mg protein/h)	GABA-T (nmol/mg protein/h)
Liver	Cat	97		
	Human	252		
				380
				624
	Mouse	15	3.4	
			3.9	
				342
	Pig			180
	Rabbit	42.3		197
			0.79	5.7
	Rat	21.8	4.5	
		24	2.51	
		30	33.6	
		32.2		66.6
		87		1446

Despite appreciable concentrations of GABA in the liver, the source of that GABA has yet to be determined. According to the results of earlier studies, the liver is not thought to be an active site of GABA synthesis. Indeed, hepatic venous GABA concentrations are significantly lower than those present in portal venous blood [26]. This is in keeping with the relatively low hepatic concentrations of glutamate decarboxylase (GAD), the enzyme responsible for GABA synthesis in most organ systems [5]. With the exception of the thyroid gland and uterus, the lowest GAD activity reported to date is in the rabbit liver, where only 0.79 nmol/mg protein/h was measured [14]. Other species such as the mouse, pig, and rat also have low GAD concentrations in their liver, particularly when compared to organs such as the brain, kidney, and reproductive tissues. It should be stressed, however, that decarboxylation of glutamate represent only one possible pathway for GABA synthesis with the metabolism of putrescine representing another [9]. In the latter pathway, ornithine is decarboxylated to putrescine by the enzyme ornithine decarboxylase (ODC). Thereafter, putrescine is oxidized to γ-aminobutyraldehyde by diamine oxidase (DAO) followed by dehydrogenation to GABA via aldehyde dehydrogenase. The monoamine oxidase system might also contribute to GABA synthesis from putrescine if putrescine is acetylated by *N*-acetyl transferase. An additional reason for questioning the concept that the liver is not an active site for GABA synthesis is the possibility that hepatic GABA synthesis is intermittent, i.e., active only at certain times during hepatic development or function. Recent data from our laboratory suggests that this is the case in that GABA synthesis was significantly increased by the remnant liver following partial hepatectomy in the rat as compared to GABA synthesis in the resting (nonregenerating) liver (Fig. 1). It is of interest to note that all previous studies consisted of documenting GABA concentrations only in resting livers.

Unlike GABA synthesis, GABA metabolism is thought to be under the control of only one enzymatic pathway and is active in the liver at all times. The enzyme primarily responsible for GABA metabolism in the liver

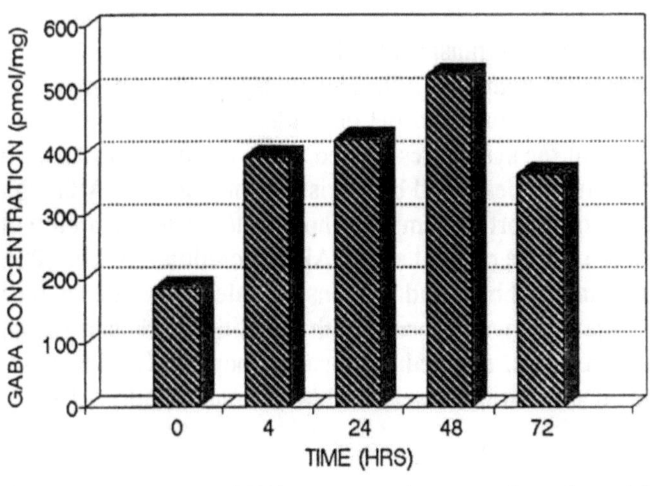

Fig. 1. Hepatic tissue GABA concentrations at various times following a 70% partial hepatetcomy in rats

is GABA transaminase (GABA-T), a mitochondrial enzyme which deaminates GABA to succinyl semialdehyde from which CO_2 and water is ultimately formed. Regardless of the animal species, the liver consistently possesses the highest of GABA-T outside the central nervous system [33]. Nonetheless, high concentrations and acitivity of GABA-T in the liver do not preclude the presence of an alternative pathway for GABA metabolism such as the shunt recently described by Fussi et al. in which GABA is converted directly to spermidine, a higher polyamine [10].

GABA Transport in the Liver

The identification and characterization of a GABA transport system in the liver was the result of testing a hypothesis which implicated GABA in the pathogenesis of hepatic encephalopathy [27]. That hypothesis stated that under normal conditions the liver is responsible for clearing GABA from the systemic circulation. In the presence of liver disease, blood GABA concentrations would be expected to rise. Elevated blood GABA concentrations then contribute to the inhibition of neuronal activity that occurs in the brains of patients with hepatic encephalopathy. The first step in testing the hypothesis consisted of demonstrating the existence of a specific GABA transport system on the surface of at least one of the various liver cell populations. This was accomplished in 1984 when, following in situ collagenase perfusion of the rat liver, isolated hepatocates but not endothelial or Kupffer cells, were reported to transport [^{3}H]GABA into the cell in a time and temperature dependent manner [24]. The transport system was also shown to be saturable and sodium and pH dependent. Competitive inhibition studies indicated that the system resembled the A-amino acid transport system of the liver originally described by Christiansen et al. [2]. Although not directly tested, the transport system was thought to be responsible not only for GABA uptake into the cell but also GABA secretion into bile [22]. Subsequent studies in rats, rabbits, and humans revealed that interference with this transport system, as would occur with hepatitis, cirrhosis, partial hepatectomy, or certain drugs, resulted in impaired hepatic GABA clearance and/or significant elevations in serum GABA concentrations [6, 36, 25, 23, 16] (Fig.2).

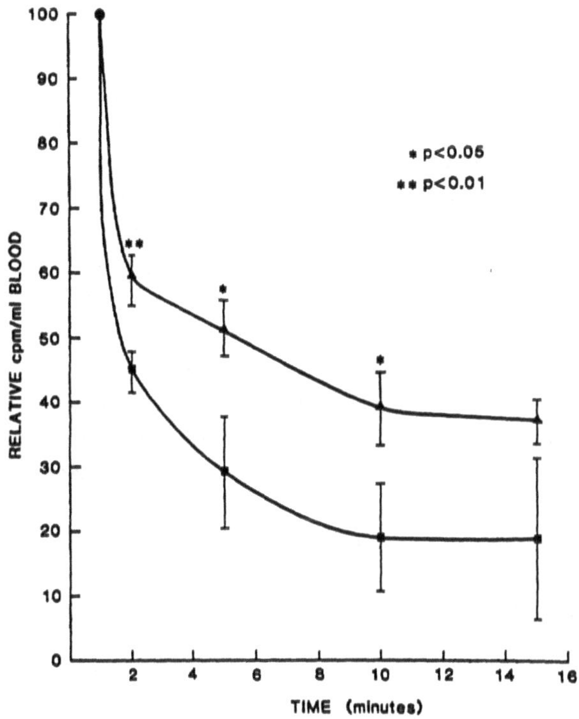

Fig. 2 Clearance of [^{3}H]GABA (0.5 μmol/kg) from the systemic circulation of adult rats (n = 4/group) pretreated 15 min earlier with either diazepam (0.2 mg/kg) (▲—▲) or an equivalent volume of saline (■—■). *Vertical bars* represent standard deviations from the mean

Clinical Implications of Impaired Hepatic GABA Clearance

Hepatic Encephalopathy

Hepatic encephalopathy is a complex disorder in which numerous factors present in the circulation in excess amount, contribute to the suppression of normal neuronal activity in patients with advanced liver disease. As discussed above, one such factor may be GABA. In support of this possibility is the finding that GABA is produced in large amount by bacterial flora which when eradicated by antibiotics results in a prompt improvement in the encepalopathic state [21, 3]. GABA concentrations are also higher in encepalopathic patients than patients without encepalopathy who have a similar degree of hepatic dysfunction [25]. Moreover, factors that are known to precipitate hepatic encepalopathy such as hyponatremia and alkalosis interfere with GABA transport and presumably clearance by the liver [24] (Fig. 2).

Hypotension

Another complication of advanced liver disease that could conceivably be worsened by impaired hepatic GABA clearance is systemic hypotension [20]. Previously, hypotension in patients with liver disease had been ascribed to a combination of anatomical shunting of blood between the arterial and venous systems and the presence of humoral vasodilators, many of which remain to be identified. The possibility that GABA might be one of these compounds was raised by a series of experiments in which infusions of GABA into the systemic circulation of dogs, at concentrations that achieved serum levels similar to those reported in animals and humans with chronic liver disease, resulted in a prompt and significant decrease in systemic, systolic, and diastolic blood pressures [20]. The drop in pressures was maintained throughout the infusion period and returned to normal only after the infusion had been discontinued. It has also been demonstrated that systemic GABA concentrations were significantly higher in patients with liver disease and hypotension than in patients with liver disease and normal blood pressures.

Bacterial Sepsis

The integrity of the hepatic GABA transport system was also stressed in a study designed to elucidate the pathogenesis of the decreased levels of consciousness obeserved in patients with bacterial sepsis [34]. In a preliminary study, eight common bacterial pathogens were found to synthesize GABA in large amount when cultured in vitro [17]. In a subsequent study, systemic serum (but not CSF) GABA concentrations were found to be significantly higher in experimental animals and patients with bacterial sepsis and decreased levels of consciousness than in septic patients with normal levels of consciousness (Fig. 3) [34]. It was proposed that during bacterial sepsis enough GABA is produced to overwhelm the hepatic GABA transport system or that a product of bacterial growth might interfere with the mechanism of action of the transport system.

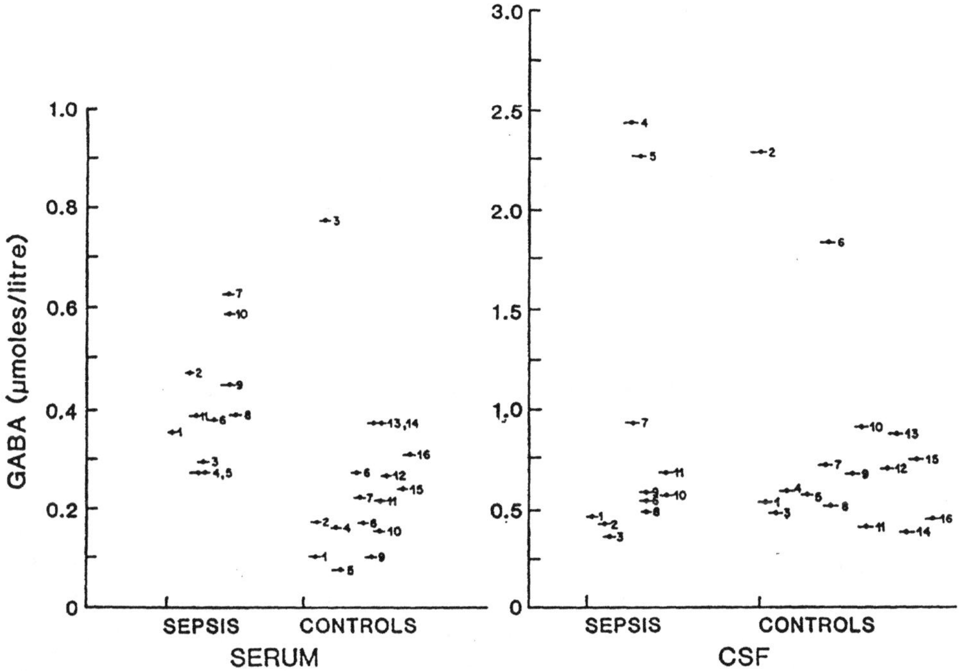

Fig. 3. Serum and cerebral spinal fluid GABA concentration in 11 septic patients with decreased levels of consciousness and 16 control patients with normal levels of consciousness. The difference in serum GABA concentrations between septic and control patients was significant at a p value less than 0.05

GABA Receptors in the Liver

During the process of identifying and characterizing the GABA transport system in the liver it was noted that bicuculline, a $GABA_A$ receptor anta gonist, caused a significant decrease in specific binding of GABA to isolated rat hepatocytes. This finding raised the possiblity that the liver, like the brain, might possess both a GABA transport and a GABA receptor system. A series of experiments were then carried out to determine whether this was the case. The results of those experiments were positive in that specific binding of GABA to hepatocytes occurred under receptor binding conditions, i.e., the absence of sodium, and that the binding could be blocked by the presence of bicuculline [19]. Moreover, the physiologic effect of activating this receptor system was identified when hepatocyte transmembrane potential was shown to decrease by approximately 5 mV when GABA was added to a suspension of isolated cells. The hyperpolarizing effect was even more striking when muscimol, a less rapidly metabolized $GABA_A$ receptor agonist, caused a 22 mV decrease in hepatocyte transmembrane potential. Finally, the specifity of this effect was confirmed by demonstrating that bicuculline both prevented and reversed the muscimol induced effect.

Clinical Implications of Hepatic GABA Receptor Activity

The importance of the hepatic GABA receptor system in health and disease states is the ongoing focus of research being carried out in our investigative unit. Initially, we designed experiments to screen for the effects of GABA on the liver based on the known effects of other neurotransmitter agents on hepatobiliary function. These include alterations in hepatic blood and bile flow, glucose homeostasis, and hepatic regeneration. With respect to hepatic blood flow, we found that intravenous infusions of GABA into anesthetized dogs caused a slight decrease in portal venous blood flow but not of statistically significant magnitude [20]. Similarly, intravenous GABA infusions into rats cause a mild increase in hepatic bile flow but only at concentrations of GABA that far exceeded normal physiologic levels [22]. Hepatic venous glucose concentrations increased following portal venous infusions of GABA but the increase never exceeded 15% of baseline values [18].

The possibility that GABA might play a role in regulating hepatic regenerative activity was intriguing for a number of reasons. First, it had been shown previously by other investigators that GABA concentrations increase significantly as the development of certain tissues in utero approached completion [12, 13, 29, 28]. This had been interpreted as indicating that GABA was responsible or at least contributed to the suppression of rapid growth by these tissues. Second, exogenous GABA had been reported to completely inhibit the growth of murine squamous cell carcinoma and Hela cell lines [1]. This data further supported the concept that GABA possesses growth regulatory properties. Third, as mentioned previously, GABA is a derivative of putrescine, a compound that is essential to hepatic regeneration [15]. Thus, it is conceivable that a negative feedback loop exists whereby excess GABA concentrations serve to inhibit further putrescine production.

In order to test the hypothesis that GABA is involved in regulating hepatic regenerative activity, adult male Sprague-Dawley rats that had undergone either 70% partial hepatectomy or sham surgery received 1 ml of intraperitoneal GABA (500 μg/g body weight) or an equal volume of normal saline. The dose of GABA had been predetermined to achieve serum concentrations that approximated those reported in patients with liver failure. GABA administration was begun 16 hours prior to partial hepatectomy and continued twice daily thereafter until the animals were killed 24, 48, or 72 h post partial hepatectomy. Serum and hepatic tissue polyamine and GABA concentrations were documented at the time of partial hepatectomy and/or the time of death. Hepatic regenerative activity was determined by calculating restitution of liver mass, [^{3}H]thymidine incorporation into DNA, and [^{14}H]leucine incorporation into protein by the liver. The results of the study revealed that exogenous GABA significantly decreased the amount of restituted liver mass and the rate of protein synthesis by the liver following partial hepatectomy. DNA synthesis, however, remained unaltered in GABA treated animals when compared to controls. The effect of GABA on hepatic polyamine concentrations provided some insight as to how GABA achieved the inhibitory effect on liver growth. Specifically, GABA attenuated by 64%-100% the marked rise in hepatic putrescine levels that occurs 24, 48, and 72 h following partial hepatectomy. Currently, efforts are underway to determine the mechanism whereby GABA prevents the increase in hepatic putrescine concentrations. To that end, the effects of GABA on hepatic uptake, synthesis, and metabolism of putrescine are being documented.

Future Directions

In our laboratory, the concept that the liver merely serves as a repository and disposal site for GABA derived from other organ systems is steadily giving ground to the belief that GABA plays an integral role in modifying hepatobi-liary growth and development. If correct, numerous opportunities exist for manipulating the GABA transport and receptor systems to the advantage of the patient with liver disease. For example, interfering with GABA synthesis or uptake by the liver should result in stimulated hepatic regeneration for patients with hepatitis or those who have recently undergone partial hepatectomy. Alternatively, the GABA system might be brought into play if hepatic regenerative growth is excessive such as with benign or malignant tumors of the liver. Here, augmentation of GABA synthesis or inhibition of its metabolism would be appropriate therapeutic considerations.

In conclusion, over the course of the past 40 years we have gone from documenting GABA concentrations and the activity of its relevant enzymes in the liver to very exciting studies in which the hepatic GABA transport and receptor systems are being manipulated in an effort to provide a better understanding and therapeutic approach to patients with liver disease.

Acknowledgement. The authors wish to thank Ms. D. Byron for her prompt and accurate typing of this manuscript.

References

1. Boggust WA, Al - Nakib (1986) Promotion and suppression of tumor growth and cell proliferation by acetylputrescine and putrescine and their oxidation products acetyl-GABA and GABA. IRCS Med Sci 14: 174 - 175
2. Christensen HN, Cespedes C, Hadlogten ME, Ronquist G (1973) Energization of amino acid transport studied for the Ehrlich ascites tumor cell. Biochim Biophys Acta 300: 487-522
3. Crossley IR, Williams R (1984) Progress in the treatment of chronic porta-systemic encephalopathy. GUT 25: 89-98
4. Drummond RY, Phillips AT (1974) L-glutamic acid decarboxylase in non neural tissues of the mouse. J Neurochem 23: 1207-1213
5. Erdö SL, Kiss B (1986) Presence of GABA, glutamate decarboxylase and GABA transaminase in peripheral tissues: a collection in quantative data. In: Erdö SL, Bowery NG (eds) GABAergic mechanisms in the mammalian periphery. Raven, New York, pp 5-17
6. Ferenci P, Covell D, Schafer DF, Waggoner JG, Schrager R, Jones AE (1983) Metabolism of the inhibitory neurotransmitter γ-aminobutyric acid in a rabbit model of fulminant hepatic failure. Hepatology 3: 507-512
7. Ferenci P, Ebner J, Zimmermann C (1988) Overestimation of serum concentrations of γ-aminobutyric acid in patients with hepatic encephalopathy by the γ-aminobutyric acid-radioreceptor assay. Hepatology 8: 69-72
8. Ferenci P, Schafter DF, Kleinberger G, Hoofnagle JH, Jones EA (1983) Serum levels of γ-amonobutyric acid-like activity in acute and chronic hepatocellular disease. Lancet 1: 811-814
9. Fogel WA (1986) GABA and polyamine metabolism in peripheral tissues. In: Erdö SL, Bowery NG (eds) Raven, New York, pp 35-56.

10. Fussi F, Savoldi F, Curti M (1988) Identification of N-carboxyethyl-γ-aminobutyric acid in brain: a key for a postulated pathway from spermidine to GABA? In: Perin A, Scalabrino G, Sessa A, Ferioli ME (eds) Perspectives in polyamine research. Wichtig, Milano
11. Gerber JC, Hare TA (1979) γ-Aminobutyric acid in peripheral tissue, with emphasison the endocrine pancreas. Diabetes 28: 1073-1076
12. Gilon P, Remacle C, Janssens de Varebeke Ph, Pauwels G, Hoet JJ (1987) GABA content and localisation of high-affinity GABA uptake during the development of the rat pancreas. Cell Mol Biol 33: 573-585
13. Gilon P, Reusens-Billen B, Remacle C, Janssens de Varebeke Ph, Pauwels G, Hoet JJ (1987) Localisation of high-affinity GABA uptake and GABA content in the rat duodenum during development. Cell-tissue Res 249: 593-600
14. Haber B, Kuriama K, Roberts E (1970) An anion stimulated L-glutamic acid decarboxylase in non-neural tissues. Occurence and subcellular localization in mouse kidney and developing chick embryo brain. Biochem Pharmacol 19: 1119-1136
15. Luk GD (1986) Essential role of polyamine metabolism in hepatic regeneration: inhibition of deoxribonucleic acid and protein synthesis and tissue regeneration by difluoromethylornithine in the rat. Gastroenterol 90: 1261-1267
16. Maddison JE, Dodd PR, Johnston GA, Farrell GC (1987) Brain γ-aminobutyric acid receptor binding is normal in rats with thioacetamide-induced hepatic encephalopathy despite elevated plasma γ-aminobutyric acid-like activity. Gastroenterol 93: 1026-1068
17. Minuk GY (1986) γ-Aminobutyric acid (GABA) production by eight common bacterial pathogens. Scand J infect Dis 18: 465-467
18. Minuk GY (1986) GABAergic mechanism and their functional importance in the liver. In: Erdö SL, Bowery NG (eds) GABAergic mechanisms in the mammalian periphery, Raven, New York, pp 325-337
19. Minuk GY, Bear CE, Seargent EJ (1987) Sodium independent bicuculline sensitive, [^{3}H] GABA binding to isolated rat hepatocytes. Am J Physiol 252: 642-647
20. Minuk GY, Mac Cannell KL (1988) It is the hypotension of cirrhosis a GABA-mediated process? Hepatology 8: 73-77
21. Minuk GY, Sarjeant EJ (1988) The effect of (a) neomycin and lactulose treatment on systemic and portal serum GABA levels in rats and (b) pH changes on [^{3}H]-GABA binding to isolated rat hepatocytes. Clin Invest Med 95: 1356-1363
22. Minuk GY, Sarjeant EJ (1984) γ-Aminobutyric acid (GABA) and hepatic bile flow in the adult rat. Clin Invest Med 7: 193-197
23. Minuk GY, Sarjeant EJ, Buchan K (1985) Elevated serum γ-aminobutyric acid (GABA) levels in children with Reyes syndrome. J Pediatr Gastroenterol Nutr 4: 528-531
24. Minuk GY, Vergalla J, Ferenci P, Jones EA (1984) Identification of an receptor system for γ-aminobutyric acid on isolated rat hepatocytes. Hepatology 4: 180-185
25. Minuk GY, Winder A, Burgess ED, Sarjeant EJ (1985) Serum γ-aminobutyric acid (GABA) levels in patients with hepatic encephalopathy. Hepatogastroenterology 37: 171-174
26. Minuk GY, Winder A, Burgess ED, Sarjeant EA (1981) Colonic bacteria: a source of γ-aminobutyric acid in blood. Prod Soc Exp Biol Med 167: 301-303
27. Schafer DF, Fowler JM, Jones EA (1982) Hepatic encephalopathy and the γ-aminobutyric acid neurotransmitter system. Lancet 1: 18-19
28. Seiler N, Bink G, Grove J (1980) Relationships between GABA and polyamines in developing rat brain. Neuropharmacology 19: 251-258

29. Sobue K, Nakajima T (1977) Metabolism of putrescine in neuroblastoma and glioma cells during culture. J Biochem 82: 1121-1126
30. Tallan HH, Moore S, Stein WH (1954) Studies on the free acids and related compounds in the tissues of the cat. J Biol Chem 211: 927-939
31. Tsuiji M, Nakajima T (1978) Studies on the formation of γ-aminobutyric acid from putrescine in rat organs and purification of this sythetic enzyme from rat intestine. J Biochem 83: 1407-1412
32. Waniewski RA, Suria A (1977) Alterations in γ-aminobutyric acid content in the rat superior cervical ganglion and pineal gland. Life Sci 21: 1129-1142
33. White HL, Sato TL (1978) GABA-transaminases of human brain and peripheral tissues-kinetic and molecular properties. J Neurochem 31: 41-47
34. Winder TR, Minuk GY, Sarjeant TP (1988) γ-Aminobutyric acid (GABA) and sepsis-related encyphalopathy. Can J Neuro Sci 15: 23-25
35. Zachmann M, Tocci P, Nyhan WL (1966) The occurence of γ-aminobutyric acid in human tissues other than brain. J Biol Chem 241: 1355-1358
36. Zeneroli ML, Iuliano E, Racagni G, Baraldi M (1982) Metabolism and brain uptake of γ-aminobutric acid in galactosamine-induced hepatic encephalopathy in rats. J Neurochem 38: 1219-1222

GABA-Immunoreactive Structures in Rat Kidney[1]

E. Dobó, Á. Párducz, J.R. Wolff, and S.L. Erdö

Introduction

A large body of evidence obtained from biochemical and pharmacological studies shows that each element of the γ-aminobutyric acid (GABA)ergic system is present in the mammalian kidney. The GABA content in the kidney of several species, including humans, was found to be as high as 1/20-1/5 of that in brain [3]. The activity of the GABA synthesizing enzyme in kidney is about 1/6 that of the brain glutamic acid decarboxylase (GAD) [3]. Moreover, the kidney exhibits nearly half as high an activity of GABA transaminase (GABA-T) as the brain does [26].

The occurence of binding sites of muscimol and baclofen, known to label $GABA_A$ and $GABA_B$ receptors, respectively, has been proved [1, 2]. Recently, a high-affinity transport system for GABA has also been identified [9, 16].

The distribution of makers for the GABA system has also been studied. The activity of the catabolizing enzyme of GABA was localized by means of histochemistry in the juxtamedullary zone of renal cortex [22]. In line with this, immunohistochemical findings, using an antiserum raised against mouse GABA-T, showed that the apical parts of the convoluted tubules were labeled, while the medulla was not found to be immunoreactive. Light microscopic autoradiographic studies using radiolabeled muscimol or baclofen indicate that $GABA_A$ and $GABA_B$ receptor sites, respectively, are located in the renal cortex [1, 2]. On the basis of these histochemical

[1] This study was supported by the Deutsche Forschungsgemeinschaft (Wo 279/6-5 and 8-3).

findings, GABA might be expected to influence renal function in the kidney cortex rather than in the medulla, most probably in the proximal convoluted tubules.

Although excellent GAD antisera [15, 24] have been widely utilized to reveal this enzyme in the nervous system, the distribution of GAD has not yet been reported in the kidney. This is probably due to the fact that brain GAD and kidney GAD display weak immunological cross-reactivity [23, 25], which makes the immunohistochemical demonstration of the renal GAD difficult. In addition, tissue preparation further decreases the chance of detecting the rate limiting enzyme of GABA biosynthesis.

It is presumable that the sites of GAD and GABA-T may not necessarily correspond with the location of GABA in different cellular components of various tissues. Immunohistochemical techniques are available which allow the visualization of small molecules such as amino acid transmitters. Since the sensivity and reliability of amino acid immunohistochemistry are also superior to enzyme immunohistochemistry [18], we attempted to localize GABA itself rather than any other markers of the GABA system in the rat kidney.

GABA Immunohistochemistry

Antisera

Two rabbit antisera to GABA, designed as no. 483a[2] or no. 7 (gifts from Drs. P. Petrusz and P. Somogyi, respectively) were used in this study. These antisera were found to be highly specific for GABA conjugated to proteins via glutaraldehyde [10, 12]. Apart from the GABA antibodies, antisera against other amino acids such as glycine, glutamic acid, and taurine (No. 507a, 428a[2], and 508b, respectively; kindly provided by Dr. P. Petrusz) were also used to screen renal sections. Each antiserum was tested in a dilution range from 1:2 000 to 1:18 000 in order to find its optimum concentration. In addition, the sera were pretreated with glutaraldehyde-reacted bovine serum albumin, which was prepared according to the method of Storm-Mathisen and coworkers [20].

Tissue Preparation

CD male rats (weighing 300-350 g), with free access to drinking water and normal laboratory animal feed, were used. The animals were anesthetized with ether and perfused transcardially, first with physiological saline (PBS) then with fixative containing a high concentration (5%) of glutaraldehyde. Efforts were made to perfuse the kidney as thoroughly as possible. Tissue was only processed if the entire organ was completely free of blood. The kidneys on both sides were cut transversely, at 75 μm thickness, in a cryostat.

Pre-embedding Immunohistochemistry

For light microscopy, sections were treated with methanolic H_2O_2 to facilitate the penetration of antibodies and to suppress the peroxidase-like activity of erythrocytes, which might still be present within the blood vessels. Free-floating sections were then processed for immunohistochemistry using Sternberger's peroxidase anti-peroxidase (PAP) method [19]. Incubations were carried out at 37°C in the following sequence: (a) in normal goat serum; (b) in primary antibody overnight; (c) in goat anti-rabbit IgG; (d) in rabbit PAP.

The immunoreactions were visualized by 3,3-diaminobenzidine-4 HCl (0.05%) and H_2O_2 (0.005%) in the presence of $NiCl_2$ (0.1%) for 10-15 min. For electron microscopy, selected specimes were treated with OsO_4, dehydrated in a graded series of ethanol, and flat-embedded in Araldite.

Postembedding Immunohistochemistry

The frozen sections, prepared as above, were osmicated, dehydrated, and flat-embedded in Durcupan ACM (Fluka). Semithin sections (1-2 μm) were cut and then air-dried onto glass slides. The samples were etched according to Maxwell's method [14]. Tissue-bound osmium was removed with $NaIO_4$ [18]. Afterwards, immunoreagents were overlayered onto the sections in the above-mentioned order. It should be noted that the postembedding staining was only performed with the GABA antibody no.7 diluted 1:2 000-1:6 000. The immunohistochemical protocol, apart from the dilution of primary antibody, was the same as described above.

Control Experiments

Two sets of experiments were performed. Primary antibodies were (1) omitted or replaced by normal rabbit serum at the same dilution or (2) preincubated with the corresponding amino acids coupled to bovine serum albumin via glutaraldehyde at 37°C for 8 h, then applied to the tissue [11].

Light Microscopic Observations

In pilot experiments, the optimal fixation were sought for GABA immunohistochemistry. Higher concentrations (up to 5%) of glutaraldehyde yielded more reproducible immunostaining. Therefore, all subsequent results were obtained with this high glutaraldehyde concentration.

Intense immunoreactivity was encountered in the kidney using either of the GABA antisera (Fig. 1). In the cortex, the connecting and collecting tubules stood out from the background (Fig. 2, 3). Their walls displayed a certain mosaic-like pattern (Fig. 4) in which one subpopulation of their constituent cells was immunoreactive for GABA, while the other was negative (Fig. 5). The staining intensity of the positive cells varied considerably. The nuclei of the immunoreactive cells were also often labeled by reaction end product. The cortical collecting tubules were still well stained with, e.g., GABA antiserum no. 7 at a dilution of 1:10 000, while a 1:6 000 dilution of this antiserum produced strong staining.

The renal corpuscels, including the glomeruli and the Bowman's capsules and the proximal convoluted tubules were not labeled at all. The distal convoluted tubules, again, were generally found to be negative. However, increasing the titer of the antibody, e.g., up to 1:2 000 using antiserum no. 7 resulted in a more pronounced staining in the distal convoluted tubules but raised nonspecific staining, especially in the glomeruli.

Strikingly, the most intense immunoreactivity was encountered in the medulla, especially in its outer zone (Fig. 1). At a light microscopical level,

Fig. 1-4. Light micrographs showing the distribution of GABA immunoreactivity in kidney. **1** GABA-immunoreactive tubular structures are located both in the cortex and in the medulla. Note that the latter displays much more intense immunoreactivity. For comparison see also Fig. 5. *Scale bar*, 200 μm. **2** In the cortex, immunostaining is confined to the collecting *(c)* and connecting *(c)* tubules. *Scale bar*, 60 μm. **3** A merging of a collecting *(c)* and a connecting *(c)* tubule is presented. Note that the glomeruli *(g)* are not immunolabeled. *Scale bar*, 40 μm. **4** The immunostaining of the cortial collecting tubules displays a checkered pattern at higher magnification. *Scale bar*, 10 μm

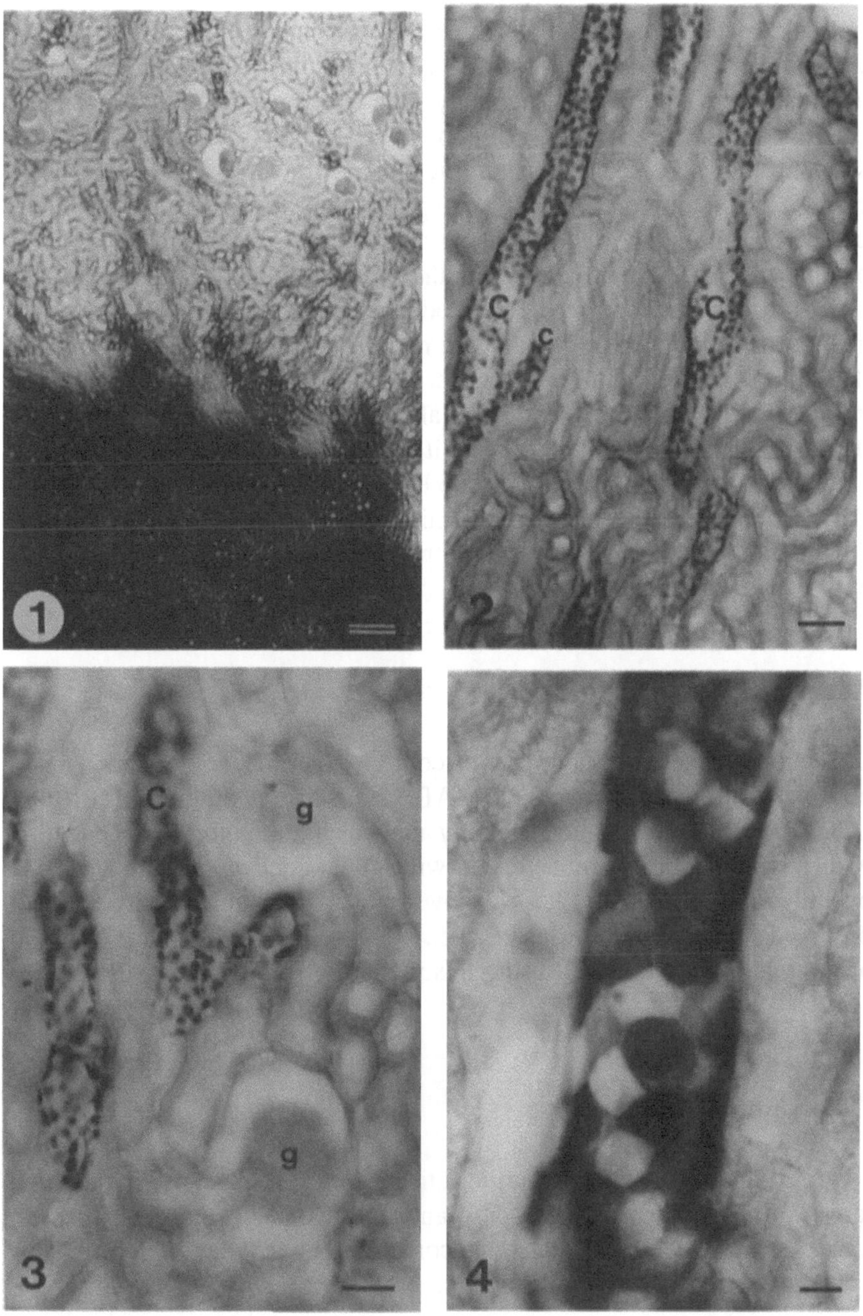
1
2
C
c
C
3
C
g
g
4

in thick frozen sections, it was not possible to distiguish the constituents of the tubular epithelia. This overall staining of the medulla could not be attenuated by reducing the antibody titer. The medulla was still labeled with the antiserum no. 7 at a dilution of 1:18 000. Therefore, it was reasonable to apply a postembedding technique for GABA. This technique made it possible to differentiate among the tubules (Fig. 6). The medullary collecting tubules showed the heaviest staining (at 1:6000-1:2000 dilution of antiserum no. 7). In contrast to the cortical collecting tubules, which were of checkered appearence, the medullary tubules exhibited more homogenous staining, in the sense that the majority of the epithelial cells lining these tubules were immunopositive for GABA. Their staining intensity also varied greatly. Immunolabeling was also found in the bulging cells of the thin limb of Henle's loop. Although it is somewhat difficult to distinguish this type of epithelium from the endothelium in the capillaries in the semithin sections, some segments of the thin limb of Henle's loop were clearly devoid of staining for GABA. It is worth mentioning that nerve fibers were not labeled at all.

Antisera Specifity

In sections made from both sets of control experiments, no staining was observed. The absorption of the GABA antisera (at dilutions 1:10 000-1:2000) with GABA conjugate totally abolished the staining. Preincubation with bovine serum albumin treated with glutaraldehyde also reduced the overall staining intensity, but selective loss of immunoreactivity was not observed, In fact, the best results, i.e., strongly labeled structures with no background, were obtained using this method of pretreatment of the antisera.

Other Amino Acid Antisera

Glycine, glutamic acid, and taurine antisera were also screened in the same range of dilutions as the GABA antisera. None of these resulted in immunoreactivity either in the renal cortex or the medulla. Therefore, their specificity was not tested thoroughly.

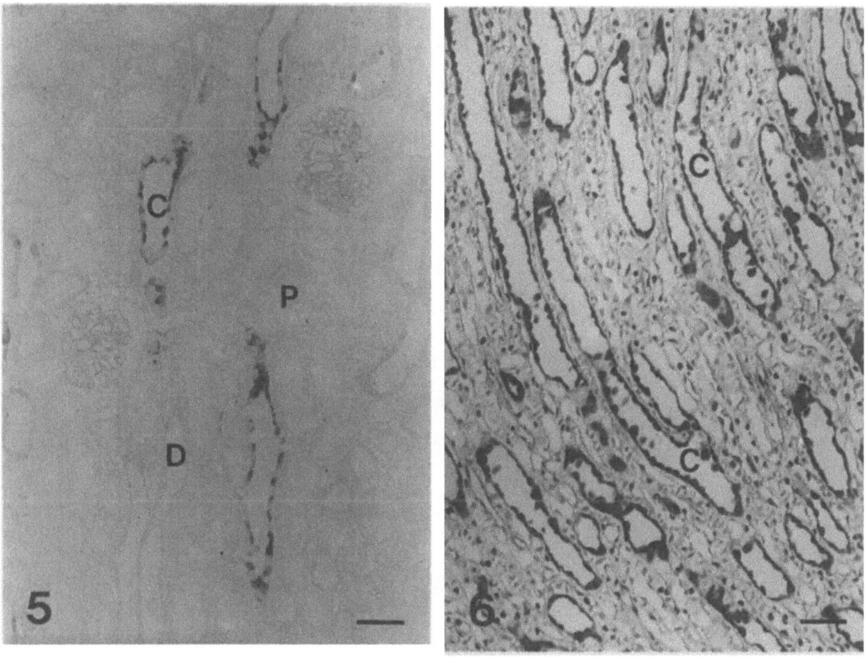

Fig. 5-6. Light micrographs demonstrating GABA immunoreactivity in semi-thin sections of kidney. **5** Note the checkered staining in the collecting tubules *(C)* and that the proximal *(P)* and distal convoluted tubules are not immunolabeled. *Scale bar*, 75 μm. **6** The postembedding application of the GABA anitbodies revealed that the majority of the epithelial cells lining the medullary collecting tubules *(C)* are positive. *Scale bar*, 50 μm

Electron Microscopical Observations

Under the electron microscope, our light microscopic observations were confirmed. Renal corpuscles and proximal and distal convoluted tubules were free of reaction endproduct. Immunoreactivity was found in the epithelial cells constituting the straight part and some segments of the thin limb of Henle's loop. About half of the squamous cells lining the thin limb were immunopositive. In electron micrographs, the negative and positive cells were located frequently in alternate succession (Fig. 7). No structural

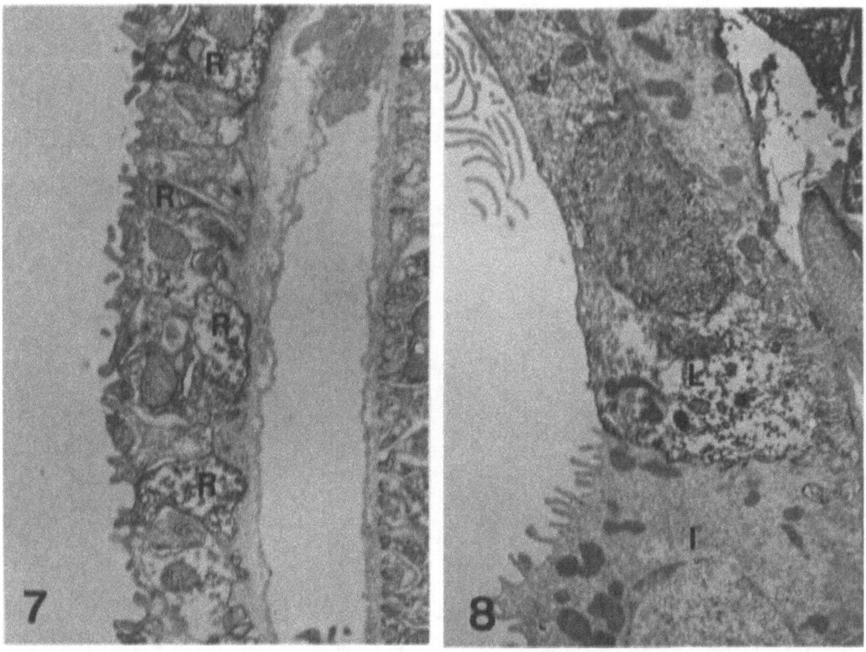

Fig. 7-8. Electron micrographs of GABA immunoreactivity in kidney. **7** Reaction endproduct *(R)* is visible in about half of the squamous cells lining the thin limb of Henle's loop; X 20000. **8** In the collecting tubule, the principal or light cell *(L)* exhibits immunodeposits, while the intercalated cell *(I)* does not; X 11000

differences could be established between the GABA-negative and the GABA-positive cells.

The straight (i.e., the thick ascending) segment of Henle's loop, which is located in the outer zone of medulla, also displayed checkered organization. The connecting and the cortical collecting tubules of checked appearance, were also examined. Electron microscopy revealed that their immunoreactive epithelial cells showed the characteristic features of principal or light cells. This cell type is endowed with few organelles and a relatively smooth apical surface, whereas the intercalated or dark cells possess more organelles, denser cytoplasm and more and longer microvilli (Fig. 8). Occasionally, some endothelial cells lining capillaries also contained reaction endproduct.

Discussion

Indirectly, evidence has already been presented to indicate that the GABA content of kidney is not of neuronal origin [8]. Accordingly, we were unable to detect GABA-immunoreactive neuronal elements in this organ. Instead, application of GABA immunohistochemistry to renal tissue produced intensely stained tubular structures in both the cortical and medullary areas. Our findings concerning cortical labeling are partly in agreement with earlier biochemical results which indicated the presence of GABA and its synthesizing enzyme in the cortex [8]. These results that the glomerular fraction was not enriched in GAD activity, coincide with our observation that the renal corpuscles are not immunoreactive for GABA. Moreover, both Amenta and coworkers [1] and Erdö [2], using radioligand autoradiography, did not find $GABA_A$ and $GABA_B$ receptor sites in the glomeruli. Taken together, the data indicate that the glomeruli neither produce GABA nor contain sites sensitive to GABA. Consequently, the GABA system does not seem to have any specific function in the renal corpuscule.

Using different biochemical techniques, Goodyer and associates [8] detected the majority of GAD activity in the fraction of proximal tubules which were obtained from renal cortex by differential centrifugation methods. They speculated that GABA may have a function in the proximal tubular cells. This supposition seems to be in line with the histochemical and immunocytochemical distribution of GABA-T in the kidney of rats [22, 25], even though GABA-T is considered to be a poor marker of the GABA system. However, our immunohistochemical observations do not correspond with their supposition, since the GABA antibodies did not label the convoluted and the straight parts of the proximal tubule.

Since previous biochemical studies on the renal GABA system were restricted to the entire kidney or the cortex, biochemical data on the medullary GABA system have not yet been available. Strikingly, our present results show that the densest GABA immunoreactivity can be found in the medulla. In addition to the connecting and the cortical collecting tubules, immunolabeling could be observed both in the thin limb and its ascending part of Henle's loop and in the medullary collecting duct.

The epithelial cells lining a certain segment of the thin limb of Henle's loop and the straight part of the distal tubule are considered to be of identical origin and morphology. In spite of this, the application of GABA antisera revealed some heterogeneity in these tubular cells. These observations may raise the question whether GABA immunoreactivity in approximately every second epithelial cell reflects any functional heterogeneity or not.

The distribution of the immunostaining was also heterogeneous within the walls of the connecting and the collecting tubules. In contrast to the former segments of the nephron, using electron microscopy we succeeded in differentiating between the immunopositive and the immunonegative profiles. According to their morphological features, the labeled cells correspond to principal cells.

In fact, the remarkable immunostaining within the medulla was unexpected. The autoradiographic demonstration of GABA receptor sites and the immunocytochemical localization of GABA-T suggested that, if anything, the cortex rather than the medulla would be involved in the renal GABA function. Nevertheless, the tubular segments, namely, the connecting and the collecting tubules within the cortex, the ascending part of Henle's loop, and the collecting duct within the medulla, that were found to exhibit GABA immunoreactivity, have recently been reported to possess vasopressin binding sites [7]. This coincidence may suggest a possible interaction between GABA and vasopressin.

The renal segments listed above, apart from the collecting ducts of the inner medulla, contain (Na^+, K^+) ATPase [6,13,17]. This led to the assumption that, through ATPase inhibitors, GABA may be released from kidney. Our preliminary experiments provided evidence that remarkable amounts of both the cortical and the medullary GABA content could be released by 2.5 m*M* ouabain, a selective inhibitor of the (Na^+, K^+)ATPase, or, alternatively, by 56 m*M* K^+ [4]. It would be useful to know whether GABA is released into the blood vessels or directly into the urine. Interestingly, it was found that the urine of cats normally contains two orders of magnitude more GABA than the blood of the same species [21]. The high GABA content of urine might be derived, in part, from the kidney via its functional release. The most intense GABA immunoreactivity was observed in those tubules which are known to be responsible for conserving water, thus influencing the final volume of urine. Since the urinary bladder and its sphincter are endowed with high-affinity GABA receptors, with K_d values around 50 n*M* [5], we cannot exclude that the urinary GABA, in the μM range, might be a humoral agent which modulates the motility of the bladder. As this mediation of GABA between the kidney and the lower urinary tract is quite speculatve, it is evident that much further work is needed to elucidate the possible role of GABA in the urinary system.

Acknowledgements. We gratefully acknowledge the expert technical help of Ms. H. Tytko and A. Wolff, and thank Prof. Drs. R. Herken (Göttingen) and F. Joó (Szeged) for their valuable advice. The generous donation of GABA antiserum No. 7 by Dr. P. Somogyi (Oxford) is acknowledged. We are particularly grateful to Dr. P. Petrusz (North Carolina) for providing us with antisera against GABA (no 483a[2]), glycine, glutamic acid, and taurine.

References

1. Amenta F, Cavallotti C, Iacopino L, Erdö SL (1988) Autoradiographic localization of the $GABA_A$ receptor agonist [^{3}H]-muscimol within rat kidney. Pharmacology 36: 390-395
2. Erdö SL (1990) Baclofen binding to $GABA_B$ receptor sites in rat kidney. Eur J Pharmacol 184: 305-309
3. Erdö SL, Kiss B (1986) Presence of GABA, glutamate decarboxylase, and GABA transaminase in peripheral tissues: a collection of quantitative data. In: Erdö SL, Bowery NG (eds) GABAergic mechanisms in the mammalian periphery. Raven, New York, pp 9-17
4. Erdö SL, Dobó E, Párducz A, Wolff RJ (1990) Releasable GABA in tubular epithelium of rat kidney. Experientia 47: 227-229
5. Erdö SL, Mione CM , Amenta F, Wolff JR (1989) Binding of (^{3}H)-muscimol to $GABA_A$ sites in the guinea-pig urinary bladder: biochemical assay and autoradiography. Br J Pharmacol 96: 313-318
6. Ernst SA (1975) Transport ATPase cytochemistry: ultrastructural localization of potassium-dependent and potassium-independent phosphatase activities in rat kidney cortex. J Cell Biol 66: 586-608
7. Fadool, MJ, Aggarwal KS (1990) Immunohistochemical demonstration of vasopressin binding in rat kidney. J Histochem Cytochem 38: 7-12
8. Goodyer PR, Mills M, Scriver CR (1982) Properties of gamma-aminobutyric acid synthesis by renal cortex. Biochim Biophys Acta 716: 348-357
9. Goodyer PR, Rozen R, Scriver CR (1985) A gamma-aminobutyric acid-specific transport mechanism in mammalian kidney. Biochim Biophys Acta 818: 45-54
10. Hepler JR, Petrusz P, Rustioni A (1986) Antisera to GABA, glutamate and aspartate: characterization by immunoabsorption and immunocytochemistry. J Histochem Cytochem 34: 110
11. Hepler JR, Toomim CS, McCarthy KD, Conti F, Battaglia G, Rustioni A, Petrusz P (1988) Characterization of antisera to glutamate and aspartate. J Histochem Cytochem 36: 13-22
12. Hodgson JA, Penke B, Erdei A, Chubb WI Somogyi P (1985) Antisera to gamma-aminobutyric acid. I. Production and characetrization using a new model system. J Histochem Cytochem 33: 229-239
13. Kashgarian M, Biemesderfer D, Caplan M, Forbush B (1985) Monoclonal antibody to NA, K-ATPase: immunocytochemical localization along nephron segments. Kidney Int 28: 899-913

14. Maxwell HM (1978) Two rapid and simple methods used for the removal of resins from 1.0 μm thick epoxy sections. J Microsc 122: 253-255
15. Oertel WH, Mugnaini E, Schmechel DE, Tappaz ML, Kopin IJ (1982) The immunocytochemical demonstration of gamma-aminobutyric acid-ergic neurons: methods and application. In: Chan-Palay V, Palay SL (eds) Cytochemical methods in neuroanatomy. Liss, New York, pp 297-329
16. Sidhu HS, Wood JD (1989) Gamma-aminobutyric acid uptake by rat kidney brush-border membrane vesicles. Eperientia 45: 726-728
17. Siegel GJ, Holm C, Schreiber JH, Desmond T, Ernst SA (1984) Purification of mouse brain (Na-K)-ATPase catalytic unit, characterization of antiserum and immunohistochemical localization in cerebellum, chorioid plexus, and kidney. J Histochem Cytochem 32: 1309-1318
18. Somogyi P, Hodgson JA, Chubb WI, Penke B, Erdei A (1985) Antisera to gamma-aminobutyric acid. II. Immunocytochemical application to the central nervous system. J Histochem Cytochem 33: 240: 248
19. Sternberger LA, Hardy PH, Cuculis JJ, Meyer HG (1970) The unlabeled antibody-enzyme method of immunohistochemistry. Preparation and properties of soluble antigen antibody complex (horseradish peroxidase-antihorseradish peroxidase) and its use in identification of spirochetes. J Histochem Cytochem 18: 315-333
20. Storm-Mathisen J, Leknes AK, Bore At, Vaaland JL, Edminson P, Haug F-MS, Ottersen OP (1983) First visualization of glutamate and GABA in neurones by immunocytochemistry. Nature, 301: 517-520
21. Tallan HH, Moore S, Stein WH (1954) Studies on the free amino acids and related compounds in the tissues of the cat. J Biol Chem 221: 927-939
22. Van Gelder NM (1967) A possible enzyme barrier for gamma-aminobutyric acid in the central nervous system. Prog Brain Res 29: 259-268
23. Wu J-Y, Chude O, Wein J, Roberts E (1978) Glutamate decarboxylase form neural and non-neural tissues. J Neurochem 30: 849-857
24. Wu J-Y, Lin C-T, Brandon C, Chan TS, Mohler H, Richards JG (1982) Regulation and immunocytochemical characterization of glutamic acid decarboxylase. In: Chan-Palay V, Palay SL (eds) Cytochemical methods in Neuroanatomy. Liss, New York, pp 279-296
25. Wu J-Y, Lin C-T, Lin H, Xu Y, Liu J-W, Hwang HB, Wei SC (1986) Immunohistochemical characterization and immunohistochemical localization of glutamate decarboxylase and GABA transaminase in peripheral tissues. In: Erdö SL, Bowery NG (eds) GABAergic mechanisms in the mammalian periphery. Raven, New York, pp 19-34
26. Wu J-Y, Moss LG, Chude O (1978) Distribution and tissue specifity of GABA-2-ketoglutarate transaminase. A new assay method. Neurochem Res 3: 207-213

GABAergic Mechanisms in Bovine Adrenal Chromaffin Cells: Their Role in the Regulation of Catecholamine Secretion[1]

M.J. Oset-Gasque, E. Castro, and M.P. Gonzalez

Introduction

γ-Aminobutyric acid (GABA), the major inhibitory neurotransmitter in the central nervous system (CNS), is also present in several peripheral tissues, where it could have a functional role in the regulation of muscle contraction or hormonal secretion (for a review see [7]).

Chromaffin cells from adrenal medulla are considered to be typical postsynaptic catecholaminergic neurons, because they have several embryological, structural, and functional properties in common with catecholaminergic neurons from brain. Like neurons, these cells originate from the neural crest, a tissue of ectodermic origin, and possess secretory vesicles, i.e., the chromaffin granules. These granules synthesize, store, and release catecholamines (CAs) and a number of neuropeptides to the blood by a calcium-dependent exocytic mechanism.

CA secretion is a process physiologically stimulated by acetylcholine (Ach), released from splanchnic nerves which supply the adrenal medulla. When Ach binds to nicotinic receptors located on the plasma membrane of chromaffin cells, it induces entry of Na^+ through receptor-associated Na^+ channels which depolarizes the cells and allows the activation of voltage-dependent Ca^{2+} channels. The subsequent Ca^{2+} entry triggers the extrusion of the CA granule contents to the extracellular space [6].

However, the recent identification of several neurotransmitters in chromaffin cells, such as opioid peptides, substance P, somatostatin, vasoactive

---

[1] This work was supported by CICYT grant pm88-0021.

intestinal peptide, histamine, angiotensin, neuropeptide Y, atrial natriuretic peptide, and GABA [18-23, 26, 8-10, 13], suggests a more complex picture of CA secretion by chromaffin cells, since, together with nicotinic Ach receptors, these neurotransmitters could act as physiological modulators of CA secretion.

In this paper we present a review of recent results concerning GABA mechanisms in adrenal medulla and their role in the control of CA secretion, using isolated and cultured chromaffin cells as a model.

GABA Metabolism in Adrenal Medulla

Adrenal medulla possesses significant quantities of GABA in a concentration, determined by gaschromatography, of about 0.5 nmol/10^6 cells (unpublished results). When we examined the possibility of adrenal medullary cells being able to synthesize and catabolize GABA, we found that adrenal medulla exhibits significant activities of the main enzymes related to GABA metabolism in the brain. Glutamate-α-decarboxylase (GAD), GABA-α-oxoglutarate trans-aminase (GABA-T), and succinic semialdehyde dehydrogenase (SSADH) were detected in this tissue [8-10].

These enzymes show many similarities to their homologues in the brain with respect to their subcellular localization (cytosol for GAD and mitochondria for GABA-T and SSADH), cofactor requirements (PLP for GAD and GABA-T and NAD for SSADH), K_m values for different substrates, optimal pH, intracellular effectors, extracellular inhibitors, etc. However, in all cases the maximal enzymatic activities were lower than those of the brain enzymes [8-10].

The presence of GABA, GAD, and GABA-T in adrenal chromaffin cells has also been demonstrated by Kataoka et al. [13] by histochemical and biochemical methods. Using specific antibodies, these authors showed that only 40% of the chromaffin cells in culture contained GAD activity, while GABA-T activity was detected in more than 70% of these cells. On the basis of this difference it is possible, as these authors suggested, that one cell population could be specialized in storing and releasing GABA, whereas the other could be implicated in the inactivation of GABA, as is the case in the brain.

Uptake of [^{3}H]GABA by Chromaffin Cells in Culture

In confirmation of our previous results on adrenal medulla slices [28], we showed that bovine adrenal medulla chromaffin cells maintained in primary culture are able to accumulate extracellular [^{3}H]GABA. This accumulation is linear for the first hour but changes thereafter. Our results on the influence of the extracellular concentration of Na^+ and Cl^- ions and the effect of inhibitors of the energetic pathways and on the Na^+/K^+-ATPase indicate that this uptake is an active process coupled to the Na^+ gradient maintained by the Na^+/K^+-ATPase, being necessary for the cotransport of two Na^+ ions and one Cl^- ion for each molecule of GABA transported [28].

Table 1. Kinetic constants of GABA uptake by chromaffin cells in culture in the presence of GABA uptake inhibitors (from [28])

Compound	K_m (μM)	V_{max} (pmol/h/5x10^6 cells)
Site 1	142 ± 34	639 ± 154
Control	373 ± 139**	590 ± 143
ß-alanine	266 ± 59**	620 ± 174
Taurine		
Site 2		
Control	10.8 ± 3.2	86.9 ± 20
Nipecotic acid	23.8 ± 5.2***	91.0 ± 9
L-DABA	17.0 ± 2.0**	85.3 ± 6.2

Chromaffin cells in culture were incubated with different concentrations of [^{3}H]GABA (between 0.5 μM and 1 mM) in the absence or presence of different compounds at a concentration of $10^{-4}M$. Kinetic constants were calculated by Eadie-Hofstee analysis of data and are means (± SEM) of four experiments, each performed in duplicate. **$p < 0.01$; ***$p < 0.001$. [28]

This uptake system presents two affinity sites for GABA:
(1) a high affinity site with a K_m = 10.8 μM and V_{max} = 86.9 pmol/h/5x10^5 cells, competitively inhibited by nipecotic acid and L-2,4-diaminobutyric acid and (2) another site of low affinity with a K_m = 142 μM and V_{max} = 639 pmol/h/5x10^5 cells, which is competitively inhibited by β-alanine and taurine (Table 1).

On the basis of the above results, we can conclude that chromaffin cells possess two differrent transport systems for GABA: one of high affinity showing the characteristics of neural GABA uptake and the other, probably common to other β-amino acids, with low affinity and chemical sensitivities similar to those found for GABA uptake by glial cells in different central and peripheral tissues. The cellular localization of this system is not known, although the presence of endothelial cells as normal contaminants of this type of culture suggests that at least a part of this uptake component could be attributed to this population of non-neural cells.

Both processes could play an important role in the regulation of GABA levels in the extracellular space surrounding chromaffin cells and thus in the regulation of CA secretion by GABA.

[^{3}H]GABA Release by Chromaffin Cells in Culture

An important question arising from studies on the role of GABA in chromaffin cells concerns the origin of the putatively regulatory GABA. GABAergic nerve fibers have been described in canine and bovine adrenals [11, 14], and GABA immunoreactivity has been localized inside chromaffin cells from bovine adrenal medulla [13] (Oset-Gasque et al., unpublished results).

Our results on GABA release indicate that chromaffin cells in primary culture are able to release [^{3}H]GABA in a dose-dependent fashion when stimulated by secretagogues known to induce release, such as nicotine, high KCl, veratridine, and calcium ionophores [29]. In general, there was a parallel release of both [^{3}H]GABA and [^{3}H]norepinephrine (NE), although in all circumstances the amount of the former was lower. However, GABA release was, in all cases, a specific process, since it could be blocked by specific antagonists of the secretagogues used, i.e., hexamethonium blocked the [^{3}H]GABA secretion evoked by nicotine, and tetrodotoxin (TTX) blocked, the release induced by veratridine but not that caused by 56 m*M* KCl (Table 2).

The nicotine and high KCl-evoked [^{3}H]GABA release was absolutely Ca^{2+}-dependent thus indicating the existence of an exocytic-like mechanism. In the veratridine-induced release, a Ca^{2+}-independent but Na^+-dependent component was also detected, since this component was eliminated in the absence of Na^+ in the extracellular medium. Moreover, given that the latter component was enhanced by ouabain and inhibited by known GABA

Table 2. Specificity of [^{3}H]GABA release evoked by various secretagogues in chromaffin cells (from [29]).

Treatment	Enhancement over basal release (%)
Cholinergic receptor agonists	
10 *μM* nicotine	29.5 ± 1.5
10 *μM nicotine* + *1 m*M hexamethonium	0.1 ± 0.5
Depolarizing agents by activating voltage-dependent cation channels	
Voltage-sensitive Na^+ channels	
10 *μM* veratridine	49.0 ± 1.2
10 *μM* veratridine + 1 *μM* TTX	23.0 ± 0.8
Voltage-sensitive Ca^{++} channels	
56 m*M* KCl	36.8 ± 2.1
56 m*M* KCl + 1 *μM* TTX	38.2 ± 1.9

Results are means ± SEM of two experiments, each performed in duplicate.
* $p < 0.001$ [29]

uptake inhibitors, we can conclude that the mechanism of this GABA release is mediated by reversion of the electrogenic GABA carrier in chromaffin cells [29].

The existence of two mechanisms of GABA release seems to be supported by our results on the subcellular localization of [^{3}H]GABA taken up by chromaffin cells. Our studies show that about 85% of the GABA taken up by chromaffin cells is localized in the cytosol and the remaining 15% in particulate fractions [29].

In conclusion, our data show clearly that cultured chromaffin cells can release [^{3}H]GABA together with CAs. This ability to secrete GABA closes a regulatory feedback loop in the control of exocytosis of CAs in the adrenal medulla since, as presented below, GABA can modulate CA secretion.

GABA Binding Sites in Adrenal Medulla Membranes

The first evidence showing the possibility that adrenal medulla contains receptors for GABA was obtained by Kataoka et al. [13], who found binding of [^{3}H]muscimol to chromaffin cells in culture, and by Martinez et al. [24], who observed binding of [^{3}H]GABA displaced by muscimol and (-)-baclofen. These results suggested that adrenal medulla contained both $GABA_A$ and $GABA_B$ receptors. These preliminary results were confirmed by those of Castro et al. [4], who showed that specific binding of [^{3}H]GABA to adrenal medulla membranes is biphasic, as reflected in the curved Scatchard plot in Fig.1A, with a low and a high affinity component. Both types of binding were specifically blocked by treatment with muscimol (low affinity) and baclofen (high affinity), indicating the existence $GABA_A$ low affinity binding sites and $GABA_B$ high affinity binding sites. Computer-assisted analysis allowed us to combine data from blockade and nonblockade experiments giving kinetic constants of K_d = 139 ± 22 n*M* and B_{max} = 0.4 pmol/mg protein for $GABA_A$ sites of low affinity (Fig. 1B) and K_d = 41 ± 6 n*M* and B_{max} = 0.35 ± 0.26 pmol/mg protein for the $GABA_B$ sites of high affinity (Fig. 1C).

The existence of both sites was confirmed by experiments on the calcium-sensitivity of binding. Specific [^{3}H]GABA binding obtained after muscimol blockade ($GABA_B$ sites), was enhanced when Ca^{2+} was added to the incubation medium, the effect reaching its maximum level at a Ca^{2+} concentration of 1 m*M*; however, [^{3}H]GABA binding obtained after blocking with (-)-baclofen ($GABA_A$ sites) was not significantly modified by Ca^{2+} ions until 1 m*M* [4].

In conclusion, these results demonstrated for the first time that both $GABA_A$ and $GABA_B$ receptors are present in adrenal medulla. The function of these receptors is discussed below.

Function and Mechanism of Action of GABA Receptors in Bovine Chromaffin Cells

$GABA_A$ Receptors

Sangiah et al. [32] reported that GABA causes release of CAs from isolated perfused bovine adrenals and pointed out the possibility that adrenal medulla contains GABA receptors. Kitayama et al. [16] found that GABA produced CA release from isolated perfused dog adrenals, this process being Ca^{2+}-dependent and independent of external Na^{+} and Cl^{-}.

According to our results [6] on bovine chromaffin cells, GABA stimulates basal CA release from these cells in a concentration-dependent manner. This release represents about 70% of that obtained by the same doses of nicotine under similar experimental conditions. These results differ from those of Kataoka et al. [15], who observed that the maximal response evoked by GABA was about ten times smaller than that evoked by ACh, a difference that could be caused by the cell preparations; in our study fresh isolated cells were used, whereas Kataoka et al. [15] used cultured cells. The fact that GABA stimulates basal CA release and that this stimulation can be mimicked by muscimol, a $GABA_A$ agonist, and blocked by bicuculine, a $GABA_A$ antagonist, indicates that this action is mediated by $GABA_A$ receptors (Fig.2). The EC_{50} for the effect of GABA on basal CA secretion was 7.2 ± 1.1 *μM*, which is well in line with GABA concentrations necessary to stimulate the $GABA_A$ receptor in brain preparations in vitro [2, 25, 33].

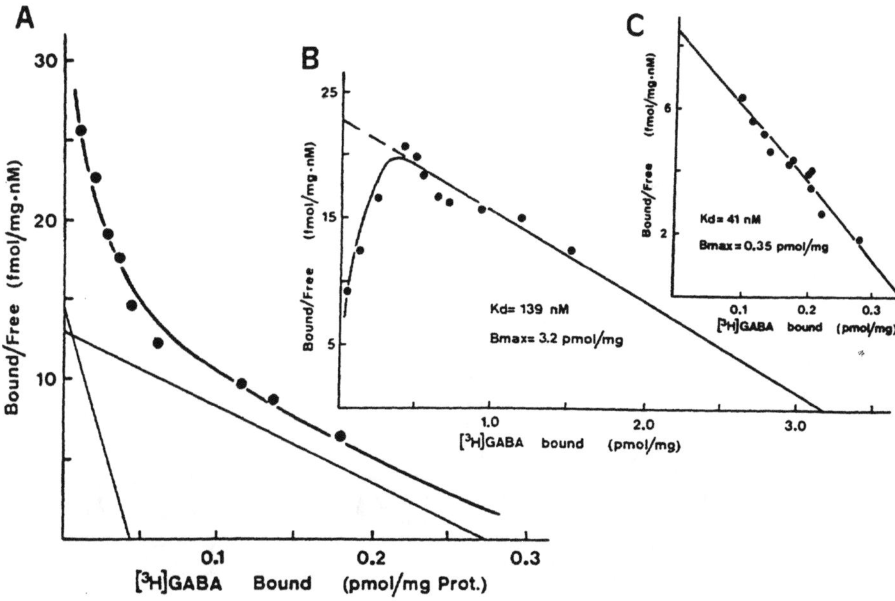

Fig. 1. A Scatchard plot of [³H]GABA binding to bovine adrenal medulla membranes in the presence of 1 m*M* $CaCl_2$ without any displacing agent. **B** Binding of [³H]GABA to adrenal medulla membranes in the presence of 100 μ*M* (-)-baclofen blocking ($GABA_A$ sites). **C** Scatchard plot of [³H]GABA binding to sites remaining after 100 μ*M* muscimol blockade ($GABA_B$ sites). (From [4]).

The CA release elicited by GABA is Ca^{2+}-dependent and is inhibited in the absence of Ca^{2+} in the extracellular medium (Fig. 2) and by both blockers of the L-type of voltage-dependent calcium channels, such as nifedipine and Cd^{2+}, but not by blockers of the T-type of voltage-dependent calcium channels, such as Ni^{2+} [5]. This pattern of inhibition is in agreement with the known pathways for Ca^{2+} entry into chromaffin cells [17]. In addition, GABA and muscimol increased the intracellular concentration of Ca^{2+}, as measured by the fluorescent probe Fura-2. This increase is specifically blocked by the $GABA_A$ antagonist bicuculline and the calcium channel blocker nifedipine.

These results suggest that GABA evokes CA secretion by producing a depolarization due to an increase in the membrane conductance to Cl^-, as occurs when the neurotransmitter binds to its specific site on the GABA receptor/Cl^- channel complex. Similarly, our more recent results on the efflux of $^{36}Cl^-$ show that muscimol induces an increase in the outflow of $^{36}Cl^-$ from chromaffin cells in culture. This effect was blocked by picrotoxin, a specific blocker of the Cl^- channel associated with the $GABA_A$ receptor (González et al., manuscript in preparation).

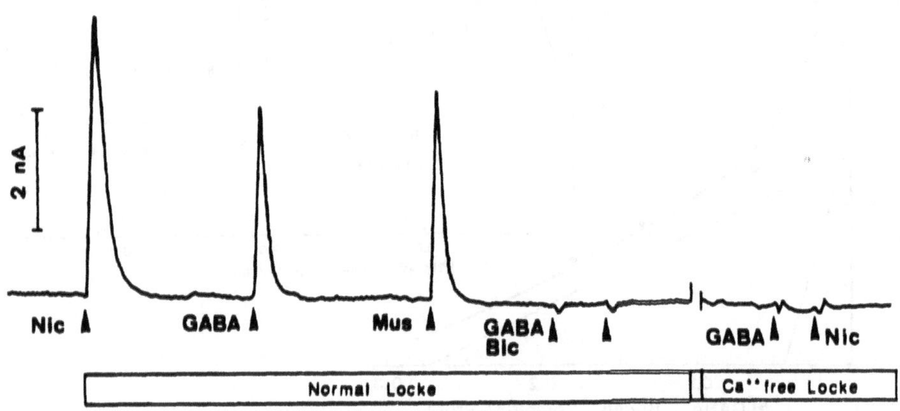

Fig. 2. Typical records of electrochemically monitored CA secretion peaks evoked by 1 m*M* nicotine, 1 m*M* GABA, 1 m*M* muscimol, and 1 m*M* GABA + 1 m*M* bicuculline. (From [5])

In addition to the GABA effect on the basal secretion of CAs, GABA, through its $GABA_A$ receptors, increases the CA release evoked by submaximal doses of nicotine but has an inhibitory effect on the nicotine-evoked secretion of CAs when nicotine is given in maximal doses [5]. This is also true in the case of KCl-evoked CA secretion by chromaffin cells. Thus at a low concentration of KCl (between 5 and 15 mM) the effect of muscimol is excitatory but at higher concentrations the effect is inhibitory (Fig. 3).

These effects could be explained by our recent results on the muscimol effect on membrane potential, measured by a fluorescent bisoxonol probe, and estimations of resting potential and equilibrium potential for Cl^- (González et al., manuscript in preparation). At membrane potentials lower than the equilibrium potential for Cl^- (in our study -50 mV), muscimol induced a depolarization of chromaffin cells and consequently, a stimulatory effect on CA secretion. Maximum stimulation was obtained when the muscimol effect was assayed at resting conditions (in our study the

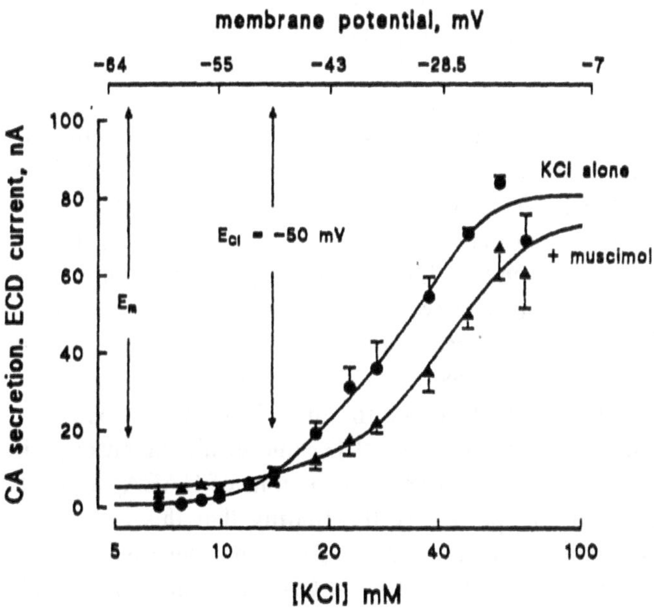

Fig. 3. Muscimol effect on the CA secretion evoked by different concentrations of KCl. (-•-•-) CA secretion evoked by KCl alone; (-▲-▲-) CA secretion evoked by KCl in presence of 100 μM muscimol. The equilibrium potential for Cl^- (E^{Cl}) was calculated by measuring the ratio between intracellular and extracellular Cl^- using a radioactive method and the application of the Nerst equation: $E^{Cl}=NR/T \ln ([Cl^i]/[Cl^e])$. Em resting potential

resting potential was about -65 mV). However, when cells were stimulated by muscimol at membrane potentials greater than the equilibrium potential for Cl- (obtained by KCl concentrations higher than 15 m*M*), the effect of the $GABA_A$ agonist was inhibitory.

Therefore, the present findings demonstrate that the effect of $GABA_A$ stimulation on CA secretion by chromaffin cells depends on the membrane potential, depolarizing excitatory or hyperpolarizing inhibitory and on the degree of change in membrane potential produced by the stimulatory agent, below or above the equilibrium potential for Cl^-. These opposite actions could be mediated by an increase in the efflux or influx of Cl^- through Cl^- channels, respectively.

$GABA_B$ Receptors

Our early results on the effect of baclofen on CA secretion seemed to indicate that this agent depressed both basal and nicotine-evoked CA release in fresh perfused cells stimulated within seconds [5]. However, when we further investi-gated the effect of $GABA_B$ agonists on CA secretion in both perfused fresh cells and cells in culture, we found that, in most cases (90%), the effect of the $GABA_B$ agonists baclofen and 3-aminopropyl phosphinic acid (3-APA) on CA secretion was stimulatory rather than inhibitory. This stimulatory effect was dose-dependent with an EC_{50} of 273 ± 38 μM and 370 ± 98 μM for 3-APA and baclofen, respectively, the baclofen effect always being smaller than the 3-APA effect (50% - 75%) (Fig.4).

With respect to the mechanism by which $GABA_B$ agonists modulate basal CA secretion, our recent results show that baclofen increases the intracellular levels of cAMP, a good correlation existing between increases in cAMP levels and amount of CA secretion (Fig. 4, insert). This result is consistent with our finding that forskolin and Cl-phenyl cAMP (PC PcAMP) increase basal CA secretion (Oset-Gasque et al., manuscript in preparation). The participation of cAMP in the actions of $GABA_B$ agonists on basal CA secretion are supported by the finding that this effect was increased by pretreatment of cells with cholera and pertussis toxins. Similarly, $GABA_B$ agonist modulation of basal CA secretion also parallels the increase in intracellular cAMP levels produced by baclofen. Moreover, these results seem to suggest the participation of G-proteins in the effect of $GABA_B$ agonists.

In contrast, $GABA_B$ agonists decrease the CA secretion evoked by nicotine in a dose-dependent manner. Since the same inhibitory effect was produced when cells were preincubated with forskolin and PC PcAMP, our results indicate that this effect of $GABA_B$ agonists may be related to increased intracellular cAMP which desensitizes, in a dose-dependent fashion, the nicotinic receptor [1, 34]. In Fig. 5 we present the dose-response curves for the effect of 3-APA, forskolin, and PC PcAMP on CA secretion evoked by 10 *μM* nicotine. In the same way, baclofen reduced the CA secretion evoked by muscimol, an effect also implicating cAMP since this is in agreement with our finding that the treatment of chromaffin cells with cholera and pertussis toxins strongly inhibited muscimol-evoked CA secretion (Oset-Gasque et al., manuscript in preparation).

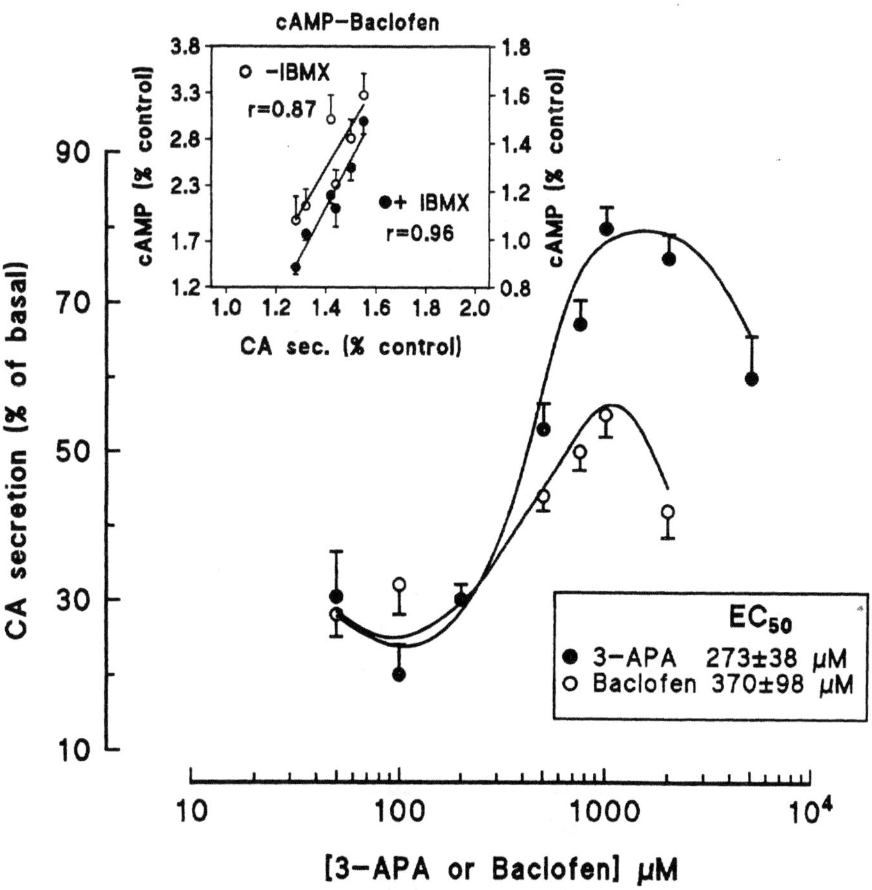

Fig. 4. Effect of $GABA_B$ agonists on basal CA secretion by chromaffin cells in culture. *Insert:* correlation between the increase in cAMP levels, measured in the absence (*r*=0,87) or presence (*r*=0,96) of IBMX, and CA secretion induced by baclofen

Hence, in our system, cAMP seems to have a modulatory role on CA secretion, increasing the basal release but reducing the evoked one by activating nicotinic ACh and $GABA_A$ receptors. It is possible that the alterations in the responses to these receptors results from the activation of cAMP-dependent protein kinase (cAMP-PK) and the subsequent receptor phosphorylation, as has been described in other systems (1, 12). Finally, these results also suggest that both $GABA_A$ and $GABA_B$ receptors in chromaffin cells can interact in order to produce a final effect on CA secretion, similar to the regulation of secretion evoked by the physiological regulator of CA secretion, ACh.

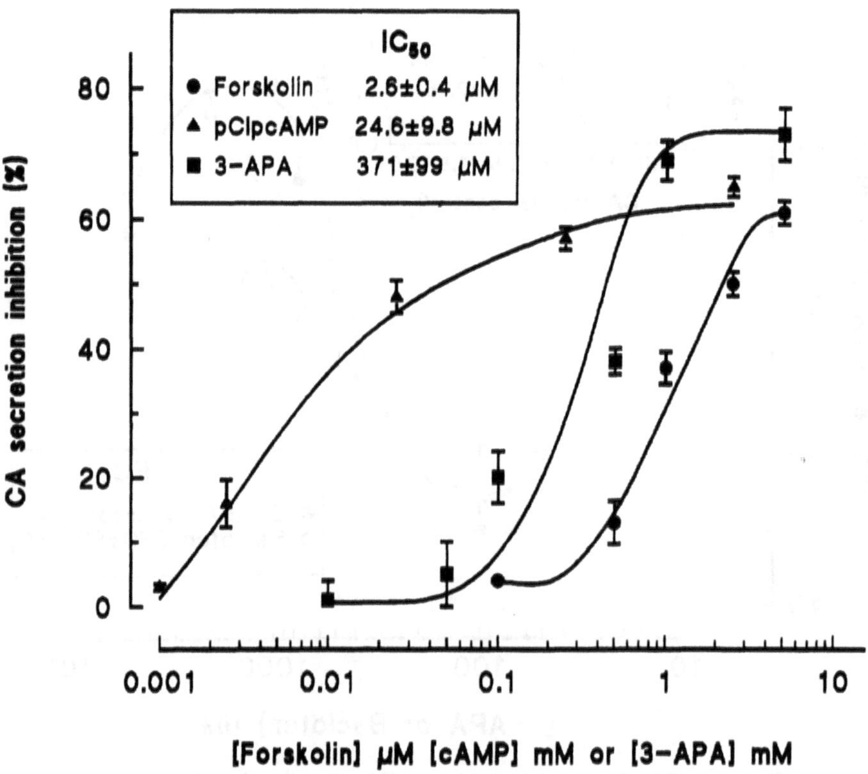

Fig. 5. Dose-dependent inhibition of CA secretion induced by several effectors; ■—■ 3-APA, ●—● forskolin, and ▲—▲ PClPcAMP

Concluding Remarks

The above results demonstrate that GABA, when bound to $GABA_A$ and $GABA_B$ receptors on the plasma membranes of chromaffin cells, can modulate both basal and evoked CA secretion, enhancing basal CA secretion and inhibiting secretagogue-evoked release. $GABA_A$ receptor stimulation evokes basal CA secretion by partial depolarization of the chromaffin cell membrane and probably by a discharge of action potentials. This leads to Ca^{2+} entry from the extracellular space but induces an inhibition of evoked CA release by clamping the membrane potential to values near the equilibrium potential for Cl^-. This mechanism is also supported by the electrophysiological findings shwoing that GABA activates Cl^--selective ion channels in bovine chromaffin cell membranes [3] and depresses the ACh-induced depolarizations in outside-out patches, while increasing the firing rate of spontaneous action potentials in intact whole cells [30]. In contrast, $GABA_B$ receptor activation produces these effects by mechanisms which are mediated by increased intracellular cAMP and probably involving G-proteins, although the nature and the exact direct or indirect mechanisms which cause this effect remain unknown. Both types of receptors could interact in order to regulate CA secretion. Regulation of effective GABA levels and thus of the resulting modulatory action of GABA could be exerted "in situ" by metabolizing enzymes, active transport, or specific release from chromaffin cells by exocytic-like and carrier reversion mechanisms. All of these results strongly suggest a physiological regulatory role of GABA in adrenal medulla.

References

1. Albuquerque EX, Deshpande SS, Aravaca Y, Alkondon M, Daly JW (1986) A possible involvement of cyclic AMP in the expression of desensitization of the nicotinic acetylcholine receptor. FEBS Lett 199: 113-120
2. Allan AM, Harris RA (1986) γ-Aminobutyric acid agonists and antagonists alter chloride flux across brain membranes. Mol Pharmacol 29: 497-505
3. Bormann J, Clapham DE (1985) γ-Aminobutryric acid receptor channels in adrenal chromaffin cells: a patch-clamp study. Proc Natl Acad Sci USA 82: 2168-2172
4. Castro E, Oset-Gasque MJ, Cañadas S, Gimenez A, González MP (1988) $GABA_A$ sites in bovine adrenal medulla membranes. J Neurosci Res 20: 241-245

5. Castro E, Oset-Gasque MJ, González MP (1989) $GABA_A$ and $GABA_B$ receptors are functionally active in the regulation of catecholamine secretion by bovine chromaffin cells. J Neurosci Res 23: 290-296
6. Douglas WW (1968) Stimulus-secretion coupling: the concept and clues from chromaffin and other cells. Br J Pharmacol 34: 451-474
7. Erdö SL (1985) Peripheral GABAergic mechanisms. Trends Pharmacol Sci 6: 205-208
8. Fernández Ramil JM, Sánchez-Prieto J, González MP (1982) Presence of glutamate decarboxylase in bovine adrenal medullary cells. Rev Esp Fisiol 38: 91-96
9. Fernández Ramil JM, Sánchez-Prieto J, Cañadas S, González MP (1983) GABA-T in bovine medulla cells: kinetic properties and comparisson with GABA-T from other tissues. Rev Esp Fisiol 39: 299-304
10. González MP, Oset-Gasque MJ, Giménez Solves A, Cañadas S (1987) Succinic semi-aldehyde dehydrogenase activity in bovine adrenal medulla and blood platelets: a comparative study with the brain enzyme. Comp Biochem Physiol 86B: 489-492
11. Guidotti A, Hanbauer I (1986) Participation of GABA/benzodiazepine receptor system in the adrenal chromaffin cell function. In: Racagni G, Donoso AO (eds) GABA and endocrine function. Raven, New York, pp 165-172
12. Harrison NL, Lambert NA (1989) Modification of $GABA_A$ receptor function by an analog of cyclic AMP. Neurosci Lett 105: 137-142
13. Kataoka Y, Gutman Y, Guidotti A, Panula P, Wroblewski J, Cosenza-Murphy D, Wu J, Costa E (1984) Intrinsic GABAergic system of adrenal chromaffin cells. Proc Natl Acad Sci USA 81: 3218-3222
14. Kataoka Y, Fujimoto M, Alho H, Guidotti A, Geffard M, Kelly GD, Hanbauer I (1986) Intrinsic gamma aminobutyric acid receptors modulate the release of catecholamines from canine adrenal gland in situ. J Pharmacol Exp Ther 239: 584-590
15. Kataoka Y, Ohara-Imaizumi M, Veki S, Kumakura K (1988) Stimulatory action of γ-aminobutyric acid on catecholamine secretion from bovine adrenal chromaffin cells measured by a real-time monitoring system. J Neurochem 50: 1765-1768
16. Kitayama S, Morita K, Dohi T, Tsujimoto A (1984) The nature of the stimulatory action of γ-aminobutyric acid in the isolated perfused dog adrenals. Naunyn Schmiedebergs Arch Pharmacol. 326: 106-110
17. Ladona MG, Aunis D, Gandía L, García A (1987) Dihydropiridyne modulation of the chromaffin cell secretory response. J Neurochem 48: 483-490
18. Leboulenger F, Leroux P, Tonon MC, Coy DH, Vaudry H, Pelletier G (1983) Coexistence of vasoactive intestinal peptide and enkephalins in the adrenal chromaffin granules of the frog. Neurosci Lett 37: 221-225
19. Linnoila RI, Diaugustine RP, Hervonen A, Miller RJ (1980) Distribution of (met5)- and (leu5)-enkephalin, vasoactive intestinal polypeptide and substance P-like immunoreactivities in human adrenal glands. Neuroscience 5: 2247-2259
20. Livett BG, Marley PD (1986) Effects of opioid peptides and morphine on histamine-induced catecholamine secretion from cultured bovine adrenal medulla chromaffin cells. Br J Pharmacol 89: 327-334
21. Lundberg JM, Hamberger B, Schulyberg M, Hockfelt T, Granberg PO, Efendie S, Terenius L, Goldtein M, Luft R (1979) Enkephalin- and somatostatin-like immunoreactivities in human adrenal medulla and pheochromocytoma. Proc Natl Acad Sci USA 76: 4079-4083
22. Majane EA, Alho H, Kataoka Y, Lee CH, Yang HYT (1985) Neuropeptide Y in bovine adrenal glands: distribution and characterization. Endocrinology 117: 1162-1168

23. Marley PD, Bunn SJ, Wan DCC, Allen AM, Mendelsohn FAO (1989) Localization of angiotensin II binding sites in the bovine adrenal medulla using a labelled specific adrenal antagonist. Neuroscience 28: 777-787
24. Martínez P, Giménez A, Castro E, Oset-Gasque MJ, González MP (1987) GABA binding in bovine adrenal medulla membranes is sensitive to baclofen. Comp Biochem Physiol 88C: 155-157
25. Nowak AM, Young AB, McDonald RL (1982) GABA and bicuculline actions on mouse spinal cord and cortical neurons in cell culture. Brain Res 244: 155-164
26. Okazaki M, Yanagihara N, Izumi F, Nakajima Y, Kuroiwa A (1989) Carbachol-induced cosecretion of immunoreactive atrial natriuretic peptides with catecholamines from cultured bovine adrenal medullary cells. J Neurochem 52: 222-228
27. Oset-Gasque MJ, Cañadas Correas S, Massó-Córdoba JM, González MP (1985) Uptake of GABA by bovine adrenal medulla slices. Naunyn Schmeidebergs Arch Pharmacol 329: 372-375
28. Oset-Gasque MJ, Aunis D (1989) γ-Aminobutyric acid uptake in bovine chromaffin cells in culture. Biochem Pharmacol 38: 2227-2232
29. Oset-Gasque MJ, Castro E, González MP (1990) Mechanisms of $[^3H]$γ-aminobutyric acid release by chromaffin cells in primary culture. J Neurosci Res 26: 181-187
30. Peters JA, Lambert JJ, Cottrell GA (1989) An electrophysiological investigation of the characteristics and function of $GABA_A$ receptors on bovine adrenomedullary chromaffin cells. Eur J Physiol 415: 95-103
31. Saito H, Saito S, Ohuchi T, Oka M, Sano T, Hosoi E (1984) Co-storage and co-secretion of somatostatin and catecholamines in bovine adrenal medulla. Neurosci Lett 52: 43-47
32. Sangiah S, Borowitz JL, Yim GKW (1974) Action of GABA, picrotoxin and bicuculline on adrenal medulla. Eur J Pharmacol 27: 130-135
33. Tanaka T, Yokohama H, Negishi M, Hayashi H, Ito S, Hayaishi O (1987) Pertussis toxin facilitates secretagogue-induced catecholamine release from cultured bovine adrenal chromaffin cells. Biochem Biophys Res Comm 144: 907-914
34. Thampy KG, Barbes EM Jr (1984) γ-Aminobutyric acid-gated chloride channels in cultured cerebral neurons. J Biol Chem 259: 1753-1757
35. Wagoner PK, Pallota BS (1988) Modulation of acetycholine receptor desensitization by forskolin is independent of cAMP. Science 240: 1655-1657

The GABAergic System in Human Female Genital Organs

Á. László, G.L. Nádasy, E. Monos, B. Zsolnai, and S.L. Erdö

Introduction

It has been recently shown that γ-aminobutyric acid (GABA) and its related enzymes, i.e., glutamate decarboxylase (GAD) and GABAtransaminase (GABA-T), are present in the female genitale tract of various mammals at higher levels than in most other organs, at least in the rat (for reviews see [13, 16, 35, 47, 60, 64]).

Specific GABA receptor sites [11, 17-19, 21, 24, 29], mediating contractile and other functional responses [14, 22, 25, 30, 38, 57] and high-affinity GABA uptake systems [14, 15] have also been demonstrated in mammalian genital organs. Moreover, it has been found that, in female sex organs, certain GABAergic markers change during gestation, both as a function of the estous cyle and with age. These findings suggested some physiological relevance of the GABA systems in the mammalian ovary, oviduct, and uterus. The characteristics and possible roles of GABA systems in the female reproductive organs of laboratory animals have been recently reviewed [14]. The present paper is an attempt to summarize the knowledge of GABA systems in the human female genital tract and to compare the data with previous findings in rodents.

GABA Content

Rat oviduct contains more than twice as much GABA as the whole brain [4, 8, 9, 20, 23, 31, 32, 46]. Especially high GABA levels, exceeding by ten fold those in the brain, are concentrated in the oviductal mucosa [49]. Moreover, GABA concentrations show a gradient along the rat oviduct, descending from the fimbriated ampulla of the uterine tube towards the uterotubal junction [42]. The distribution of GABA and GAD is similar [12], but GABA-T shows a reverse distribution profile along the tube [1].

The GABA concentration in the human Fallopian tube is considerably lower than in the respective organ of the rat (Table 1). As in rodents, there are higher GABA levels in the ampulla than in the isthmus of the human oviduct [21].

Table 1. A comparison of oviductal, ovarian, and uterine GABA concentrations in rats and humans

Organ	GABA concentrations (nmol/mg tissue)			
	Rat	Reference	Human	Reference
Oviduct	1 - 3 4 - 6	[23, 31,32] [4, 8, 20, 48]	0.09 - 0.37	[21, 40]
Ovary	0.04 - 0.10 0.50 - 0.56	[4, 46, 49, 52] [20, 31, 59]	0.12 - 0.22	[19, 49]
Uterus	0.027 - 0.03 0.4 - 0.6	[11, 46] [31, 32]	0.07	[27]

Values refer to concentration ranges given in the cited references. Measures are standardized according to the following approximation: 1 mg wet or frozen tissue = 0.1 mg protein = 0.1 mg dry tissue.

The data reported for GABA levels in the rat ovary are rather variable. Depending on assay conditions and perhaps on the rat strain used, higher [20, 31, 32, 47, 52, 59] and much lower [4, 49] ovarian GABA levels have been described. However, there is general agreement that, in rat, GABA is significantly less abundant in the ovary than in the oviduct. The GABA level in the human ovary is about half of that detected by comparable assay techniques in the rat ovary [20, 31, 59].

It appears that GABA is less abundant in the rat uterus than in the ovary and oviduct of the same species [11, 46]. However, uterine GABA levels, similar to those in the ovary, have also been demonstrated in the rat [31,

32, 42]. The GABA concentration in the human uterus [27] is in the same range as found in the uterine tissue of rodents [11].

GAD and GABA-T

Very low GAD activity, i.e., 1%-3% of the value for whole brain, could be detected in the rat ovary, while remarkably higher activities, amounting to 10%-12% of cerebral levels, were found in the rat oviduct [4, 8, 9, 23, 46, 59]. The activity of GAD was unevenly distributed in the rat oviduct, with especially high activities confined to the mucosa layer [49]. Moderate GAD activities have also been detected in the rat and rabbit uterus [11, 35].

GAD obtained from rat oviducts has been reported to exhibit biochemical [23, 46] and immunochemical [4] characteristics resembling those of brain GAD, thus differing from the so-called non-neuronal form of the enzyme [35, 64]. This finding, along with recent immunocytochemical evidence [28], indicates that the neuronal type of enzyme may be present in this organ, in epithelial cells of the mucosa [28].

In the rat oviduct, GAD may be be a major source of the GABA content, but in the ovary and the uterus, other metabolic pathways might also contribute to GABA production (reviewed in [13]). In fact, in homogenates of the ovary, uterus [2, 3], and oviduct [56], GABA may be formed from putrescine via oxidative deamination and subsequent oxidation.

The occurrence of GABA-T, the enzyme responsible for GABA degradation in the CNS, has also been demonstrated in the rat oviduct [1, 46]. As revealed by enzyme histochemistry [62], GABA-T activity is chiefly confined to the oviductal mucosa [26]. However, it has been revealed that, in rats in vivo, a potent inhibitor of cerebral GABA-T, i.e., aminooxyacetic acid (AOAA), elevates GABA levels in the brain, while decreasing GABA concentrations in the oviduct [20] and ovary [20, 47]. This difference in the behavior of GABA related enzymes is thought to be dreived from a preferential inhibition by AOAA of GABA-T in the brain and of GAD in the female reproductive tract [20, 46, 48].

In the human ovary and uterus, the activities of GABA related enzymes were found to be in the same range as those in rats and rabbits, whereas the human oviduct contained significantly lower GAD activities than the rat oviduct (Table 2). Moreover, the activities of GAD and GABA-T failed to show any correlation with age and endocrine milieu and were found to be similar in tissues obtained from control patients and patients with uterine myoma or carcinoma [27, 40].

Table 2. A comparison of GAD activities in oviducts, ovaries, and uteri of rats and humans

Organ	GAD activity (nmol/mg protein/h)			
	Rat	Reference	Human	Reference
Oviduct	20 - 30	[4, 9, 23, 46]	0.43	[40]
Ovary	0.4 - 3.8 17.6	[4, 23] [59]	0.27	[40]
Uterus	0.68	[11]	0.32	[27]

Values refer to enzyme activity ranges reported in the respective references.

Specific GABA Binding Sites

GABAergic inhibiton in the CNS is mediated by specific receptors and pharmacologically and functionally distinct GABA receptor subpopulations have been identified in the CNS [6, 7, 10, 37, 54, 61] and in peripheral tissues [5, 13]. These are: the bicuculline-sensitive $GABA_A$ receptors and the baclofen-sensitive $GABA_B$ receptors. $GABA_A$ receptors can be specifically labeled by GABA in a sodium-, calcium- and magnesium-free medium, while in the presence of calcium and/or magnesium ions GABA labels both receptor populations [6, 7, 37]. For labeling of $GABA_A$ and $GABA_B$ receptor sites in a selective fashion, muscimol and baclofen, respectively, are generally used as radioligands.

It has been shown that in homogenates of the rat ovary [^{3}H]muscimol binds to a population of high-affinity binding sites, characterized by an equilibrium dissociation constant (K_d) of 15-21 nM and a maximal binding capacity (B_{max}) of 45 fmol/mg protein [59]. Similarly, a single population of bicuculline-sensitive, high-affinity (K_d = 40-52 nM) binding sites has been identified in homogenates of the rat ovary [18], oviduct [19], and uterus [11] using [^{3}H]GABA as ligand. In these organs of the rat, the B_{max} values were found to average around 100 fmol/mg protein [14]. Interestingly, in rabbit uterus, the density of high-affintiy GABA binding sites was about four times higher than that in rat uterus [11]. In each organ and species, the pharmacological specifity of GABA binding to various displacers was consistent with the presence of $GABA_A$ receptors.

Bicuculline-sensitive [^{3}H]GABA binding sites have also been demonstrated in the female genital organs of humans [19, 21, 27, 40]. In freeze-thawed membranes of the human ovary, oviduct, and uterus GABA labeled single populations of binding sites, which showed affinities similar to those in the respective rat organs [11, 17, 18, 59] and in the human anterior pituitary [33]. However, the density of [^{3}H]GABA binding sites in the human sex organs exceeded, by about fivefold, that found in rats (Table 3). Moreover, in the human Fallopian tube, GABA binding sites showed an uneven distribution, with the isthmus exhibiting a slightly higher binding capacity than the ampulla [14, 21].

Table 3. A comparison of [^{3}H]GABA binding (K_d and B_{max}) to bicuculline-sensitive sites in the genital organs of female rats and humans

Organ	Rat			Human		
	K_d (n*M*)	B_{max} (fmol/mg protein)	Reference	K_d (n*M*)	B_{max} (fmol/mg protein)	Reference
Oviduct	52	115	[14, 16]	40	690	[21]
Ovary	52	130	[14, 16]	38	676	[19]
Uterus	42	131	[11]	47	690	[27]

The binding parameters of [^{3}H]GABA determined for the human genital organs failed to show correlation with the age and endocrine background of patients and did not differ in membrane preparations obtained from noncancer patients and from those with uterine myoma or carcinoma.

The specificity of [^{3}H]GABA binding to membranes of the human sex organs was characterized in displacement experiments using various agonists and antagonists of GABA receptors (Table 4). Muscimol and bicuculline proved to be potent in displacing [^{3}H]GABA binding in each organ, with IC_{50} values in nanomolar and micromolar ranges, respectively. In these experiments, bicuculline showed a somewhat weaker displacing potency than found in cerebral membranes [53]. By contrast, picrotoxin, a noncompetitive $GABA_A$ antagonist; baclofen, a selective $GABA_B$ receptor agonist and L-diaminobutyric acid (L-DABA), a specific substrate of the neuronal GABA uptake carrier, were inactive. These findings support the view that the labeled sites in the human reproductive organs are identical to $GABA_A$ type receptors.

Table 4. Displacement of specific GABA binding from membranes of the human oviduct, ovary, uterus, and placenta

Organ	Compound	IC_{50} value (μM)	Reference
Oviduct	Muscimol	0.028	[21]
	GABA	0.056	
	(+)-Bicuculline	163	
	(±)-Baclofen	> 500	
	Picrotoxin	> 500	
Ovary	Muscimol	0.022	[19]
	GABA	0.047	
	(+)-Bicuculline	110	
	(±)-Baclofen	> 200	
	Picrotoxin	> 200	
Uterus	Muscimol	0.019	[27]
	GABA	0.057	
	(+)-Bicuculline	18.5	
	(±)-Baclofen	> 300	
	L-DABA	> 300	
Placenta	Muscimol	0.008	[24]
	GABA	0.056	
	(+)-Bicuculline	21.0	
	(±)-Baclofen	> 200	

IC_{50} values are defined as displacer concentrations producing 50% reduction of specific binding.
L-DABA, L-diaminobutyric acid

Functional Responses to GABA

It is well documented that GABA, its specific agonists, and its antagonists evoke contractile responses in smooth muscle preparations obtained from different segments of the female genital tract [14, 22, 25, 30, 38, 57]. In the rat oviduct, GABA alone did not influence spontaneous contractile activity, but enhanced contractions induced by cholinergic stimulation [30]. However, GABA and Baclofen increased the spontaneous contractility of the isolated rabbit oviduct [14, 30] and uterus [14, 57], suggesting a $GABA_B$ receptor-mediated stimulatory effect. The uterus was about ten times more sensitve to GABAergic stimulation than the oviduct. The effects of GABA and baclofen were not affected by blockers of adrenergic and cholinergic receptors and were insensitive to tetrodotoxin and indomethacin [14, 30, 57]. These findings support the view that $GABA_B$ receptors in the rabbit oviduct and uterus are located on the smooth muscle cells, rather than on neuronal elements.

In the rabbit oviduct and uterus, $GABA_A$ receptor-mediated contractile responses have also been demonstrated [14, 30, 57]. Although the spontaneous motility was not affected by $GABA_A$ receptor agonists, bicuculline, a specific antagonist of $GABA_A$ receptor binding, elicited contractions in both organs. As these effects of bicuculline could be eliminated by muscimol, the above responses of the isolated oviduct and uterus were proposed to be mediated via $GABA_A$ receptors [14, 30, 57].

GABAenergic contractile responses have also been described in ovarian tissue. In muscle preparations obtained from bovine ovarian follicles GABA inhibited alpha-adrenergic contraction in a bicuculline-insensitve fashion [38]. As the effect of GABA was mimicked by baclofen but not by muscimol, these results are consistent with the presence of inhibitory $GABA_B$ receptors on noradrenergic nerve terminals of the follicular wall and suggest a role for GABA in the regulation of follicular rupture during ovulation.

Functional responses to GABA, other than changes in muscles contractility, have been demonstrated in anesthesized, pseudopregnant rats [25]. After administration of GABA to the surface of the rat ovary, ovarian blood flow and estradiol secretion increased, while the secretion of progesterone decreased in a significant manner [25]. Although the pharmacological specificity of these effects has not been tested, the above findings, along with the appearance of specific GABA binding sites in ovarian granulosa cells [59], are consistent with a role for GABA in the regulation of endocrine functions in the ovary.

The effects of GABA and related compounds on spontaneous contractile activity have been examined in ampullar segments of the isolated human Fallopian tube [41]. In these experiments, the motility of circular and longitudinal muscle was simultaneously studied [39, 50]. In both muscle layers, GABA increased the frequency, but decreased the amplitude, of spontaneous contractions (Fig. 1). In accordance with earlier findings in the rabbit oviduct and uterus [14, 22, 57], the effect of GABA could be mimicked by baclofen but not by muscimol, suggesting the involvement of $GABA_B$ receptors. In the longitudinal muscle layer, the effect of GABA was independent of the endocrine milieu, while in the circular muscle layer the amplitude of contraction was reduced by GABA only in the follicular phase and in postmenopause, but not in the luteal phase [41]. The periodic changes in the spontaneous contractile activity of the circular and longitudinal layers proved to be mostly parallel, indicating that motilities of both muscle layers are under the control of a common pacemaker center.

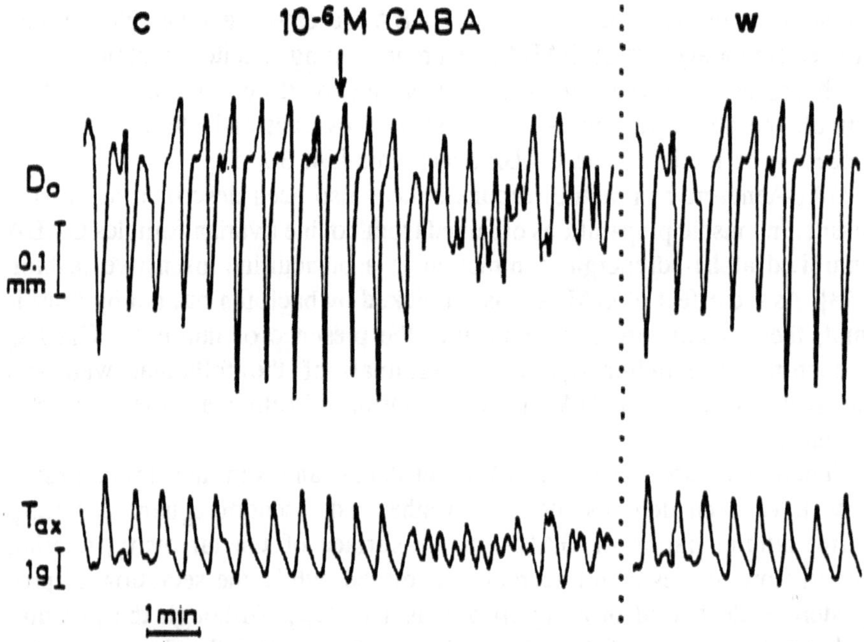

Fig. 1. The effect of GABA on the spontaneous contractile activity of isolated human Fallopian tube ampullar segment. D_o outer diameter, and T_{ax}, axial tension, refer to circular and longitudinal muscular activity, respectively; *C*, control; *W*, washout

The contractile activity of the human Fallopian tube is regulated by ovarian steroid hormones (for references see [41]). Furthermore, recent lines of evidence show that ovarian steroids may be involved in both up- and down-regulation of GABA receptors [36, 43-45]. These findings may provide an explanation for the hormone dependence of contractile responses to GABA in the human oviduct and are compatible with the view that, in the human oviduct, GABA may be involved in the regulation of motily and ovum transport.

Cyclic and Gestation-Related Chances

Many lines of evidence have acclumulated to suggest that the hormonal milieu may play a role in the regulation of GABA systems in female reproductive organs [14]: (1) GABA content has been found to change during the estrous cycle. (2) In rat ovary [42] and uterus [31, 32], the highest GABA levels were measured on the day of estrus, whereas, in rat oviduct, maximal values were found during late diestrus or proestrus [14, 42, 46]. (3) Transiently high GAD activity was demonstrated in the mucosa of the rat oviduct during estrus [49]. In other studies, with less time points around proestrus and estrus, ovulation related changes could not be recognized [9, 31, 32, 42, 49, 63]. (4) Treatment with diethystilbestrol decreased the levels of GABA and GAD in the rat ovary [59]. (5) Both hypophysectomy and ovariectomy induced a marked decrease in GABA concentration and GAD activity in the rat oviduct in vivo which could be restored by gonadotropin, estradiol, and progesterone treatment [8, 9]. In the uteri of ovariectomized rats, GABA and GAD levels were found to be in the same range as in diestrus (i.e., lower than in estrus) and increased after estradiol administraion [32]. (6) In the rat oviduct, the GABA receptor-mediated inhibitory effects on cholinergic contractions [30] and the GABA-induced stimulation of ribosomal protein synthesis [55] were found to be maximal in diestrus.

The mechanisms of hormonally dependent changes in GABA systems in the female sex organs is not yet understood. Partial information has been presented about the mechanisms of steroid regulation of GABA receptors. It has been shown that various estrogen and progesterone metabolites may acutely regulate the function of GABA-gated chloride channels, i.e. $GABA_A$ receptors [10], via allosteric sites and may also influence the development of $GABA_A$ type receptors in brain tissue [36, 43-54]. Such an obvious relationship has not been revealed between steroids and $GABA_B$ receptors, which are known to interact with cation channels and second messenger systems [10, 51]. In human genital organs, the biochemical characteristics of the GABA system have been found to exhibit high quantitative variability in different individuals. Consequently, no statistically significant correlation could be found between GABAergic parameters and serum

Table 5. Biochemical characteristics of the GABA systems in the oviduct, ovary and uterus of nonpregnant and pregnant women

Organ	Parameter	Nonpregnant	Pregnant	Reference
Oviduct	GABA concentration (nmol/g tissue)	161	110*	[40]
	GAD activity (nmol/mg protein/h)	0.43	nd	
	GABA-T acitvity (nmol/mg protein/h)	3.5	5.9**	
	[^{3}H]GABA binding (fmol/mg protein)	116	192**	
Ovary	GABA concentration (nmol/g tissue)	128	160***	[40]
	GAD activity (nmol/mg protein/h)	0.27	nd	
	GABA-T acitvity (nmol/mg protein/h)	1.97	1.78	
	[^{3}H]GABA binding (fmol/ mg protein)	87	93	
Uterus	GABA concentration (nmol/g tissue)	72	41*	[27]
	GAD activity (nmol/mg protein/h)	0.32	0.39	
	GABA-T acitvity (nmol/mg protein/h)	3.3	6.1**	
	[^{3}H]GABA binding (fmol/mg protein)	317	681**	

*$p < 0.05$; **$p < 0.01$; ***$0.05 < p < 0.1$ vs nonpregnant; *nd*, not determined

levels of estradiol, progesterone, LH, and FSH in nonpregnat patients [27, 40].

By contrast, in pregnancy, biochemical markers of the GABA system exhibited significant quantitative changes in the human reproductive tract (Table 5). In the oviduct and uterus, GABA levels were decreased, while the activity of GABA-T and the binding of [^{3}H]GABA significantly increased during gestation [27, 40]. In the ovary, pregnancy related changes were restricted to a minor increase in GABA levels, without concomitant changes in enzyme activities and receptor binding. The pregnancy related decrease in GABA levels in the oviduct and uterus seems to be derived form an increase in GABA catabolic (GABA-T) capacity rather than from a reduction of biosynthetic (GAD) activity. Similar gestational changes in GABA levels and GABA-related biochemical markers have been demonstrated in the rat oviduct [1, 12, 32].

The increase in ^{3}H-GABA binding in the pregnant oviduct may reflect an increase in the density and/or affinity of $GABA_A$ receptors [40], while in the uterus it seems to be associated with an enhancement of receptor affinity [27]. Since $GABA_A$ receptors are known to inhibit both the contractility of oviducts and uteri in rodents [14, 22, 30, 57] and the motility of the human Fallopian tube [41], the increased amount of GABA receptor sites may provide a more effective target for the pysiological suppression of uterotubal motility during gestation. Moreover, because of the close anatomical connection between the Fallopian tube and the ovary, an intracanalicular action of ovarian GABA on oviductal or even uterine GABA receptors can be postulated [14, 40]. The elevation of ovarian GABA levels during gestation [12, 31, 32] (see also Table 5) might also be implicated by this hypothesis.

It has been shown that, in pregnant rats, GABA levels are reduced in the implantation region of the uterus, suggesting that GABA might have a function during implantation of the blastocyst [32]. Interestingly, the human term placenta has also been found to contain GABA at a much higher concentration than those in the blood [29] and amniotic fluid [34], a moderate GAD activity, and a high density of $GABA_A$ type receptor binding sites [24]. In displacement studies, placental GABA binding sites showed the characteristics of inhibitory $GABA_A$ receptors [24] (Tabel 4). Although the release of GABA from placental tissue has not yet been examined, one may speculate whether GABA of placental origin might also contribute to suppression of uterotubal motility during pregnancy.

Summary and Conclusion

As in other mammals, GABA and its related enzymes and receptors are present in the genital organs of the human female. Major species differences occur: the levels of GABA and GAD are much lower in humans than in rats, and GABA receptor sites are more abundant in the human ovary, oviduct, and uterus than in the respective organs of the rat. Thus, as in rodents [26, 28], the GABA systems in the human sex organs seem to be non-neuronal in nature, even though histochemical studies with human tissue are still awaited.

The predominant contractile responses to GABA are similar in the uterotubal tissue of rabbits and humans. GABA stimulates the frequency of spontaneous contractions via $GABA_B$ receptors. The physiological significance of the $GABA_A$ receptor-mediated inhibition of uterotubal motility is not yet clear, but it appears to be restricted to certain situations, such as pregnancy.

The individual variability of GABAergic biochemical markers has proven to be higher in the genital tract of humans than in the sex organs of rodents. This might also explain the lack of statistical correlation between GABAergic markers of the human reproductive tract and the hormonal background of patients.

In both rodents and humans, the GABA system of the reproductive organs seems to play a major role during pregnancy. GABA may be involved in the modulation (suppression) of uterotubal motility. As a source of endogenous GABA, the placenta may also be involved in the GABAergic regulation of uterine, and even embryonic, functions.

In conclusion, GABA may play an important role in the regulation of uterotubal and ovarian functions not only in rodents, but also in humans. Clinical studies with GABAergic drugs (e.g., baclofen, progabide, vigabatrin) may clarify for obstetricians and gynecologists the practical significance of the genital GABA system.

References

1. Amenta F, Cavallotti C, Mione MC, Erdö SL (1986) Segmental distribution and gestational changes of GABA-transaminase activity in rat oviduct. J Reprod Fertil 78: 593-599
2. Andersson AC, Henningsson S (1980) In vitro metabolism of putrescine by diamine oxidase in tissues of the pregnant rat. Agents Actions 10: 104-106
3. Andersson AC, Henningsson S, von Scheele C (1980) Formation of GABA via oxidative deamination of putrescine in ovaries of immature rats after gonadotrophin stimulation. Agents Actions 10: 106-107
4. Apud JA, Tappaz ML, Celotti F, Negri-Cesi P, Masotto C, Racagni G (1984) Biochemical and immunochemical studies on the GABAergic system in the rat fallopian tube and ovary. J Neurochem 43: 120-125
5. Bowery NG, Doble A, Hill DR, Hudson AI, Shaw JS, Turnbull MJ, Warrington R (1981) Bicuculline-insensitive GABA receptors on peripheral autonomic nerve terminals. Eur J Pharmacol 71: 53-70

6. Bowery NG, HillDR, Hudson AL (1983) Characteristics of $GABA_B$ receptor binding sites on rat whole brain synaptic membranes. Br J Pharmacol 78: 191-206
7. Bowery NG, Price GW, Hudson AL, Hill DR, Wilkin GP, Turnbull MJ (1984) GABA receptor multiplicity. Visualization of different receptor types in the mammalian CNS. Neuropharmacology 23: 219-231
8. Celotti F, Apud JA, Rovescalli AC, Melcangi RC, Negri-Cesi P, Racagni G (1986) The GABAergic extrinsic innervation of the rat follopian tubes: biochemical evidence and endocrine modulation. Adv Biochem Psychopharmacol 42: 251-264
9. Celotti F, Apud JA, Rovescalli AC, Negri-Cesi P, Racagni G (1987) Possible involvement of ovarian mechanisms other than estrogen-progesterone secretion in the regulation of glutamic acid decarboxylase activity of the rat fallopian tubes. Endocrinology 120: 700-706
10. Eldefrawi AT, Eldefrawi ME (1987) Receptors for γ-aminobutyric acid and voltage dependent chloride channels as targets for drugs and toxicants.
11. Erdö SL (1984) Identification of GABA receptor binding sites in rat and rabbit uterus. Biochem Biophys Res Commun 125: 18-24
12. Erdö SL (1984) Alteration of GABA levels in ovary and fallopian tube of the pregnant rat. Life Sci 34: 1879-1884
13. Erdö SL (1985) Peripheral GABAergic mechanisms. Trends Pharmacol Sci 6: 205-208
14. Erdö SL (1986) GABAergic mechanisms and their possible role in the oviduct and uterus. In: Erdö SL, Bowery NG (eds) GABAergic mechanisms in the mammalian periphery. Raven, New York, pp 205-217
15. Erdö SL, Amenta F (1986) Characterization and localization of high-affinity GABA uptake in slices of the rabbit oviduct. Eur J Pharmacol 130: 287-294
16. Erdö SL, Bowery NG (eds) (1986) GABAergic mechanisms in the mammalian periphery. Raven, New York
17. Erdö SL, Lapis E (1982) Presence of GABA receptors in rat oviduct. Neurosci Lett 33: 257-279
18. Erdö SL, Lapis E (1982) Bicuculline-sensitive GABA receptors in rat ovary. Eur J Pharmacol 85: 242-246
19. Erdö SL, László Á (1984) High specific gamma-aminobutyric acid binding to membranes of the human ovary. J. Neuorchem 42: 1464-1467
20. Erdö SL, Rosdy B, Szporny L (1982) Higher GABA concentrations in fallopian tube than in brain of the rat. J Neurochem 38: 1174-1176
21. Erdö SL, László Á, Szporny L, Zsolnai B (1983) High density of specific GABA binding sites in the human fallopian tube. Neurosci Lett 42: 155-160
22. Erdö SL, Riesz M, Kárpáti M, Szporny L (1984) $GABA_B$-receptor mediated stimulation of the contractility of isolated rabbit oviduct. Eur J Pharmacol 99: 333-336
23. Erdö SL, Kiss B, Szporny L (1984) Comparative characterization of glutamate decarboxylase in crude homogenates of oviduct, ovary, and hypothalamus. J Neurochem 43: 1532-1537
24. Erdö SL, László Á, Kiss B, Zsolnai B (1985) Presence of gamma-aminobutyric acid and its specific receptor binding sites in the human term placenta. Gynecol Obstet Invest 20: 199-203
25. Erdö SL, Varga B, Horváth E (1985) Effect of local GABA administration on rat ovarian blood flow, and on progesterone and estradiol secretion. Eur J Pharmacol 11: 397-400

26. Erdö SL, Somogyi J, Hámori J, Amenta F (1986) Light- and electronmicroscopic visualization of gama-aminobutyric acid and GABA-transaminase in the oviduct of rats. Predominant occurence in epithelium. Cell Tissue Res 244: 621-626
27. Erdö SL, Villányi P, László Á (1989) Gestational changes of GABA levels and GABA binding in the human uterus. Life Sci 44: 2009-2014
28. Erdö SL, Joó F, Wolff JR (1989) Immunohistochemical localization of glutamate decarboxylase in the rat oviduct and ovary: further evidence for non-neuronal GABA systems. Cell Tissue Res 255: 431-434
29. Ferkany JW, Smith LA, Seifert WE, Caprioli RM, Enna SJ (1978) Measurement of gamma-aminobutyric acid (GABA) in blood. Life Sci 22: 2121-2128
30. Fernandez I, Orensanz LM, de Ceballos ML (1984) GABA modulation of cholinergic transmission in rat oviduct. Life Sci 35: 357-364
31. Fernandez-Pardal J, Gimeno MF, Gimeno AL (1984) The presence of GABA in ovarian, tubal and uterine rat tissue and the influence of estrogens. IRCS Med Sci 12: 706
32. Gimeno MF, Fernandez-Pardal J, Viggiano M, Pezzot MC, Gimeno AL (1986) On the presence of GABA in ovarian, tubal and uterine rat tissues. Modification at different stages of the estrous cycle and during pregnancy. Adv Biochem Psychopharmacol 42: 275-282
33. Grandison L, Cavagnini F, Schmid R, Invitti C, Guidotti A (1982) γ-Aminobutyric acid- and benzodiazepine-binding sites in human anterior pituitary tissue. J Clin Endocrinol Metab 54: 597-601
34. Grossman MH, Hare TA, Manyam NVB (1979) Measurement of γ-aminobutyric acid in human whole blood and amniotic fluid. Fed Proc 38: 375
35. Haber B, Kuriyama K, Roberts E (1970) An anion stimulated L-glutamic acid decarboxylase in non-neuronal tissues. Biochem Pharmacol 19: 1119-1136
36. Harrison NL, Majewska MD, Harrington JW, Barker JL (1987) Structure-activity relationships for steroid interaction with the $GABA_A$ receptor complex. J Pharmacol Exp. Ther 241: 346-353
37. Hill DR, Bowery NG (1981) [^{3}H]GABA bind to bicuculline-insensitive $GABA_B$ sites in rat brain. Nature 290: 149-152
38. Kannisto P, Owan Ch, Schmidt G, Walles B (1986) Evidence for prejunctional $GABA_B$ receptors mediating inhibition of ovarian follicle contraction induced by nerve stimulation. Eur J Pharmacol 122: 123-129
39. László Á, Nádasy GL, Monos E, Zsolnai B (1988) Effect of pharmacological agents on the activity of the circular and longitudinal smooth muscle layers of human Fallopian tube ampullar segments. Act Physiol Hung 72: 123-133
40. László Á, Villányi P, Zsolnai B, Erdö SL (1989) Gamma-aminobutyric acid, its related enzymes and receptor-binding sites in the human ovary and fallopian tube. Gynecol Obstet Invest 28: 94-97
41. Lásló Á, Nádasy GL, Erdö SL, Monos E, Siklósi Gy, Zsolnai B (1990) Effect of GABA on the spontaneous muscular activity of the human Fallopian tube ampullar segments in vitro. Acta Physiol Hung (to be published)
42. Louzan P, Gallardo MG, Tramezzani JH (1986) Gamma-aminobutyric acid in the genital tract of the rat during the oestrous cycle. J Reprod Fertil 77: 499-504
43. Majewska MD, Harrison NL, Schwartz RD, Barker JL, Paul SM (1986) Steroid hormone metabolites are barbiturate-like modulators of the GABA receptor. Science 232: 1004-1007
44. Majewska MD (1987) Antagonist-type interaction of glucocorticoids with the GABA receptor-coupled chloride channel. Brain Res 418: 377-382

45. Majewska MD, Schwartz RD (1987) Pregnenolone-sulfate: an endogenous antagonist of the γ-aminobutyric acid receptor complex in brain. Brain Res 404: 355-360
46. Martin del Rio R (1981) γ-Aminobutyric acid system in rat oviduct. J Biol Chem 256: 9816-9819
47. Martin del Rio R, Latorre Caballero A (1980) Presence of γ-aminobutyric acid in rat ovary. J Neurochem 34: 1584-1586
48. Martin del Rio R, Sierra Lopez M (1983) Effects of aminooxyacetic acid on in vivo γ-aminobutyric system of rat oviduct. Gen Pharmacol 14: 281-283
49. Murashima YL, Kato T (1986) Distribution of gamma-aminobutyric acid and glutamate decarboxylase in the layers of rat oviduct. J Neurochem 46: 166-172
50. Nádasy GL, László Á, Monos E, Zsolnai B (1988) Spontaneous periodic contraction of the ampullar segment of the human Fallopian tube in vitro. Acta Physiol Hung 72: 13-21
51. Nicoll RA (1988) The coupling of neurotransmitter receptors to ion channels in the brain. Science 241: 545-551
52. Okada Y, Taniguchi H, Baba S (1982) High concentration of GABA in pancreatic islets with special emphasis on B cells. In: Okada Y, Roberts E (eds) Problems in GABA research from brain to bacteria. Excerpta Medica, Amsterdam, pp 379-386
53. Olsen RW, Greenlee D, Van Ness P, Ticku MK (1978) Stucies on gamma-aminobutyric acid receptor/ionophore proteins in mammalian brain. In: Fonnum F (ed) Amino acids as chemical transmitters. Plenum, New York, pp 467-486
54. Olsen RW, Bergman MO, Van Ness PC, Lummis SC, Wathins AE, Napias C, Greenlee D (1981) γ-aminobutyric acid receptor binding in mammalian brain: heterogeneity of binding sites. Mol Pharmacol 19: 217-227
55. Orensanz LM, Azuara C, Fernandez I (1985) Lack of efect of GABA on [^{3}H]leucine incorporation into a rat oviduct ribosomal system. Neurochem Res 10: 789-796
56. Orensanz LM, Fernandez I, Martin del Rio R, Storm-Mathisen J (1986) Gamma-aminobutyric acid system in the rat oviduct. Adv Biochem Psychopharmacol 42: 265-274
57. Riesz M, Erdö SL (1985) $GABA_B$ receptors in the rabbit uterus may medicate contractile responses. Eur J Pharmacol 119: 199-204
58. Roberts E, Chase TN, Tower DB (1976) GABA in nervous system function. Raven, New York
59. Schaeffer JM, Hsueh AJW (1982) Identification of gamma-aminobutyric acid and its binding sites in the rat ovary. Life Sci 30: 1599-1604
60. Tanaka C (1985) γ-aminobutyric acid in peripheral tissue. Life Sci 37: 2221-2235
61. Turner AJ, Whittle SR (1983) Biochemical dissection of the gamma-aminobutyrate synapse. Biochem J 209: 29-41
62. Vincent SR, Kimura H, McGeer EG (1980) The pharmacohistochemical demonstration of GABA-transaminase. Neurosci Lett 16: 354-348
63. Wolf R, Meier-Fleitmann A, Duker EM, Wuttke W (1986) Intraovarian secretion of catecholamines, oxytocin, beta-endorphin, and gamma-aminobutyric acid in freely moving rats: development of a push-pull tubing method. Biol Reprod 35: 599-607
64. Wu JY (1982) Characterization of L-glutamate decarboxylase in neural and non-neural tissues. In: Okada Y, Roberts E (eds) Problems in GABA research from brain to bacteria. Excerpta Medica, Amsterdam, pp 40-54

Neurotransmitter Amino Acids as Modulators of Biological Processes of Spermatozoa

H.K. Boldizsár, L. Wekerle, E. Vén, P. Sarlós, and J. Barna

It is well known that the characteristics of spermatozoa such as motility, transmembrane potential, agglutination type, decapacitation and capacitation specifically change within the male and female genital tracts. These features are generally thought to be regulated by two major endogenous factors, one of them being a protein and the other an unidentified, heat-stable, dialyzable component of low molecular weight [1, 18]. Furthermore, it has also been recognized that there are significant differences in the composition of electrolytes and amino acids in the seminal plasma, in the secretions of the male accessory glands, and in the oviduct and uterus. Interestingly, the most abundant amino acids in these tissue fluids are those which are known to function as neurotransmitters in the central nervous system (CNS).

The present study was an attempt to elucidate whether and how amino acids, such as aspartic acid (Asp), glutamic acid (Glu), glycine (Gly), and γ-aminobutyric acid (GABA), may be involved in the regulation/modulation of selected biological functions of the spermatozoon. To address this question the following studies were performed: (a) examination of free amino acid concentrations in the seminal plasma and in secretions of the female genital tract; (b) studies on the effects of amino acids on the pattern of sperm agglutination; (c) examination of the influence of excitatory and inhibitory amino acids on the the motility of spermatozoa; (d) experiments on the influence of GABA and Gly on the uptake of radiolabeled chloride by spermatozoa; (e) demonstration of Glu binding sites on seminal membranes using two different radioligands.

Free Amino Acid Concentrations

Determination of Amino Acids

Amino acid concentrations were determined with an automatic amino acid analyzer (one-column system filled with Varionchrome 9 resin) using norleucine as an internal standard. Semen samples from boars were centrifuged at 1000 g, and the seminal plasma was immediately deproteinized and stored at -20°C until analysis. The amino acid secretions of oviduct and uterus were determined in 12-14 week old New Zealand white rabbits of about 3 kg body mass. To induce ovulation, rabbits were treated with 100 IU pregnant mare serum gonadotropin (PMSG) (Gonadophyl, Phylaxia, Budapest) intramuscularly and 3 days later with 100 IU human chorionic gonadotropin HCG (Chorio-gonin, G.Richter, Budapest) intravenously. Groups of six animals were killed at the time of the second injection (zero time), 4 h later, and every 24 h until day 6 after ovulation. The oviducts and uterine horns were washed from the ampullar end of the tube with 1 ml physiological saline and the washings were collected for the assay of free amino acids. To confirm the incidence of ovula-tion, ovarian follicles, hemorrhages, and corpora lutea were checked, and the ova present in the washing fluid counted.

Amino Acids in Seminal Plasma

The concentrations of free amino acids in boar seminal plasma were compared with the respective values detected in blood. On the basis of seminal/blood concentration ratios, three characteristic groups of amino acids could be distinguished: (1) those which were several times more abundant in the seminal plasma than in the blood (e.g., Asp, Glu); (2) those showing nearly equal concentrations in both tissue fluids (e.g., Leucine); and (3) those exhibiting significantly higher levels in blood than in seminal plasma (e.g., Lysine). The results obtained for several representative amino acids are shown in Fig. 1. Based on their concentration ratios in seminal plasma and blood, similar groups of amino acids could be distinguished in bulls [11].

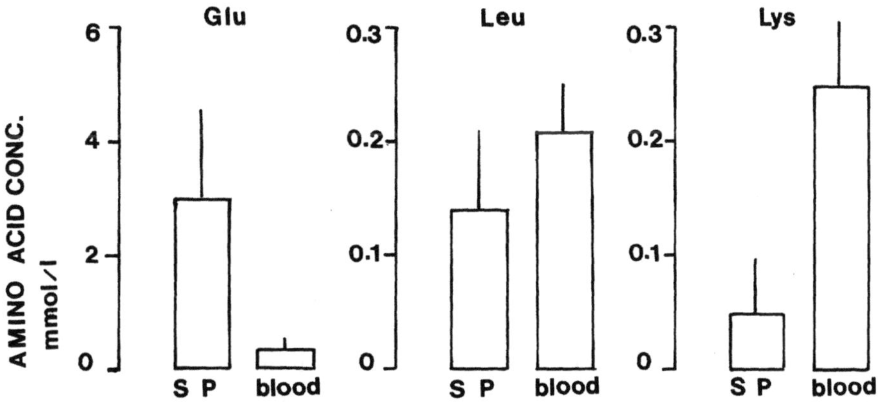

Fig. 1. Free amino acid concentrations in seminal plasma (*SP*) and blood of the boar. *Columns* and *vertical bars* represent mean ± SD, respectively (n = 27-33 for SP and 6 for blood)

Amino Acids in Oviductal and Uterine Secretions

Data reported by Fahning et al. [9] show that the major amino acid components of the uterine fluid of the cow are the same as those found to be predominant in the seminal plasma (Table 1).

Table 1. A comparison of free amino acid concentrations (mmol/l) in blood and uterine fluid of the cow (modified from [9])

	Blood	Uterine fluid days of cycle	
		7-10	17-18
Asp	0.01	0.60	0.70
Glu	0.08	1.34	1.70
Tau	0.05	1.57	0.53
Gly	0.40	4.42	1.70
GABA	trace	0.40	0.60
Tyr	0.05	0.22	0.22
Lys	0.08	trace	trace

As shown in Fig. 1 and Table 1, a group of amino acids, including Asp, Glu, Gly, Tau, and GABA, is much more abundant in the seminal plasma and in the uterine fluid than in the blood. This is consistent with an active secretion of such amino acids into both tissue fluids. Several lines of evidence indicate that the secretion of Glu and Asp into the seminal plasma is controlled by testoste-rone [2, 14], while in the female genital tract the amount of secreted Gly [9], Tau [16], and GABA [5, 6] may be under the regulation of ovarian hormones.

Hormonal Dependence of Amino Acid Secretion

Hormone-dependent changes in amino acid secretion were examined in the uterotubal fluid of rabbits at various times after hormonally induced ovulation. The most pronounced alteration was a significant increase in the Gly concentrations on days 3 and 4 after ovulation, with a concomitant decrease of Glu levels (Table 2). A similar increase in Gly secretion was observed in sows 2 days after hormonally induced ovulation (Boldizsár and Wekerle, unpublished data).

Table 2. Secretion of amino acids into the uterotubal fluid of rabbits after hormonally induced ovulation.

Hours after	Free amino acids (nmol)				
ovulation	Tau	Ser	Glu	Gly	Ala
0	134	200	255	668	301
4	138	148	287	885	238
24	74	107	122	685	144
48	110	148	153	1291	109
72	72	118	109	1600	106
96	96	100	163	1699	137
120	59	57	139	1405	85
144	nd	138	110	1168	195

Values are mean (n=6); nd, not determined

Other amino acids showed highly variable levels (e.g., Tau, Ser), without any visible correlation with postovulatory hormonal changes. In virtual contrast to the high GABA concentration demonstrated in the rat oviductal wall [8], the uterotubal secretion of the rabbit contained trace amounts of

GABA. However, in cow, Fahning et al. [9] demonstrated a slight elevation of GABA in the uterine fluid on days 17 and 18 of the estrous cycle, indicating that the secretion of GABA may also be under the control of estrogens and/or gestagens, at least in certain species.

Effects on Sperm Agglutination and Motility

Sperm Agglutination and Motility Assays

Sperm was obtained from 2-year-old roosters (each weighing 3 kg), diluted tenfold with a modified Kiev extender, centrifuged, and resuspended in the same extender to double the volume. Siliconized slides were used to inspect sperm motility and agglutination in the presence and absence of the above-mentioned amino acids. For morphological examination, stained smears were also prepared.

The motility of spermatozoa was visually evaluated by trained individuals and scored on a tenfold scale from 0 to 100%. The mean values obtained by three independent investigators were taken as an index of sperm motility.

Effects on Agglutination Pattern

In an undiluted state, rooster spermatozoa were arranged in an irregular fashion and showed agglutination-like nodules with head-to-head configuration (not illustrated). Upon 2-, 4- and 8-fold dilution with standard extender, the motility of cells increased, and thus the number of nodules was reduced. In extenders containing Asp (1-6 mM) or Glu (2-7 mM) an irregular arrangement of sper-matozoa was observed (not shown). By contrast, a regular helical or braid-like, elongated agglutination pattern occurred in GABA-containing (1-8 mM) and Gly-containing (7 mM) media within 30 sec after dilution (Fig. 2).

Amino acid concentrations (m*M*) were as follows: Asp 6, Glu 7, Gly 7, GABA 8. Cell motility was examined immediately after treatment. Values are expressed in percentage of total cell number and represent the mean of three independent observations. Control values for samples diluted freshly with amino acid-free extender varied around 80%.

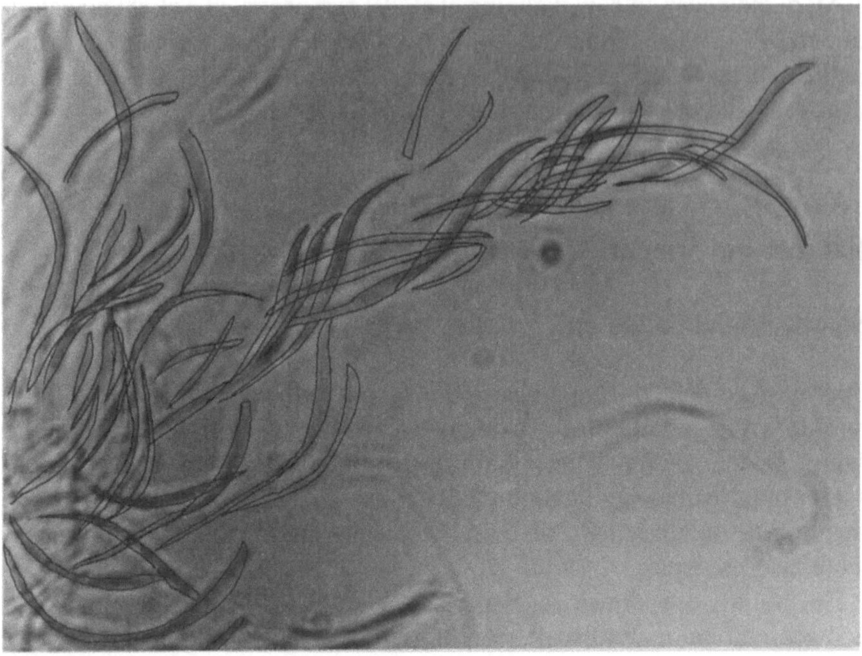

Fig. 2. Characteristic, braid-like agglutination pattern of rooster spermatozoa shown in GABA-containing extender. A similar pattern was observed in the presence of Gly

Effects on the Motility of Spermatozoa

Increasing concentrations of Glu (2-7 m*M*) immobilized the spermatozoa of roosters, while Asp (1-6 m*M*) produced a minor reduction in motility. Gly (7 m*M*) and GABA (1-8 m*M*) did not influence the spontaneous motility, but they were potent in revitalizing the cells previously immobilized by Glu (Table 3). After revitalization with 8 m*M* GABA, 20% of the cells showed vibrant motility, while 60% stationary vibration. The effects of both GABA and Gly were immediate.

Our results show that amino acid concentrations are of crucial importance for sperm motility. The Glu concentrations in the seminal plasma of roosters and in the usual dilutors are several times higher than those used in our studies, suggesting that Glu may be a potent physiological suppressor of sperm motility and that the immobilization of spermatozoa by Glu may not be due to nonspecific changes in osmolarity or pH. By contrast, Gly, and/or GABA (depending on species) may well contribute to the physiological mobilization of spermatozoa within the female reproductive tract. The rapid manifestation of the effects of Glu, Gly and GABA on sperm agglutination and motility suggests that these responses may be related to changes in the transmembrane potential of spermatocytes.

Table 3. Motility of rooster spermatozoa in extenders of different amino acid content

Asp	Glu	Gly	GABA	Mobile cells (%)
+				60
+		+		90
	+			0
	+	+		80
	+		+	70
+	+			0
+	+	+		70
		+[a]		60[a]
			+[a]	20[a] (60)[b]

[a]Effects on immobile cells. [b]Value in parentheses indicates percentage showing stationary vibration.

The present findings, along with the well known electrolyte and amino acid gradients along the genital tract, are consistent with the hypothesis that the physiological functions of spermatozoa may be regulated by neurotransmitter-type amino acids. In fact, such amino acids seem to meet all criteria established earlier for undefined, endogenous, organic compounds of low molecular weight, which have been proposed as playing a major role in the regulation of spermatozoal functions [1, 18].

Amino Acid Receptors on Spermatozoa: Receptors for Inhibitory Amino Acids

In the CNS, GABA and Gly are known to increase chloride conductance through receptor-coupled chloride channels [3]. These effects are mediated via bicuculline-sensitive $GABA_A$ receptors and strychnine-sensitive glycine receptors, respectively [3]. In the present study, the effect of GABA on the influx (uptake) of radiolabeled chloride into spermatozoa was examined.

Measurement of Chloride Uptake

The effect of GABA on $^{36}Cl^-$ influx into boar spermatozoa was examined by the method of Kardos [12], with minor modifications. Briefly, after centrifugation, the semen was diluted 1:5 with Meleagros extender, then 1:1 with a HEPES buffer (10 m*M*, pH 7.4) containing 77 m*M* Cl^-. Spermatozoa were incubated with 2.5 μCi/ml $^{36}Cl^-$ (16 mCi/g) at 30°C for 7 s in the presence or absence of 1 m*M* GABA. Chloride uptake was terminated by dilution of the samples with an ice-cold buffer containing 1 m*M* picrotoxin or 1 m*M* furose-mide. Cells were separated by filtration (Whatman GF/B), and the radioactivity accumulated in spermatozoa was counted.

Effect of GABA on Chloride Uptake

When applied at a concentration of 1 m*M*, GABA stimulated the uptake of chloride into spermatozoa in a significant manner. This increase was completely eliminated by the additional presence of a selective inhibitor of $GABA_A$ receptors, i.e., bicuculline methiodide (Table 4).

The above results indicate that bicuculline-sensitive GABA receptors ($GABA_A$) are present in seminal membranes and may regulate chloride fluxes and thus transmembrane potentials. This is in line with recent data of Erdö and Wekerle [7] showing the presence of specific ^{3}H-GABA binding sites of the $GABA_A$ type on membranes of boar spermatozoa.

Table 4. The effect of GABA and bicuculline methiodide on the influx of $^{36}Cl^-$ into boar spermatozoa

Treatment	$^{36}Cl^-$ uptake (ng/mg protein)	*n*
None	490 ± 20	5
1 m*M* GABA	662 ± 17	3
1 m*M* GABA + 1 m*M* bicuculline methiodide	488 ± 16	3

Receptors for Excitatory Amino Acids

Asp and Glu are known to be potent excitatory amino acids (EAAs) which activate common receptor populations in the brain (see [10]). In the CNS, Glu may label three major receptor populations, which are referred to, after their selective agonists, as *N*-methyl-D-aspartate (NMDA), quisqualate, and kainate receptors [10]. In the absence of chloride ions, Glu is known to chiefly label the NMDA-sensitive receptor class (chloride-independent binding). In the the presence of chloride (and calcium), Glu binds to a number of other binding sites (chloride-dependent binding), with uptake / reuptake carriers as the predominant components [10]. The present study was carried out to demon-strate both chloride-independent (NMDAtype) receptor sites and chloride-dependent binding sites on seminal membranes using [^{3}H]-Glu as the radioligand. Moreover, the occurrence of NMDA-sensitive receptors was also examined using a specific, noncompetitive NMDA-receptor antagonist, i.e., [^{3}H]*N*-(1,2-thienyl)-cyclohexyl piperidine ([^{3}H]TCP).

Chloride-Independent [^{3}H]Glu Binding Assay

Chloride-independent Glu binding was investigated in extensively washed whole membranes of boar spermatozoa. Membranes were incubated in 50 m*M* Tris-acetate buffer (pH 7.4) in the presence of 5 nM [^{3}H]Glu at 37°C for 60 min. The reaction was terminated by centrifugation at 40000 g and 4°C for 20 min. The supernatant was discarded and the radioactivity bound to the pellet was measured by a conventional liquid scintillation technique. Nonspecific binding was determined in the additional presence of 1 m*M* Glu, 1 m*M* NMDA, or 1 m*M* D-2-amino-5-phosphono-valerate (D-AP5).

Chloride-Dependent [^{3}H]-Glu Binding Assay

Chloride-dependent binding to whole membranes of boar spermatozoa was examined. Briefly, previously washed membranes were incubated in a 50 m*M* Tris-HCl buffer (pH 7.4) containing 5 n*M* [^{3}H]Glu, at 25°C for 30 min. The reaction was terminated by centrifugation (40000 g, 4°C, 20 min) and the supernatant aspirated. The radioactivity bound to the membrane pellets was measured by standard liquid scintillation spectrometry.

Nonspecific binding was measured in the additional presence of 1 m*M* unlabeled Glu.

[^{3}H]TCP Binding Assay

Boar seminal pellets were homogenized in distilled water. Membranes were centrifuged and resuspended in a 50 m*M* Tris-HCl buffer (pH 7.4) containing 0.05% Triton X-100 and preincubated at 22°C for 30 min in order to remove endogenous Glu and polyamines. The detergent was removed by three cycles of washing, i.e., suspension in fresh buffer and centrifugation (40000 g, 4°C, 20 min). Incubations were performed in the presence of 10 n*M* [^{3}H]TCP at 22°C for 30 min. The reaction was terminated by filtration. Nonspecific binding was measured in the presence of 22 m*M* (+)-ketamine.

Binding Sites on Spermatozoa

The chloride-independent binding of [^{3}H]Glu was reduced by approximately 20% in the presence of 1 m*M* NMDA or 1 m*M* D-AP5, indicating that NMDA-sensitive receptors may be responsible only for a minor component of Glu binding to seminal membranes (not shown).

The NMDA-selective radioligand [^{3}H]TCP, at a concentration of 3.5 nM, showed moderate specific binding to seminal membranes amounting to 79 fmol/mg protein. Although both results are preliminary in nature, they suggest that NMDA-sensitive EAA receptors are present on membranes of spermatozoa. Furthermore, it should be noted that, in contrast with previous findings for cerebral membranes, 10 μM Glu did not significantly enhance the binding of TCP to seminal membranes.

Although the above studies did not provide information about the distribution of Glu receptor sites on spermatozoa, one can speculate whether they are enriched in particular portions (head, tail) of gametocytes. Indeed, selectively decapitated spermatozoa may represent a suitable tool for the clarification of such a question.

Chloride-dependent Glu binding was also demonstrated in membranes of boar spermatozoa. The binding of 5 n*M* [^{3}H]Glu could be displaced by increasing concentrations of unlabeled Glu (Table 5), with 50% displacement being found at 300 n*M* Glu (IC_{50}). This value resembles those reported for chloride/calcium-dependent Glu binding sites on specific transport carriers of cerebral membranes [13, 15], suggesting that similar Glu transporters

Table 5. Displacement of chloride-dependent [^{3}H]Glu (5 n*M*) binding from membranes of boar spermatozoa

Unlabeled Glu (nmol/l)	[^{3}H]Glu specific binding (%)
-	100
10	86 ± 8
30	98 ± 4
100	71 ± 5
300	50 ± 8
1000	28 ± 7

Values are mean ± SD; *n* = 3-5

also be present on spermatozoa. Further research is needed to identify the putative Glu uptake mechanisms in spermatozoa.

It has been shown that a portion of chloride-dependent calcium-stimulated Glu binding is confined to calcium-dependent enzymes in brain membranes [17, 19]. Accordingly, a contribution of such enzymes to chloride-dependent Glu binding in seminal membranes seems plausible. However, it is worth mentioning that the stimulating effect of calcium on Glu binding to boar spermatozoa was not evident in our experiments (not shown). This resembles previous findings for Glu binding to cerebrocortical astrocytes [4].

Summary and Conclusions

The results summarized above provide evidence that neurotransmitter amino acids may also be important in the local regulation of seminal functions, such as motility, agglutination, etc. Thus, Glu, Gly, and GABA were most potent in influencing spermatocytes. The effects of these amino acids seem to be mediated via specific receptors located on seminal cell membranes that show basic characteristics resembling those of the respective receptors in the brain. Although our findings indicate that Glu, Gly, and GABA may play key roles in basic reproductive phenomena, much further work is needed to elucidate the exact mechanisms through which they regulate the functions of spermatozoa in different species.

Acknowledgements. The authors wish to thank J. Kardos, I. Kovács, and G. Maksay (Central Research Institute for Chemistry, Budapest) for their help in receptor binding assays.

References

1. Austin CR (1975) Membrane fusion events in fertilization. J Reprod Fertil 44: 155-166
2. Boldizsár H, Wekerle L, Fehér T (1988) Actual concentrations of testosterone, estrone, estradiol and cortisone in blood and of testosterone and amino acids in seminal plasma in swine (in Hungarian). Proceedings of the Andrology Meeting, Herceghalom, Hungary
3. Bormann J, Hamill OP, Sakmann B (1987) Mechanisms of ion permeation through channels gated by glycine and γ-aminobutyric acid (GABA) in mouse cultured spinal neurons. J Physiol (Lond) 385: 243-286
4. Bridges RJ, Nieto-Sampedro M, Kadri M, Cotman CW (1987) A novel chloride dependent L-^{3}H-glutamate binding site in astrocyte membranes. J Neurochem 48: 1709-1715
5. Del Rio RM (1981) γ-Aminobutyric acid system in rat oviduct. J Biol Chem 256: 9816-9819
6. Erdö SL (1984) Alteration of GABA levels in ovary and Fallopian tube of the pregnant rat. Life Sci 34: 1879-1884
7. Erdö SL, Wekerle L (1990) $GABA_A$ type binding sites on membranes of spermatozoa. Life Sci 47: 1147-1151
8. Erdö SL, Rosdy B, Szporny L (1982) Higher GABA concentrations in Fallopian tube than in brain of the rat. J Neurochem 38: 1174-1176
9. Fahning ML, Schultz RH, Graham EF (1967) The free amino acid content of uterine fluids and serum in the cow. J Reprod Fertil 13: 229-236
10. Hansen JJ, Krogsgaard-Larsen P (1990) Structural, conformational, and stereochemical requirements of central excitatory amino acid receptors. Med Res Rev 10: 55-94
11. Ibrahim MAR, Boldizsár H (1981) Studies on free amino acid content in seminal plasma of A.I. bulls of different performance. Acta Vet Acad Sci Hung 29: 263-269
12. Kardos J (1989) $^{36}Cl^-$ flux measurements on $GABA_A$ receptor-activated chloride exchange. Biochem Pharmacol 38: 2578-2591
13. Kessler M, Bandry M, Lynch G (1987) Use of cysteine to distinguish glutamate binding from glutamate sequestration. Neurosci Lett 81: 378-381
14. Kochakian CD (1973) Hypotaurine: regulation of production in seminal vesicles and prostate of guinea pig by testosterone. Nature 241: 202-203
15. Mena EE, Fagg GE, Cotman CW (1982) Chloride ion enhance L-glutamate binding to rat brain synaptic membrane. Brain Res 243: 378-381
16. Van der Horst CJG, Brand A (1971) The occurrence of hypotaurine and inositol in the reproductive tract of the ewe and its regulation by pregnenolone and progesterone. Nature 223: 67-68
17. Vargas F, Greenbaum L, Costa E (1980) Participation of cysteineproteinase in the affinity of Ca^{2+}-dependent binding of glutamate to hippocampal synaptic membranes. Neuropharmacology 19: 791-794

18. Yanagimachi R (1970) In vitro capacitation of golden hamster spermatozoa by homologous blood sera. J Reprod Fertil 3: 147-153
19. Yoneda Y, Ogita K, Yamada K, Kouda T, Ohgaki T, Uchida S, Megura H (1989) High concentrations of calmodulin antagonists inhibit accumulation and binding activities of ^{3}H-glutamate in rat brain. Neurochem Int 15: 17-23

Pharmacology and GABA Receptors

GABA Receptors Inside and Outside the Brain

N.G. Bowery

Introduction

Establishment of any substance as a neurotransmitter or neurohormone requires the fulfillment of defined criteria [56]; this is often easier outside than inside the brain. However, the reverse appears to be true for γ-aminobutyric acid (GABA). Evidence for its localized presence and discrete release from neuronal processes has been demonstrated in brain tissue and inhibitory neuronal responses to exogenous GABA compare favourably with the synaptic response to stimulation of inhibitory pathways [45, 24]. The significance of GABA in these mechanisms is substantiated by the use of selective antagonists such as bicuculline and 2-hydroxysaclofen or phaclofen [22, 32, 38], which substantially block synaptically mediated responses including that to applied GABA.

In the periphery the evidence is not so conclusive. GABA and its forming enzyme glutamic acid decarboxylase (GAD) are present, although in lesser amounts than in the brain [35] and its stimulated release from myenteric neurones has been demonstrated [41]. However, substantial evidence for the presence of GABAergic neurones outside the enteric nervous system is not so obvious and, although they appear to be present in sympathetic ganglia (see Wolff et al. and Häppölä et al., this volume), the significance of GABA as a peripheral neurotransmitter is a matter for debate.

One undeniable and consistent feature of the GABA system outside the brain is the existence of receptors on many peripheral organs [29, 54, 47, 34]. These receptors appear to be comparable to their CNS counterparts although the possibility that there are subtle distinctions cannot be ignored. It seems possible that the receptors have no physiological role in many cases, but this does not mean that they cannot be used for pharmacological manipulation, a point I will return to later.

Historical Aspects

The significance of GABA within the mammalian brain became established in the mid-1960s after considerable debate during the preceding 10 years [4, 50, 25, 6, 20]. The crucial evidence came from electrophysiological studies showing that GABA mimicked synaptic inhibition within higher centres [46].

An understanding of the action of GABA outside the brain had been pursued at the same time although with much less emphasis on its potential as a neurotransmitter. Among the earliest reports was that by Hobbiger [40], which showed that the isolated intestine of the guinea pig could be contracted and relaxed by GABA. However, the significance of this biphasic effect was not realized at the time and was not interpreted as an activation of separate receptor subtypes. Subsequent studies involving neuronal (and cardiovascular) preparations showed that GABA can elicit effects analogous to those reported in the CNS. It transpired that GABA receptors gate Cl^- channels in peripheral neurones in a similar manner to those in central neurones and this effect can be mimicked and antagonised by the same agonists and antagonists as in the CNS [28, 9].

The superior cervical ganglion has provided an ideal in vitro system to examine this point. Activation of GABA receptors in the ganglion increases neuronal Cl^- conductance which depolarizes the cells (Fig. 1) since the equilibrium potential for Cl^- is in the region of -40 mV (compared with 70 mV in most CNS neurones) [1]. If the distribution of Cl^- across the membrane is altered such that the internal $[Cl^-]$ is reduced relative to the external concentration then a hyperpolarization can be elicited in response to GABA. The depolarization produced under physiological conditions is generally insufficient to produce neuronal firing and is thus analogous to depolarizing inhibition in the CNS.

The receptor responsible for this action of GABA we would now classify as a $GABA_A$ site in view of the agonist and antagonist selectivity [34]. Moreover, barbiturates, benzodiazepines and inverse benzodiazepines receptor ligands all modulate the response to GABA in a manner consistent with $GABA_A$ receptor involvement [36, 14, 48].

Within the CNS a series of protein subunits have been identified comprising the GABA receptor complex [53], indicating that $GABA_A$ receptors are heterogeneous although none of the $GABA_A$ recognition ligands are able to differentiate these sites at present. The molecular nature of the $GABA_A$ site in peripheral ganglion neurones or other peripheral sites is unknown, but whether all of the sequences so far described in the CNS are present remains an interesting possibility. The number of receptors per unit area is much lower than on central neurones and this may suggest that the full complement of site sequences is not expressed.

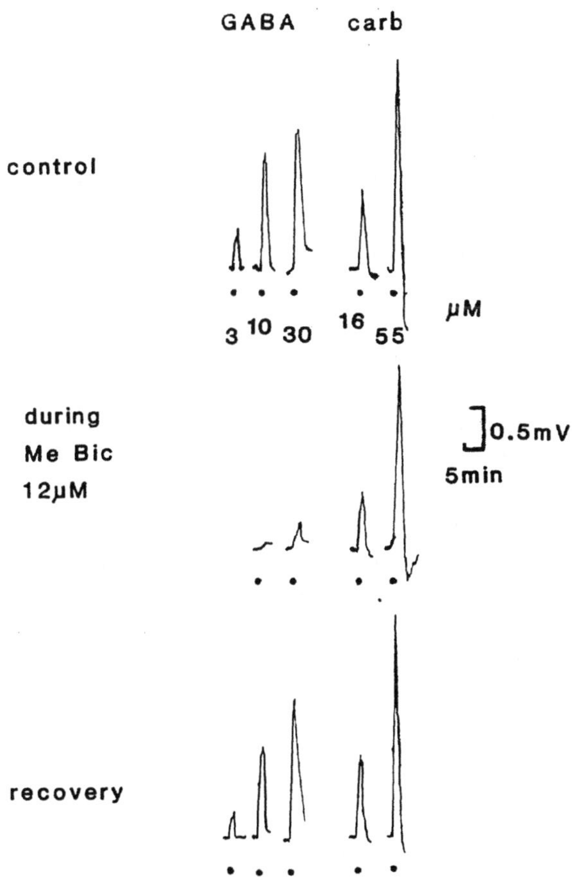

Fig. 1. Depolarizing responses to GABA and carbachol in a rat isolated superior cervical ganglion obtained by surface recordings. The ganglion was superfused with Krebs-Henseleit solution (1 ml/min) at room temperature and the voltage between $Ag^{+}/AgCl$ electrodes placed on the ganglion cell body and postganglionic trunk monitored continuously. GABA (3, 10 of 30 *μM*) or carbachol (*carb*, 16 or 55 *μM*) was added to the superfusion solution for 2 min periods indicated by the *black dots*. The *middle panel* responses were obtained 20 min after the addition of bicuculline methobromide (*MeBic*, 12 *μM*) to the superfusion solution. The *lowest panel* was obtained 30 min after washing out the bicuculline. GABA or carbachol were administered at 15 min intervals. Atropine (10 *μM*) was present throughout

Are GABA Receptors in Ganglia Physiologically Important?

Whilst receptors for GABA have been adequately demonstrated in sympathetic and parasympathetic ganglia their activation by endogenous ligands has been difficult to demonstrate. The major source of GABA within ganglia appears to be in the glial cells [8, 63, 5] although a neuronal location has been reported [43]. Glial cells actively accumulate GABA [8, 12] and this can be released by, for example, an excess of the carrier substrate ß-alanine or by excess [K^+] [18, 11]. Normally this accumulated GABA would be degraded by GABA transaminase present in glia, but in the presence of an enzyme inhibitor, such as amino oxyacetic acid, the amount available for release is sufficient to activate the neuronal GABA receptors to produce a depolarization [10]. Whether this reflects a physiological release mechanism inside or outside the brain is unknown but does indicate that glial cells play an active role in GABA function. The possibility of a neuronal release of GABA in ganglia has been proposed but this may be restricted, physiologically, to the enteric nervous system.

The major evidence for GABA release from peripheral neurones stems from the work of Jessen and colleagues [41, 42] within the enteric nervous system. GABAergic neurones have been clearly demonstrated in the intestine and these appear to project to motor fibres innervating the intestinal muscle. The GABA receptors on which these fibres impinge are presumably those responsible for the effects that Hobbiger first observed in 1958 [40].

$GABA_B$ Receptors

The presence of bicuculline-sensitive GABA receptors on the soma of sympathetic neurones led to the postulate that they would also be present on sympathetic terminals arising from these neurones [17]. In this case their activation might be expected to depolarize the nerve terminal in a manner analogous to that thought to occur at primary afferent terminals in the spinal cord [21]. This would have provided a model for presynaptic depolarization of inhibition mediated via GABA within the cord. It has long been considered that activation of bicuculline-sensitive GABA receptors on primary afferent terminals within the dorsal horn of the cord decreases transmitter release (for references see [21]).

Since depolarization of sympathetic terminals cannot be readily measured I decided to use the direct approach of monitoring the evoked release of tritiated norepinephrine following its accumulation by sympathetic terminals of rat isolateral atria. The outcome of these experiments was as predicted in that GABA inhibited release in a dose-dependent manner. What was not predicted, however, was the nature of the receptor responsible for the action of GABA.

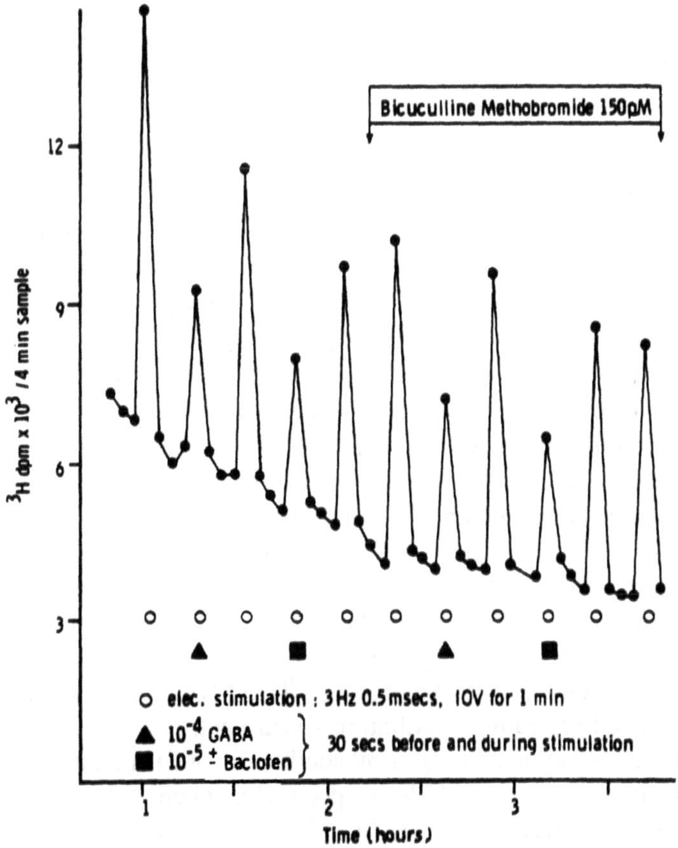

Fig. 2. Depression of the evoked release of [^{3}H]norepinephrine from rat isolated atria by GABA and baclofen: insensitivity to bicuculline. The atria from one rat were incubated in 0.4 μM [^{3}H]noradrenaline for 40 min and then superfused with Krebs-Henseleit solution (0.45 ml/min). The tritium content of consecutive 4 min superfusate samples is represented by the *closed circles*. *Open circles* below the record indicate periods of transmural stimulation. GABA (*triangles*) or baclofen (*squares*) were added to the superfusion solution where indicated. Bicuculline methobromide (150 μM) was present throughout the period indicated by the *horizontal bar*

The effect of GABA was not mimicked by many GABA analogues normally considered to be receptor agonists, such as isoguvacine, and was not blocked by the accepted antagonists, such as bicuculline (Fig. 2) or picrotoxin. The effect of GABA appeared to be mediated through a novel receptor which was not coupled to Cl^- channels and it seemed unlikely that any resulting alteration in transmitter release was due to terminal depolarization as first proposed. Instead, we now believe that the diminution in transmitter release results from a reduction in [Ca^{++}] conductance produced by activation of the novel receptor [13, 30, 31, 19], although the possibility of an increase in [K^+] conductance cannot be excluded [52].

In attempting to establish the pharmacological characteristics of this apparently novel receptor the affects of numerous GABA analogues were studied. Among these was the substance ß-p-chlorophenyl-GABA (baclofen). This analogue had already been in clinical use as a muscle relaxant for at least 8 years although its mode of action was unknown. Baclofen mimicked GABA in inhibiting the evoked release of noradrenaline and was stereospecifically active [13]. By contrast it was inactive at bicuculline-sensitive chloride-dependent GABA receptors in the superior cervical ganglion preparation (personal observations) and had already been shown to be inactive at GABA receptors coupled to Cl^- channels within the brain [27, 23]. Thus (-)-baclofen provided a selective agonist for studying the significance of these novel receptors which were designated $GABA_B$ to differentiate them from classical $GABA_A$ receptors [39].

Further evidence for the existence of novel receptor binding sites for GABA in rat atria was obtained using 10-μm cyrostat sections. Each section was incubated in sodium-free Tris-HCl buffer containing 50 n*M* [^{3}H]GABA. In initial experiments sections obtained from six rats (six sections per rat) were incubated with [^{3}H]GABA alone or together with an excess (100 μM) of the unlabelled receptor ligands isoguvacine, bicuculline methobromide or (±)-baclofen. The amount of [^{3}H]GABA bound to the sections was determined by liquid scintillation spectrometry.

Under these conditions isoguvacine and bicuculline reduced the amount bound by approximately 50% and 40%, respectively, whereas baclofen produced no significant effect (Fig. 3). In subsequent experiments, in which Ca^{++} (2.5 m*M*) and isoguvacine (40 μM) were added to the incubation buffer, (±)-baclofen and GABA (100 μM) but not bicuculline reduced the amount bound (Fig. 4). These data are the same as obtained in experiments using rat brain synaptic membranes [22, 38] and indicate the presence of both Ca^{++}-dependent, isoguvacine- and bicuculline-insensitive $GABA_B$ binding sites and $GABA_A$ sites.

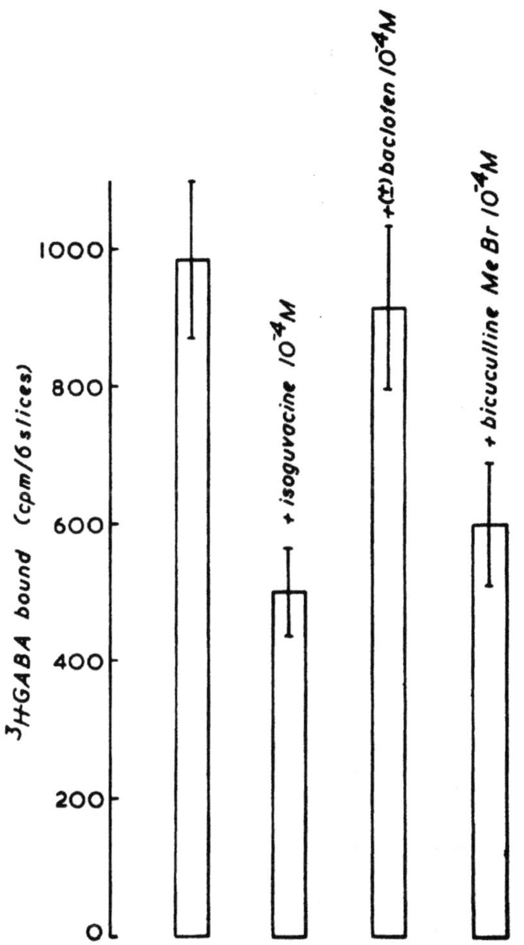

Fig. 3. Presence of isoguvacine- and bicuculline-displaceable [^{3}H]GABA binding sites in 10 μm sections of rat atria. Atria from 6 rats were removed and immersed in 0.1% paraformaldehyde for 30 min prior to freezing in isopentane in liquid nitrogen; 10 μm sections were prepared at -20°C, mounted on glass slides and stored frozen at -20°C. Thawed sections were incubated in Tris-HCl buffer solution containing [^{3}H]GABA (100 n*M*) for 20 min at 20°C with or without unlabelled displacers. *Vertical bars* indicate SEM (Data from Bowery and Hudson, unpublished)

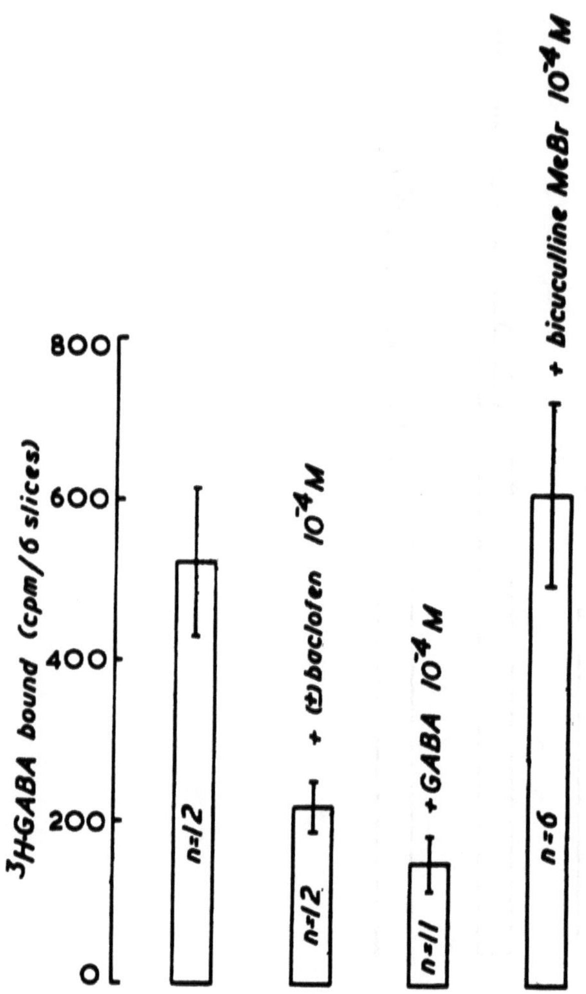

Fig. 4. Presence of baclofen-sensitive [^{3}H]GABA binding sites in 10 μm rat atrial sections. The sections from 6-12 rat atria were prepared as indicated in Fig. 3 but incubated in Tris-HCl buffer solution containing Ca^{++} (2.5 m*M*) and isoguvacine (40 μ*M*) to prevent binding to $GABA_A$ sites (see [49]). *Vertical bars* indicate SEM (data from Bowery and Hudson, unpublished)

These sites ($GABA_A$ and $GABA_B$) appear to be present mainly on sympathetic nerve fibres since pretreatment with 6-hydroxydopamine (6-OHd) reduced the binding of [^{3}H]GABA to both sites. Rats were injected with 6-OHd on four occasions before killing and the atria excised on day 13. The amount of [^{3}H]GABA bound to 10-μm sections was determined as before. $GABA_A$ and $GABA_B$ binding was significantly reduced as illustrated in Fig. 5 which shows the mean level of binding in single atria expressed as percent of the total [^{3}H]GABA associated with the tissue.

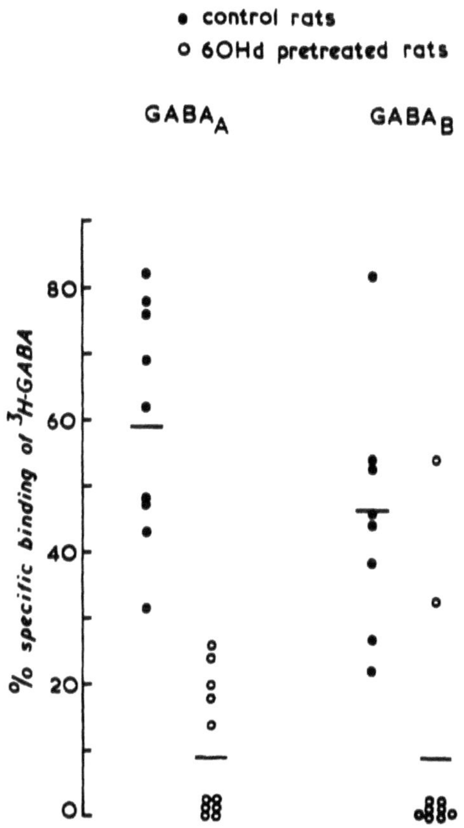

Fig. 5. Influence of 6-hydroxydopamine (6-OHd) pretreatment on the binding of [^{3}H]GABA to $GABA_A$ and $GABA_B$ sites in 10 μm rat atrial sections. Rats were injected with 6-OHd on days 1 and 2 (50 mg/kg) and days 8 and 9 (100 mg/kg) and the atria excised on day 13. [^{3}H]GABA binding to sections in the absence ($GABA_A$) and presence ($GABA_B$) of Ca^{++} and isoguvacine was determined and mean values from at least 6 sections per rat determined. These are indicated by the *dots* (control) or *open circles* (6-OHd). Each dot or *circle* represents a single rat. Group means are indicated by the *horizontal lines* (data from Bowery and Hudson, unpublished)

Atria from six of the treated and control animals were incubated in [^{3}H]norepinephrine to assess their capacity to accumulate the amine. Pretreatment with 6-OHd reduced the uptake by >80% (Fig. 6) supporting the notion that the sympathetic fibres had degenerated and that the reduction in GABA binding may be due to a loss of the sympathetic fibres.

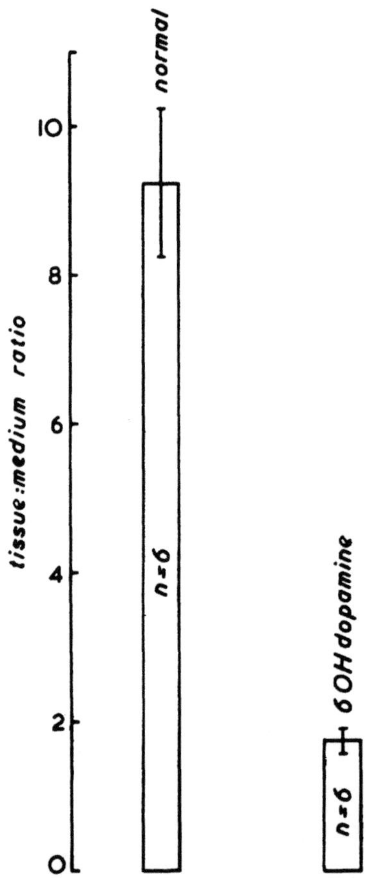

Fig. 6. Effect of 6-OHd treatment on [^{3}H]norepinephrine accumulation by isolated atria. Atria from 6 control and 6-OHd-treated rats (regimen as for Fig. 5) were incubated at 32°C for 40 min in Krebs-Henseleit solution containing [^{3}H]norepinephrine (0.4 μM) (plus pargyline 3 μM and ascorbic acid 100 μM). After incubation the tissues were rinsed briefly and immersed in Soluene for 24 h prior to determining the tissue: medium ratios from the tritium levels in the atria and incubation solution. (data from Bowery and Hudson, unpublished)

$GABA_B$-mediated inhibition of transmitter release manifests as a depression in neurally evoked contractions of isolated preparations of smooth muscle and this may reflect a physiological role, at least within the intestine. In this tissue $GABA_B$ site activation can produce a decrease in peristalsis and muscle relaxation [44, 55], and the presence of GABA neurones within the enteric nervous system could provide the source of natural agonist.

Subsequent studies have demonstrated the presence of $GABA_B$ sites within the brain [52, 15, 16, 7], but it was particularly rewarding to know that it was the data obtained in the initial studies in peripheral tissues which led to their discovery in higher centres.

It is now apparent that $GABA_B$ sites play a significant physiological role in the brain and this is of greater importance than the original effects observed in the periphery. However, whilst the majority of the peripheral actions may not be physiologically relevant the $GABA_B$ receptor still provides a site for pharmacological manipulation. Among the peripheral actions attributed to $GABA_B$ receptor activation are those listed in Table 1. In most organs in which $GABA_B$ receptors have been demonstrated, $GABA_A$ sites have also been shown and frequently their activation has an opposite effect. For example, in the intestine $GABA_A$ receptor activation

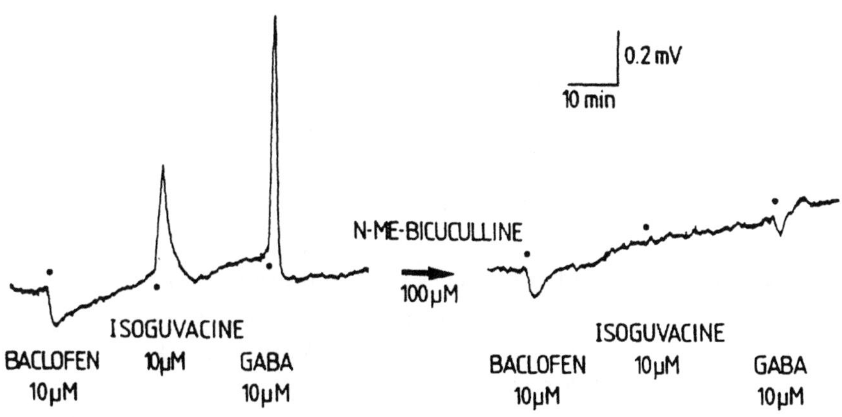

Fig. 7. Bicuculline-sensitive and -insensitive actions of GABA recorded extracellularly in a rat isolated superior cervical ganglion. In the absence of bicuculline, GABA depolarized the ganglion, whereas in the presence of the antagonist GABA produced a hyperpolarizing response. Each agonist was applied for 1 min at the points indicated by the *dots above and below* the recording. Note that baclofen and isoguvacine mimicked the two effects of GABA in the presence and absence of antagonist, respectively. From [52] with permission

increases tone, whereas $GABA_B$ site stimulation decreases both tone and peristaltic activity [44, 37]. Another recent example has been demonstrated in the rat superior cervical ganglion [51] (Fig. 7) where $GABA_A$ sites mediate depolarization though an increase in neuronal chloride ion conductance and $GABA_B$ site activation produces neuronal hyperpolarization presumably through an increase in $[K^+]$ conductance although this has not been established. These comparative results obtained in the same preparation were of particular note because as pointed out earlier, it was extrapolation from data obtained in this ganglion preparation which led to the detection of $GABA_B$ sites on terminals arising from this tissue. When we obtained these original data we never considered the converse, that $GABA_B$ sites would also be present on the cell bodies.

Table 1. Responses to GABA receptor activation in periphery

Tissue	$GABA_A$	$GABA_B$
Intestine (varies with region)	Contractile activity ↑	Peristalsis ↓
Lung	ACh release ↑ Bronchial hyperreactivity ↓	ACh release ↓ Bronchial hyperreactivity ↓
Trachea	ACh release ↓	
Gall bladder	Contraction	Contraction ?
Urinary bladder	Motility ↓	Motility ↓
Blood vessels	Constriction - PG mediated Dilation - ACh mediated	Dilation - neuronally mediated
Sex organs (species dependent)	Inhibition (oviduct, fallopian tube)	Stimulation (oviduct, uterus)
Endocrines	Somatostatin release ↓ Prolactin release ↓ MSH release ↑	Glucagon ↑ plasma MSH release ↓
Adrenal medulla	Catecholamine release ↑	Catecholamine release ↓

G-Protein-Involvement

The ganglion hyperpolarization produced by baclofen and GABA (in the presence of bicuculline) can be prevented by pretreatment of the tissue with pertussis toxin [51], which ADP-ribosylates membrane G-proteins associated with inhibition of adenylyl cyclase. This not only reduces the functional responses which are G-protein coupled but also ligand binding to associated receptors. Treatment of synaptic membranes prepared from rat cerebral cortex with pertussis toxin reduces $GABA_B$ binding [2] and $GABA_B$-mediated inhibition of forskolin-sitmulated adenylyl cyclase activity in cortical slices and cerebral granule cells [64]. Pertussis toxin also appears to selectively reduce post- but not pre-synaptic $GABA_B$ events recorded in rat hippocampus [38, 33]. thus ganglionic $GABA_B$ responses may be analogous to the post-synaptic responses recorded in the CNS.

In view of the insensitivity of presynaptic $GABA_B$ effects to pertussis toxin in hippocampus it has been proposed that $GABA_B$ sites are heterogeneous. However, much more data are required to establish this possibility, although it may even transpire that peripheral $GABA_B$ sites differ from their central counterparts.

A further source of receptor differentiation within the CNS may arise from the two distinct effects of baclofen (and GABA) on adenylyl cyclase. Inhibition and stimulation of cAMP formation have been reported for both of these agonists in a manner consistent with a $GABA_B$ receptor involvement in each case. However, the pharmacological profile may be different for these two effects [58, 61]. For example, 3-aminopropylphosphinic acid (3-APA) is equiactive with baclofen in inhibiting forskolin-stimulated cAMP production, whereas it appears to be significantly less active in potentiating norepinephrine-induced cAMP formation. Furthermore the $GABA_B$ antagonist, CGP 35348 (p-[3-aminopropyl]-p-diethoxymethylphosphinic acid) prevents baclofen-mediated inhibition much more effectively than the potentiation produced by baclofen (Pratt and Bowery, personal observations).

$GABA_B$ effects on adenylyl cyclase activity in peripheral tissues have not been studied but it seems likely that they would be detected in view of the ganglionic sensitivity to pertussis toxin [51]. Whether both the inhibitory and potentiating activity can be demonstrated is an interesting question. If peripheral $GABA_B$ receptor characteristics do differ from their CNS counterparts andenylyl cyclase studies may provide important information.

Therapeutic Use of GABA Receptor Agonists and Antagonists

The potential use of GABA receptor ligands in modulating peripheral function has yet to be fully explored. However, it seems unlikely that a $GABA_A$ antagonist will be of any value particularly in view of the potentially lethal effects on gaining access to the CNS. $GABA_A$ receptor agonists may have some potential use within the periphery particularly in the control of endocrine release mechanisms such as for somatostatin and prolactin [59, 60].

By contrast both $GABA_B$ agonists and antagonists may be useful therapeutic agents both outside and within the brain. For example, present indications suggest that a selective $GABA_B$ agonist will be (a) an effective bronchiorelaxant free of the acetylcholine releasing properties of a $GABA_A$ agonist and (b) an intestinal muscle relaxant decreasing peristaltic activity. This latter effect may be of significance in conditions of excessive intestinal activity whereas the former may be important as a treatment for asthma.

As pointed out earlier the evidence supporting a physiological role for $GABA_B$ receptors in the periphery is low. Without this knowledge predictions of the actions of any antagonist in vivo are impossible since its activity should be dependent on endogeneous agonist activation of the receptor. At present there appear to be no indications for using a $GABA_B$ antagonist within the periphery. More likely is its therapeutic potential within the CNS. Predictions that such a drug will be antidepressant, anticonvulsant, a cognition improver [7] and anxiolytic are based on the knowledge of $GABA_B$-mediated events within in the brain. For example, $GABA_B$ receptor activation reduces transmitter amine release from brain slices. If the monoamine theory of depression [62] is valid then an increase in synaptic amine function may be valuable in the treatment of depression. This provides the hypothetical basis for the therapeutic activity of tricyclic antidepressants and monoamine oxidase inhibitors. A $GABA_B$ antagonist might be expected to prevent any suppression of monoamine release exerted by an endogeneous agonist.

It has been reported that baclofen can decrease the retention and consolidation of memory, thus, an agonist might reverse this effect.

The potential antiepileptic properties could arise in particular with an antagonist selective for presynaptic $GABA_B$ autoreceptors. It has been reported that autoreceptors on GABAergic terminals are of the $GABA_B$ class [57, 3] and these are present, for example, on GABA inhibitory processes to pyramidal cells within the hippocampus [26]. A $GABA_B$ autoreceptor antagonist should therefore increase inhibition to the pyramidal cells suppressing neuronal excitation and generation of an epileptic focus. This could be an important mechanism even with an antagonist of all $GABA_B$ receptors if the autoreceptor mechanism prevails physiologically or in pathological conditions. Whilst $GABA_B$ site activation can suppress glutamate release from hippocampal synaptosomes [49] and an antagonist might therefore be expected to increase rather than suppress excitation, this mechanism may be of less importance physiologically since no evidence for any GABAergic innervation to glutamate terminals, in particular those innervating pyramidal cells, has been described. By contrast an autoreceptor mechanism is more likely to be of physiological significance.

In conclusion, there is no doubting the importance of GABA receptor mechanisms within the CNS. Both $GABA_A$ and $GABA_B$ subtypes are of physiological significance and their pharmacological modulation is therapeutically relevant. By contrast, in the periphery their physiological significance is less clear but both receptor types are present and even if their physiological roles prove to be insignificant they can still provide sites for pharmacological manipulation. Moreover, it should be remembered that it was data from peripheral preparations which provided the lead and basis for establishing the current classification of GABA receptor subtypes in the brain.

References

1. Adams PR, Brown DA (1975) Actions of γ-aminobutyric acid on sympathetic ganglion cells. J Physiol (Lond) 250: 85-120.
2. Asano T, Ui M, Ogasawara, N (1985) Prevention of the agonist binding to γ-aminobutyric acid B receptor by guanine nucleotides and islet-activating protein, pertussis toxin, in bovine cerebral cortex. J Biol Chem 260: 12653-12658
3. Baumann PA, Wicki P, Stierlin C, Waldmeier PC (1990) Investigations on $GABA_B$ receptor-mediated autoinhibition of GABA release. Naunyn Schmiedebergs Arch Pharmacol 341: 88-93
4. Bazemore AW, Elliott KAC, Florey E (1957) Isolation of Factor I. J Neurochem 1: 334-339
5. Bertilsson L, Suria A, Costa E (1976) γ-Aminobutyric acid in rat superior cervical ganglion. Nature 260: 540-541
6. Bindman LJ, Lippold OCJ, Redfearn JWT (1962) The nonselective blocking action of γ-aminobutyric acid in the sensory cerebral cortex of the rat. J Physiol (Lond) 162: 105-120

7. Bowery NG (1989) $GABA_B$ receptors and their significance in mammalian pharmacology. Trends Pharmacol Sci 10: 401-407
8. Bowery NG, Brown DA (1972) γ-Aminobutyric acid uptake by sympathetic ganglia. Nature New Biol 238: 89-91
9. Bowery NG, Brown DA Depolarizing actions of γ-aminobutyric acid and related compounds on rat superior cervical ganglia in vitro. Br J Pharmacol 50: 205-218
10. Bowery NG, Brown DA, Collins GGS, Galvan M, Marsh S, Yamini G (1976) Indirect effects of amino acids on sympathetic ganglion cells mediated through the release of γ-aminobutyric acid from glial cells. Br J Pharmacol 57: 73-91
11. Bowery NG, Brown DA, Marsh S (1979) γ-Aminobutyric acid efflux from sympathetic glial cells: effect of depolarizing agents. J Physiol (Lond) 293: 75-101
12. Bowery NG, Brown DA, White RD, Yamini G (1979) ^{3}H-γ-aminobutyric acid uptake into neuroglial cells of rat superior cervical sympathetic ganglion. J Physiol (Lond) 293: 51-74
13. Bowery NG, Doble A, Hill DR, Hudson AL, Shaw JS, Turnbull MJ, Warrington R (1981) Bicuculline-insensitive GABA receptors on peripheral autonomic nerve terminals. Eur J Pharmacol 71: 53-70
14. Bowery NG, Dray A (1978) Reversal of the action of amino acid antagonists by barbiturates and other hypnotic drugs. Br J Pharmacol 63: 179-215
15. Bowery NG, Hill DR, Hudson AL (1983) Characteristics of $GABA_B$ receptor binding sites on rat whole brain synaptic membranes. Br J Pharmacol 78: 191-206
16. Bowery NG, Hill DR, Hudson AL, Doble A, Middlemiss DN, Shaw J, Turnbull MJ (1980) Baclofen decreases neurotransmitter release in the mammalian CNS by an action at a novel GABA receptor. Nature 283: 92-94
17. Bowery NG, Hudson AL (1979) γ-Aminobutyric acid reduces the evoked release of ^{3}H-noradrenaline from sympathetic nerve terminals. Br J Pharmacol 66: 108P
18. Bowery NG, Jones GP, Neal MJ (1976) Selective inhibition of neuronal GABA uptake by *cis*-1,3-aminocyclohexane carboxylic acid. Nature 264: 281-284
19. Cherubini E, North KA (1984) Inhibiton of calcium spikes and transmitter release by γ-aminobutyric acid in the guinea-pig myenteric plexus. Br J Pharmacol 82: 101-105
20. Crawford JM, Curtis DR (1964) The excitation and depression of mammalian cortical neurones by amino acids. Br J Pharmacol 23: 313-329
21. Curtis DR (1978) Pre- and non-synaptic activities of GABA and related amino acids in the mammalian nervous system. In: Fonnum F. (ed.) Amino acids as chemical transmitters. Plenum, New York, pp 55-86
22. Curtis DR, Duggan AW, Felix D, Johnston GAR (1970) GABA, bicuculline and central inhibition. Nature 226: 1222-1224
23. Curtis DR, Game CJA, Johnston GAR, McCulloch RM (1974) Central effects of ß-(p-chlorophenyl)-γ-aminobutyric acid. Brain Res 70: 493-499
24. Curtis DR, Johnston GAR (1974) Amino acid transmitters in the mammalian central nervous system. Ergeb. Physiol 69: 97-188
25. Curtis DR, Phillis JW, Watkins JC (1959) The depression of spinal neurones by γ-amino-*n*-butyric acid and ß-alanine. J Physiol (Lond) 146: 185-203
26. Davies CH, Collingridge GL (1990) Involvement of presynaptic $GABA_B$ receptors in paired pulse depression of GABAergic IPSPs in rat hippocampal slices. In: Bowery NG, Bittiger H, Olpe HR (eds.) $GABA_B$ receptors in mammalian function. Wiley, Chichester (to be published)
27. Davies J, Watkins JC (1974) The action of ß-phenyl-GABA derivatives on neurones of the cat cerebral cortex. Brain Res 70: 501-505

28. De Groat WC (1970) The actions of γ-aminobutyric acid and related amino acids on mammalian autonomic ganglia. J Pharmac Exp Ther 172: 384-396
29. Demeneix BA, Desaulles E, Feltz P, Loeffler JP (1984) Dual population of $GABA_A$ and $GABA_B$ receptors in rat pars intermedia demonstrated by release of MSH caused by barium ions. Br J Pharmacol 82: 183-190
30. Desarmenien M, Feltz P, Occhipinti G, Santangelo F, Schlichter R (1984) Co-existence of $GABA_A$ and $GABA_B$ receptors on Aδ and C primary afferents. Br J Pharmacol 81: 327-333
31. Dunlap K (1981) Two types of γ-aminobutyric acid receptor on embryonic sensory neurones. Br J Pharmacol 74: 570-585
32. Dutar P, Nicoll RA (1988) A physiological role for $GABA_B$ receptors in the central nervous system. Nature 332: 156-158
33. Dutar P, Nicoll RA (1988) Pre- and postsynaptic $GABA_B$ receptors in the hippocampus have different pharmacological properties. Neuron 1: 585-598
34. Erdö SL, Bowery NG (1986) GABAergic mechanisms in the mammalian periphery. Raven, New York
35. Erdö SL, Kiss B (1986) Presence of GABA, glutamate decarboxylase, and GABA transaminase in peripheral tissues: a collection of quantitative data. In: Erdö SL, Bowery NG (eds) GABAergic mechanisms in the mammalian periphery. Raven, New York, pp5-17
36. Evans RH (1976) Potentiation of the effects of GABA by pentobarbitone. Brain Res 171:113-120
37. Giotti A, Luzzi S, Spagnesi S, Zilletti L (1983) $GABA_A$ and $GABA_B$ receptor-mediated effects in guinea-pig ileum. Br J Pharmacol 78: 469-478
38. Harrison NL (1990) On the presynaptic action of baclofen at inhibitory synapses between cultured rat hippocampal neurones. J Physiol 422: 433-446
39. Hill DR, Bowery NG (1981) ^{3}H-Baclofen and ^{3}H-GABA bind to bicuculline-insensitive $GABA_B$ sites in rat brain. Nature 290: 149-152
40. Hobbiger F (1958) Effects of γ-Aminobutyric acid on the isolated mammalian ileum. J Physiol (Lond) 142: 147-164
41. Jessen KR, Hills JM, Dennison ME, Mirsky R (1983) γ-aminobutyrate as an autonomic neurotransmitter. release and uptake of ^{3}H-γ-aminobutyrate in guinea-pig large intestine and cultured enteric neurons using physiological methods and electron microscopic autoradiography. Neuroscience 10: 1427-1442
42. Jessen KR, Hills JM, Dennison ME, Burnstock G (1979) GABA may be a neurotransmitter in the vertebrate peripheral nervous system. Nature 281: 71-74
43. Jessen KR, Mirsky R, Hills JM (1986) GABAergic neurons in the vertebrate peripheral nervous system. In: Erdö SL, Bowery NG (eds) GABAergic mechanisms in the mammalian periphery. Raven, New York, pp 117-134
44. Kerr DIB, Ong J (1986), GABAergic mechanisms in the gut: their role in the regulation of gut motility. In: Erdö SL, Bowery NG (eds) GABAergic mechanisms in the mammalian periphery. Raven, New York, pp 153-174
45. Krnjević K (1974) Chemical nature of synaptic transmission in vertebrates. Physiol Rev 54: 418-540
46. Krnjević K, Schwartz S (1966) Is γ-aminobutyric acid an inhibitory transmitter? Nature 211: 1372-1374
47. Kusonoki M, Taniyama K, Tanaka C (1984) Neuronal GABA release and GABA inhibition of ACh release in guinea-pig urinary bladder. Am J Physiol 246: R502-R509

48. Little HJ (1984) The effects of benzodiazepine agonists, inverse agonists and Ro15-1788 on the responses of the superior cervical ganglion to GABA in vitro. Br J Pharmacol 83: 57-68
49. Maguire JJ, Fowler LJ, Bowery NG (1990) (-)Baclofen inhibits K^+-evoked release of endogenous aspartate, glutamate and GABA from rat hippocampal synaptosomes. proceedings IUPHAR Congress, July, Amsterdam, 268 (abstr)
50. McLennan H (1959) The identification of one active component from brain extracts containing Factor I. J Physiol 146: 358-368
51. Newberry NR, Gilbert MJ (1989) Pertussis toxin sensitivity of drug-induced potentials on the rat superior cervical ganglion. Eur J Pharmacol 163: 245-252
52. Newberry NR, Nicoll RA (1984) Direct hyperpolarizing action of baclofen on hippocampal pyramidal cells. Nature 308: 450-452
53. Olsen RW, Tobin AJ (1990) Molecular biology of $GABA_A$ receptors. FASEB J 4: 1469-1480
54. Ong J, Kerr DIB (1983) $GABA_A$ and $GABA_B$ receptor-mediated modifications of intestinal motility. Eur J Pharmacol 94: 9-17
55. Ong J, Kerr DIB (1984) evidence for a physiological role of GABA in the control of guinea-pig intestinal motility. Neurosci Lett 50: 339-343
56. Paton WDM (1958) Central and synaptic transmission in the nervous system (pharmacological aspects). Annu Rev Physiol 20: 431-470
57. Pittaluga A, Asaro D, Pellegrini G, Raiteri M (1987) Studies on [^{3}H] GABA and endogenous GABA release in rat cerebral cortex suggest the presence of autoreceptors of $GABA_B$ type. Eur J Pharmacol 144: 45-52
58. Pratt GD, Knott C, Davey R, Bowery NG (1989) Characterisation of 3-aminopropyl phospinic acid as a $GABA_B$ agonist in rat brain tissue. Br J Pharmacol 96: 141P
59. Robbins MS, Grouse LH, Sorensen RL, Elde RP (1981) Effect of muscimol on glucose-stimulated somatostatin and insulin release from the isolated perfused rat pancras. Diabetes 30: 168-171
60. Schally AV, Redding TW, Arimura A, Dupont A, Linthicum GL (1977) Isolation of gamma-aminobutyric acid from pig hypothalamus and demonstration of its prolactin release-inhibiting (PIF) activity in vivo and in vitro. Endocrinology 100: 681-691
61. Scherer RW, Ferkany JW, Enna SJ (1988) Evidence for pharmacologically distinct subsets of $GABA_B$ receptors. Brain Res Bull 21: 439-443
62. Schildkraut JJ (1965) The catecholamine hypothesis of affective disorders: a review of supporting evidence. Am J Psychiat 122: 509-522
63. Young JAC, Brown DA, Kelly JS, Schon F (1973) Autoradiographic localization of sites of ^{3}H-γ-aminobutyric acid accumulation in peripheral autonomic ganglia. Brain Res 63: 479-486
64. Xu J, Wojcik WJ (1986) Gamma aminobutyric acid B receptor-mediated inhibition of adenylate cyclase in cultured cerebellar granule cells: blockade by islet-activating protein. J Pharmacol Exp Ther 239: 568-573

Autoradiographic Localization of Peripheral $GABA_A$ Receptors

W.L. Collier, M. Carlesimo, P. Napoleone, and F. Amenta

Introduction

We are now very comfortable with the fact that γ-aminobutyric acid (GABA) is both a peripheral neurotransmitter and a major central nervous system inhibitory transmitter. In fact, in recent years more attention has been paid to the localization and identification of GABA receptor sites at the peripheral level than at the central level. We now know that in the peripheral tissues there are two types of GABA receptors. The $GABA_A$ receptor responds to the antagonist bicuculline and is considered to be the classical GABA receptor; the $GABA_B$ receptor does not respond to bicuculline but it is sensitive to stimulation by baclofen [1, 5, 13, 15, 29].

Prior to 1979, it was not possible to visualize receptor binding sites due to the lack of irreversibly binding ligands, i. e., those which would remain bound during the histological processing of study organs. In 1979, however, Young and Kuhar [33] proposed a protocol which permitted the autoradiographic visualization of receptor binding sites using reversibly binding radioligands. This procedure thus allows analysis of pharmacological characteristics and of distribution of receptor binding sites in adjacent cryotome sections. The autoradiographic visualization of a receptor binding site can demonstrate where a physiological phenomenon takes place and may actually help to understand the biological function occurring by showing the precise site of action of a given substance.

Localization and identification of peripheral GABA receptor sites has been accomplished using classical receptor binding studies and a newer autoradiographic technique. To understand fully the underlying pharmacology and microscopic anatomy, both methods need to be used.

Consequently, in this study we analyzed the biochemical characteristics and the anatomic location of $GABA_A$ receptor sites in various parts of rat peripheral organs. First, we studied $GABA_A$ receptor sites in three key arterial beds: the cerebral arteries, the renal arteries, and the mesenteric arteries. We subsequently studied $GABA_A$ receptor sites in the adrenal gland and in the kidney. We then focused on the gastrointestinal tract, specifically the stomach, duodenum, jejunum, and the ileum. We concluded by examining $GABA_A$ receptor sites in the male reproductive system, including the epididymus, seminal vesicles, and the prostate.

Methods

For binding experiments the following procedure was modified from earlier work done in our laboratory [1,3]: Male Wistar rats, weighing between 200 and 300 g, were given light ether anesthesia and decapitated. The target tissue was quickly removed, cleaned of adhering connective tissue, and then frozen with a dry-ice/acetone mixture. Serial frozen sections of approximately 8 μm thickness were cut by means of a cryotome. These sections were then mounted on microscope slides. The sections were preincubated for 20 min at 4°C in the same buffer but containing different concentrations of [^{3}H]muscimol. Under these conditions, [^{3}H]muscimol is known to bind with high affinity to the $GABA_A$ receptor sites of peripheral organs [3]. In order to determine nonspecific binding, adjacent sections were incubated as described above but in the presence of excess unlabeled GABA (0.1 mmol/l).

The specificity of the binding of [^{3}H]muscimol to peripheral receptor sites was always assessed. Specifically, sections were incubated with [^{3}H]muscimol in the presence of various concentrations of nonradioactive muscimol (0.1 μmol/l) or GABA (1 nmol/l).

After incubation, the sections were wiped and put into preweighed scintillation vials and solubilized in Soluene; total radioactivity was counted in a Beckman liquid scintillation spectrometer. Specific binding was calculated by subtracting the nonspecific binding (which occurred in the presence of 0.1 mmol/l unlabeled GABA). Equilibrium dissociation constants (K_d) and the maximum binding capacity (B_{max}) were calculated from Scatchard plots of saturation isotherms [3, 4, 9, 27]. Sample sections of each tissue were sonicated so that protein determinations could be performed against a standard of bovine serum albumin using the method proposed by Lowry et al. [21].

For autoradiographic experiments, sections not employed for binding assays were fixed by exposing them to formaldehyde vapors at 80°C for 1h. The optimal conditions for autoradiographic demonstration of [^{3}H]muscimol binding sites are [1]:

1. Quick killing of the experimental animal
2. Rapid removal and isolation of the organs to be investigated
3. Thin sections (6 - 8 μm) of tissue using a cryotome
4. Mounting the sections on gelatin coated slides
5. Preincubation of the sections for 20 min at 4°C in Tris-citrate buffer, 0.3 *M*, pH 7.1
6. Incubation of the sections with 5-20 n*M* [^{3}H]muscimol for 30-40 min at 4°C in Tris-citrate buffer
7. Evaluation of the specificity of the reaction by comparison with control sections which were incubated using 0.1 m*M* muscimol or 1 m*M* GABA
8. Washing the sections for 1 min using ice-cold Tris-citrate buffer
9. Air-drying and fixation of the sections by exposure to formaldehyde vapors for 60 min at 80°C
10. Attachment of nuclear emulsion-coated cover slips for 4-8 weeks at 4°C
11. Development of the autoradiographs, staining of the sections, and observation under a light microscope equipped with both dark- and bright-field condensers

Coverslips were coated with Ilford L4 nuclear emulsion which was diluted 1:1 with distilled water. These coverslips were then attached to the slides and exposed at 4°C for a period of up to 8 weeks. The autoradiographs were then developed in D19 Kodak developer and counterstained with toluidine blue. All sections were then studied using a light microscope which had both dark- and bright-field condensers.

The density of silver grains developed within the organs studied, after exposure to [^{3}H]muscimol, could be ascertained by reflectance photometry. Basically, a Fluoval Photometrie microphotometer was utilized; it was calibrated by considering as "zero" the background of control sections which had been incubated without radiolabeled muscimol. Reflectance measurements were made of randomly selected portions of the organs under investigation using a measuring diaphragm 10 μm in diameter. In all studies, the photometer recordings were measured by using arbitrary units proportional to the density of the developed silver grains [4].

Current Status

Rat Cerebral Arteries

[^{3}H]muscimol specifically bound to selected segments of the rat cerebral arterial tree. Napoleone et al. [27] reported that the ligand was specifically bound by sections of the circle of Willis arteries and by sections of the pia-arachnoid membrane. Scatchard analysis of the binding isotherms produced a dissociation constant (K_d) of 46 n*M* and a binding site density (B_{max}) of 0.60 pmol/mg tissue.

Autoradiographic analysis revealed that silver grains which represented specific binding sites were visualized within the wall of the circle of Willis and within the pial and arachnoid membranes. No specific labeling was observed in the common carotid arteries, the pia-arachnoid arteries, and the pia-arachnoid veins. [^{3}H]muscimol was bound by the medial layer of the arteries under investigations while the adventitia and the adventitial-medial border did not show significant binding. The greatest density of [^{3}H]muscimol binding sites were, in descending order, the basilar and posterior cerebral arteries, the main pia-arachnoid arteries, the posterior communicating artery, the middle cerebral artery, the internal carotid artery, and the anterior cerebral artery (Fig. 1).

It may be that the vasodilatory effect of GABA on cerebral vessels is endothelium independent. This is because of the lack of [^{3}H]muscimol binding sites in the intimal layer of the rat circle of Willis and the pia-arachnoid arteries. If this is indeed true, it is unlike acetylcholine, adenosine-5-triphosphate, substance P, and similar vasoactive agents which are endothelium dependent [27].

The higher density of [^{3}H]muscimol binding sites in the posterior cerebral arteries than in the anterior cerebral arteries may be the basis of the frontal cortex after muscimol challenge, as reported by Edvinsson and coworkers [7]. We must also remember that the lack of binding sites in the cerebral arterioles, which are the key vascular components supplying local cerebral blood flow [31], may explain the difficulty in pharmacologically altering cerebral tissue perfusion in the intact rat when using GABAergic agents [18,24].

Using a similar procedure and [^{3}H]GABA in the presence of isoguvacine as a ligand, no $GABA_B$ receptor sites were noticeable in the rat cerebrovascular tree (Fig. 1).

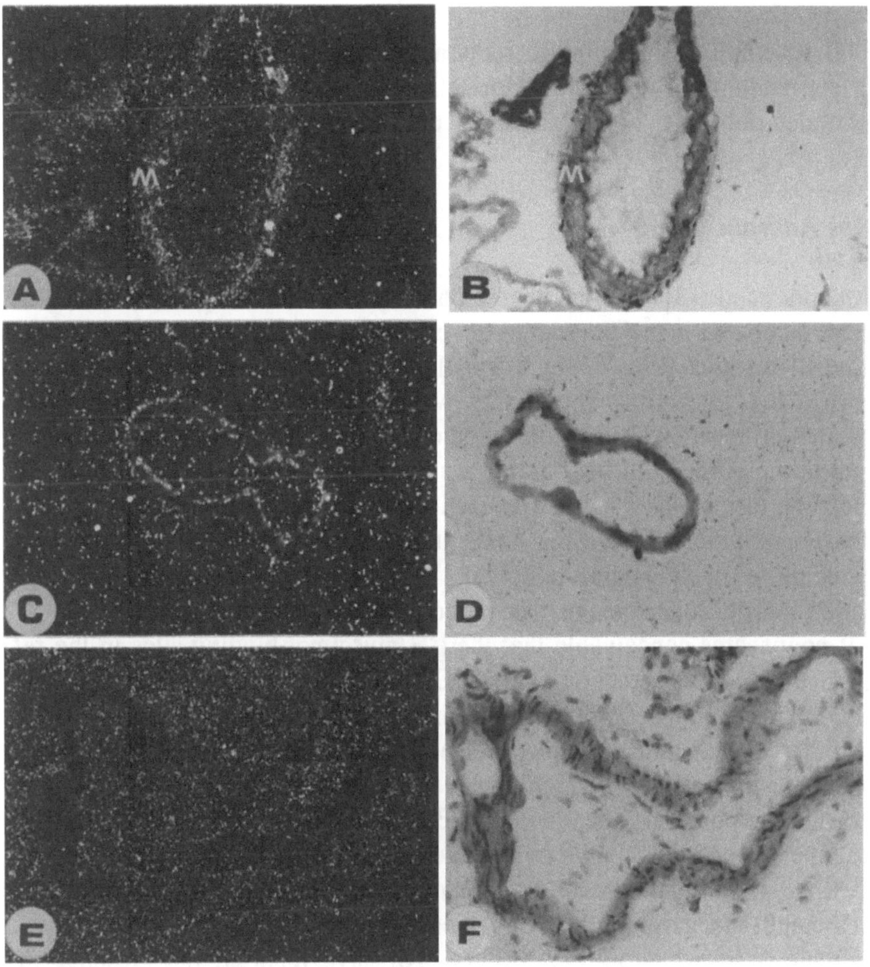

Fig. 1. A, B Dark and light field photomicrographs of a section of rat middle cerebral artery incubated with [^{3}H]muscimol (20 nmol/l) to demonstrate GABA$_A$ receptors. Silver grains are primarily within the medial layer (*M*) of the artery; **C, D** Dark and light field photomicrographs of a section of rat middle cerebral artery incubated with [^{3}H]muscimol (10 nmol/l) in the presence of isoguvacine (40 μmol/l) and $CaCl_2$ (2.5 mmol/l) to label GABA$_B$ receptors. No specific accumulation of silver grains is noticeable in the artery. **E, F** Dark and light field photomicrographs of a section of rat superior mesenteric artery branch incubated with [^{3}H]muscimol (20 nmol/l) to demonstrate GABA$_A$ receptors. Silver grains are within tissue-free areas but not within the wall of the artery. X 120

Rat Renal and Mesenteric Arteries

[^{3}H]muscimol did not bind to sections of renal artery, whether they were large, medium, or small [4]. Unpublished results showed that [^{3}H]muscimol also did not bind to sections of the mesenteric artery (Fig. 1).

Rat Adrenal Gland

Amenta et al. [3] reported that sections of rat adrenal gland which had been incubated in [^{3}H]muscimol developed silver grains chiefly within the adrenal medulla (Fig. 2), while smaller amounts were seen in the adrenal cortex. It is interesting to note that, within the medulla, silver grains were located primarily within chromaffin cells. There was binding to [^{3}H]muscimol in the cortex, but the binding was more or less of a diffuse nature. In the zona fasciculata and the zona reticularis the authors reported a greater density of grains. Increasing concentrations of GABA, isoguvacine, and bicuculline methyl iodine inhibited [^{3}H]muscimol accumulation in a concentration-dependent manner in sections of the adrenal gland.

It now seems conceivable that the [^{3}H]muscimol binding sites in the rat adrenal medulla represent functional $GABA_A$ receptors which, when activated by intravenous GABA or muscimol, cause a depolarization of the chromaffin cells and subsequent release of stored neurotransmitters. When reserpine pretreatment, which significantly reduces the excitatory cardiovascular response to intravenous GABA, is withheld, peripheral sympathetic innervation plays only a minor role in this response. This is in keeping with the notion that adrenal catecholamines are more resistant to reserpine treatment than those of the sympathetic nerves [32].

Previous investigators [17, 30, 19, 16] have reported the presence of functional $GABA_A$ receptors modulating transmitter release from chromaffin cells in bovine and canine adrenals. The existence of an intrinsic GABAergic mechanism regulating adrenal function has been proposed by Kataoka et al. [17]. Additional data indicate that endogenous GABA has complicated effects on catecholamine secretion in the adrenal medulla. These involve not only an increase in catecholamine secretion by GABA per se [19, 16], but also an inhibitory effect on catecholamine release induced by splanchnic nerve stimulation and nicotinic agents. Therefore, within the adrenal medulla, GABA may exert, by acting directly on $GABA_A$ receptor sites, a dual modulatory action on catecholamine secretion as was also previously described in the guinea pig bladder [23].

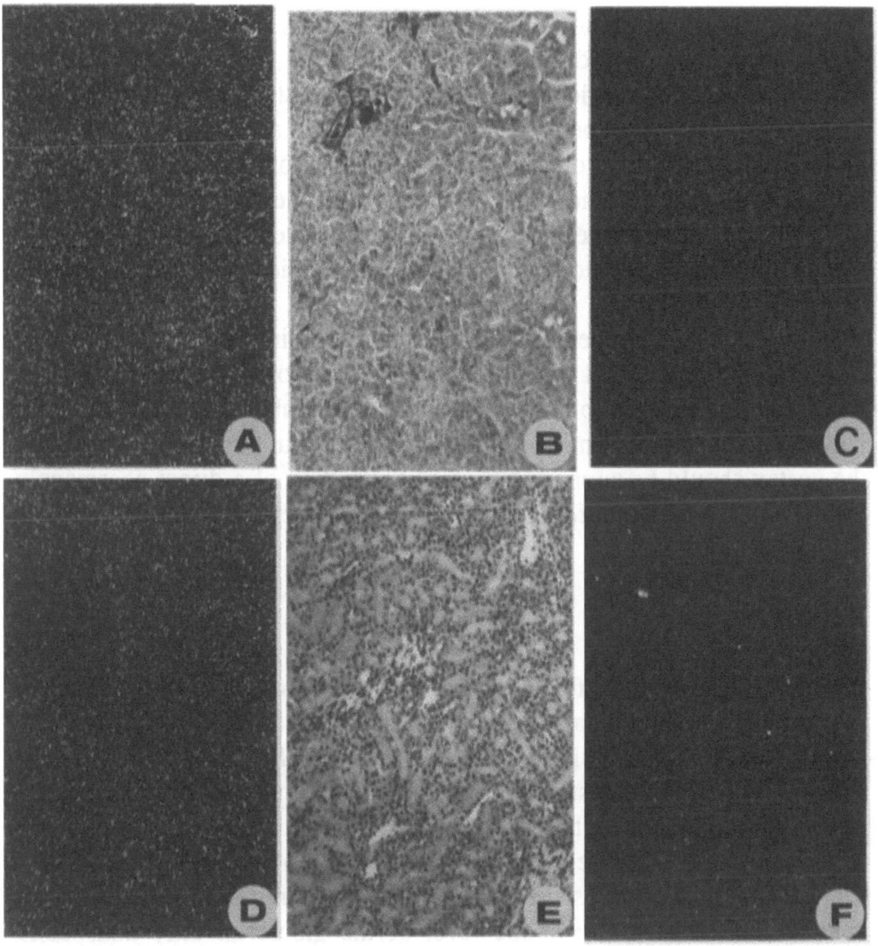

Fig. 2. Sections of rat adrenal medulla (**A** - **C**) and renal cortex (**D** - **F**) exposed to 20 nmol/l [^{3}H]muscimol to label GABA$_A$ receptors. **A, D**, dark field photomicrographs showing the accumulation of silver grains within chromaffin cells of the adrenal medulla (**A**) and renal cortex (**D**). **B, E**, bright field photomicrographs of **A** and **D**, respectively, stained with toluidine blue to verify microanatomical details. **C, F**, dark field photomicrographs of reactions adjacent to those shown in photomicrographs **A** and **D**, respectively, incubated in the presence of 0.1 mmol/l muscimol to cause nonspecific binding. X 140

It is interesting to note that the GABA induced secretion of catecholamines from dog adrenals is antagonized by bicuculline [19], as the adrenal dependent cardiovascular effects in anesthetized rats were antagonized in the study performed by Amenta et al. [7]. By contrast, in guinea pigs the adrenal dependent pressor effect of intravenous GABA or muscimol in barbital anesthetized animals was hexamethonium sensitive while being bicuculline resistant. This suggested the involvement of "atypical" GABA receptors which activate nicotinic preganglionic input to chromaffin cells [10].

The low density of specific [^{3}H]muscimol binding sites in the adrenal cortex of the rat adrenal suggests that GABA recognition sites might also be present in this area, but in a much smaller number than in the medulla. However, the receptor nature of the muscimol binding sites occurring at the cortex should be supported by the observation of a direct pharmacological response of the adrenal cortex to infusions of GABA.

Rat Kidney

[^{3}H]muscimol was specifically bound by sections of rat kidney. Scatchard analysis of the binding data showed that the dissociation constant was 23.7 nmol/l and the binding site density was 1.15 pmol/mg of tissue [2].

Autoradiographic investigation revealed that [^{3}H]muscimol binding sites were found to be concentrated in the cortex and in the medulla which bordered the cortex. Subsequent investigation by Amenta and coworkers showed that the cortical structures labeled by [^{3}H]muscimol were the convoluted tubules (Fig. 2); at the level of the glomeruli only nonspecific binding was reported. Cortical collecting tubules and tubules located in the outer portion of the medulla, which are most likely the thick tubules of the ascending limb of the loop of Henle or collecting tubules, also bind [^{3}H]muscimol.

Amenta and coworkers [2] reported that, unlike the cortical tubules where both an endogenous GABAergic system and specific muscimol binding sites are located, the collecting tubules of tha rat renal cortex and medulla appear to have a large number of muscimol binding sites only, since low GABA levels have been reported for these two structures [11]. The Amenta group concluded by suggesting that [^{3}H]muscimol binding sites on collecting tubules may have a pharmacological rather than a physiological implication [2].

Rat Gastrointestinal Tract

In the stomach, specific [^{3}H]muscimol binding occurred in the body and in the antrum. While the greatest binding density was in the muscle layers, the mucosa also contained a high concentration of silver grains. Silver grains were not bound in any appreciable manner by ganglia of the gastric wall [8] (Fig. 3).

When the binding density of the circular layer and the longitudinal layer of the stomach was calculated, it was determined that the density of silver grains in the circular layer was greater than the density in the longitudinal layer. It was also noted that more binding occurred in the muscle layer of the body than in the antrum. When binding sites for the gastric glandular surface were studied, silver grains which represented specific [^{3}H]muscimol binding were observed [8].

In the small intestine, [^{3}H]muscimol bound to selected sections of duodenum, jejunum, and ileum in a manner which is considered to be consistent with the labeling of $GABA_A$ receptor sites. Napoleone and coworkers [26] noted that binding was time and concentration dependent, reversible and belonged to a single class of high affinity binding sites, at least in the concentration ranges of radioligands that they used. Interestingly, the dissociation constant value (K_d) was virtually identical in the three sections of the small intestine. The maximum density of binding sites (B_{max}) was highest in the duodenum and lowest in the ileum.

In the rat small intestine (±)-baclofen, diaminobutyric acid (DABA), and picrotoxin were not able to displace [^{3}H]muscimol, even at high concentrations. By contrast, GABA, bicuculline methiodine, isoguvacine and muscimol, were able to displace [^{3}H]muscimol binding [26].

Autoradiographic examination of sections of rat small intestine exposed to [^{3}H]muscimol showed that specific silver grains were bound to the intestinal musculature. As was expected, nonspecific binding was noted primarily in the intestinal mucosa. The largest accumulation of silver grains was in the duodenum. In this organ the silver grains were homogeneously distributed within both the longitudinal and the circular musculature. In the jejunum, the density of [^{3}H]muscimol binding sites was similar to that observed in the duodenum, although there were fewer silver grains in the longitudinal musculature. The smallest accumulation of [^{3}H]muscimol binding sites occurred in the ileum, which had silver granules rather homogeneously distributed within both longitudinal and circular smooth muscle [26].

Microphotometric analysis showed that the density of silver granules in the longitudinal muscle of the ileum was 40% and 34% lower than values reported in the same muscular layer in the duodenum or in the jejunum, respectively. When the circular muscle layer was examined, the density of [^{3}H]muscimol binding in the ileum was 54% and 45% lower than the values observed in the circular layers of the duodenum or the jejunum, respectively. There was no difference in silver grain density between the area occupied by the mesenteric ganglia and the longitudinal or circular muscle layers [26].

The neuronal pathways and receptors by which GABA actually effects intestinal motility still remain to be clarified. The results reported are inconsistent, probably due to species differences in the animal models that were used. In the rat ileum, a $GABA_A$ dependent activation of nonadrenergic noncholinergic intramural neurons is suspected to cause the relaxing effect of GABA [6,30]. This effect on small intestine muscle is neural in nature [6,30].

In their 1990 work Napoleone and coworkers [26] confirmed the presence in the rat small intestine of a single class of saturable [^{3}H]muscimol binding sites. The density of those binding sites was found to decrease with an increase in the distance from the duodenum; this correlated very well with the $GABA_A$ receptor mediated neurogenic inhibition of rat small intestine motility [30].

It is now felt that bound [^{3}H]muscimol displacement by GABA, bicuculline, isoguvacine, and muscimol supports the concept of a specific, "classic", i.e., bicuculline sensitive, $GABA_A$ receptor site within sections of the rat duodenum, jejunum, and ileum [26]. Although autoradiography showed that rat intestinal $GABA_A$ receptor sites were mainly located within the gut muscle, GABA does not interact directly with intestinal smooth muscle; its effects, as is true in the stomach, are under neuronal mediation [30].

The high concentration of silver grains in the circular and longitudinal layers and not in the myenteric plexus area suggested a predominant localization of $GABA_A$ receptors on nerve terminals rather than on cell bodies of myenteric neurons. It cannot be excluded that the prejunctional $GABA_A$ receptor sites observed in the rat small intestine could actually be targets for GABA released from the GABAergic neurons of the myenteric plexuses, as first reported by other authors [22, 14].

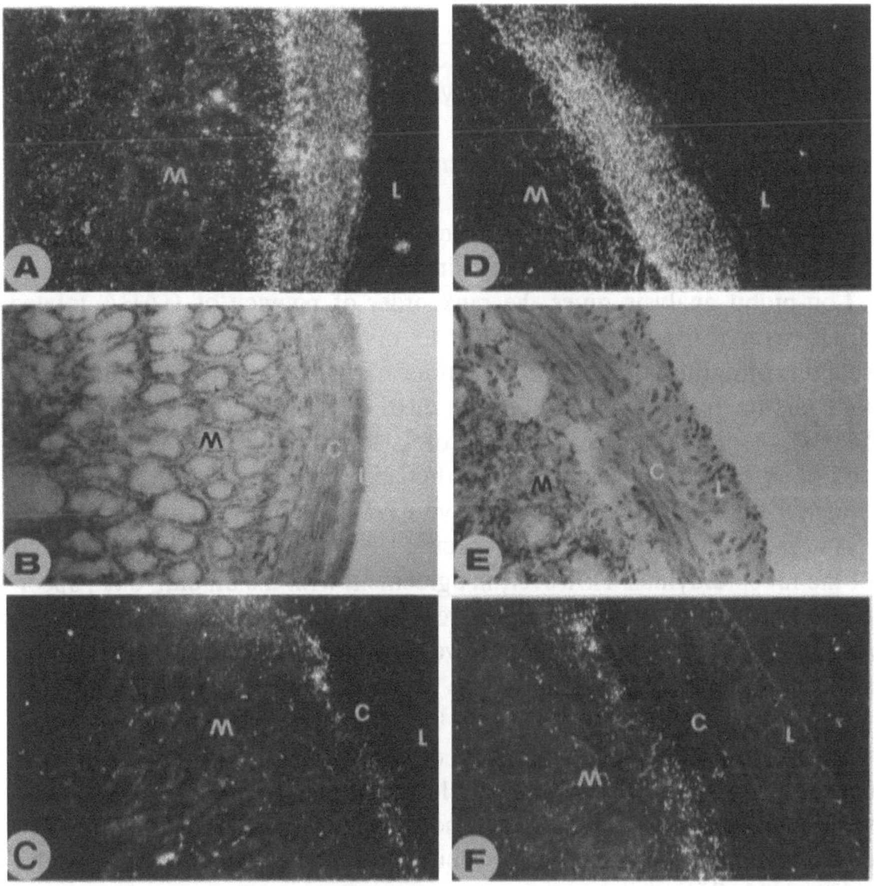

Fig. 3. Sections of rat stomach (**A** - **C**) and of rat jejunum (**D** - **F**) exposed to 20 nmol/l [^{3}H]muscimol to label GABA$_A$ receptors. *M*, mucosa; *C*, circular smooth muscle; *L*, longitudinal smooth muscle. **A**, **D**, dark field photomicrographs showing the accumulation of silver grains within the gastric and jejunal smooth muscle and within the glandular cells of the gastric mucosa. **B**, **E**, bright field photomicrographs of photomicrographs **A** and **B**, respectively, stained with toluidine blue to verify microanatomical details. **C**, **F**, dark field photomicrographs of reactions adjacent to those shown in photomicrographs **A** and **D**, respectively, incubated in the presence of 0.1 mmol/l muscimol to cause nonspecific binding. X 160

Rat Male Reproductive System

No specific [^{3}H]muscimol binding was noted in the epididymis of the rat. Instead, sections of seminal vesicles and of the lateral lobe of the prostate gland bound [^{3}H]muscimol, consistent with the labeling of $GABA_A$ receptor sites [25]. The affinity constants (K_d) were almost identical in the two tissues. The maximum binding capacity (B_{max}), however, was significantly higher in the seminal vesicles than in the lateral lobes of the prostate.

Log-probit analysis gives the rank order of potency of GABA related drugs with respect to their ability to inhibit [^{3}H]muscimol binding to sections of animal tissue. When sections of rat seminal vesicle and lateral lobe of the prostate gland were studied, it was seen that ($\pm$)-baclofen, DABA, or picrotoxin had no effect on [^{3}H]muscimol binding while bicuculline methiodide, isoguvacine, imidazoleacetic acid, and 3-aminopropanil-sulphonic acid were effective in inhibiting binding [25].

Autoradiography did not show specific [^{3}H]muscimol binding to sections of the rat epididymus. However, in agreement with the binding data, specific [^{3}H]muscimol binding sites were observed in the epithelium and in smooth muscle cells of the seminal vesicle. There were fewer silver grains in sections of the lateral lobe of the prostate gland than in sections from the seminal vesicles. As was true in the seminal vesicles, the silver grains were primarily within the epithelium of the prostatic glands and, in smaller concentrations, in the smooth muscle of the gland. In both the seminal vesicles and the prostate, numerous nonspecific silver grains were diffused onto the entire slide without specific accumulation in any particular structure [1, 25].

While we are not yet able to conclude the physiological importance of GABA in the male reproductive tract, the localization of $GABA_A$ receptor sites in the epithelium and in the smooth muscle of organs in the male reproductive system suggests a possible multifaceted role for this neurotransmitter. Since the epithelium of seminal vesicles and prostatic glandular tissue contributes to the secretion of seminal fluid [9], $GABA_A$ receptors in these structures could be involved in the regulation of secretion.

It is still unclear whether the $GABA_A$ receptors identified in the male reproductive tract represent the target for locally secreted GABA, as is described in the pacreatic islets [12] and the adrenal medulla [28].

Conclusions

The above findings suggest that the peripheral $GABA_A$ receptor system is probably involved in a variety of functions. They include the control of cerebral arterial tone, the regulation of gastrointestinal mobility and of gastric and male sex organ secretory activity, the control of renal tubular function, and the release of catecholamines from the adrenal medulla.

References

1. Amenta F (1986) Autoradiographic localization of GABA receptor sites in peripheral tissue. In: Erdö SL, Bowery NG (eds) GABAergic mechanisms in the mammalian periphery. Raven, New York, pp 99-115
2. Amenta F, Cavallotti C, Iacopino L et al. (1988) Autoradiographic localization of the $GABA_A$ receptor agonist (^{3}H)-muscimol within rat kidney. Pharmacology 32: 390-395
3. Amenta F, Collier WL, Erdö SL, Giuliani S, Maggi CA, Meli A (1988) $GABA_A$ receptor sites modulating catecholamine secretion in the rat adrenal gland: evidence from ^{3}H-muscimol autoradiography and in vivo functional studies. Pharmacology 37: 394-402
4. Amenta F, Erdö SL, Mione MC, Napoleone P (1986) ^{3}H-muscimol binding sites within guinea pig ovary. A histoautoradiographic study. Pharmacology 32: 202-207
5. Bowery NG, Doble A, Hill DR, Hudson AL, Shaw JS, Turnbull MJ, Warrington R (1981) Bicuculline-insensitive GABA receptors on peripheral autonomic nerve terminals. Eur J Pharmacol 71: 53-70
6. Davanger S, Ottersen OP, Storm-Mathisen J (1989) GABA-immunoreactive cells in the rat gastrointestinal epithelium. Anat Embryol (Berl) 179: 221
7. Edvinsson L, Larsson B, Skarby T (1980) Effects of the GABA agonist muscimol on regional cerebral blood flow in the rat. Brain Res 185: 445-458
8. Erdö SL, De Vincentis G, Amenta F (1990) Autoradiographic localization of [^{3}H]muscimol binding sites in rat stomach: evidence for mucosal $GABA_A$ receptors. Eur J Pharmacol 175: 351-354
9. Erdö SL, Német L. Szporny L (1983) The occurrence of GABA in vas deferens, prostate, epididymis, seminal vesicle and testicle of the rat. Acta Biol Hung 34: 435-547
10. Giuliani S, Maggi G, Meli CA (1986) Differences in cardiovascular responses to peripherally administered GABA as influenced by basal conditions and type of anaesthesia. Br J Pharmacol 88: 659-670
11. Goodyear PR, Mills M, Scriver CR (1982) Properties of gamma-aminobutyric acid synthesis by rat renal cortex. Biochim Biophys 716: 348-357
12. Guyton AC (1979) Textbook of physiology, 4th edn. Saunders, Philadelphia
13. Hill DR, Bowery NG (1981) ^{3}H-Baclofen and ^{3}H-GABA bind to bicuculline-insensitive $GABA_B$ sites in rat brain. Nature 290: 149-152

14. Jessen KR, Hills JM, Dennison ME et al (1983) γ-Aminobutyrate as an autonomic neurotransmitter: release and uptake of ^{3}H-γ-aminobutyrate in guinea pig large intestine and cultured enteric neurons using physiologic methods and electromicroscope autoradiography. Neuroscience 10: 1427-1442
15. Jessen KR, Hills JM, Dennison ME, Burnstock G (1979) GABA may be a new neurotransmitter in the vertebrate peripheral nervous system. Nature 281: 71-74
16. Kataoka Y, Fujimoto M, Guidotti A et al (1986) Intrinsic GABA receptors modulate the release of catecholamine from canine adrenal gland in situ. J Pharmacol Exp Ther 239: 584- 590
17. Kataoka Y, Gutman Y, Guidotti A et al (1984) Intrinsic GABAergic system of adrenal chromaffin cells. Proc Natl Acad Sci USA 81: 3218-3222
18. Kelly PAT, McCulloch J (1983) The effect of the GABA agonist muscimol upon the relationship between local cerebral blood flow and glucose utilization. Brain Res 258: 338-342
19. Kitayama S, Morita K, Dohi T et al (1984) The nature of the stimulatory action of GABA in the isolated perfused dog adrenals. Naunyn Schmeidebergs Arch Pharmacol 326: 106-110
20. Krantis A, Harding RK (1987) GABA-related actions in isolated in vitro preparations of the rat small intestine. Eur J Pharmacol 141: 291-298
21. Lowry OH, Rosenbrough NJ, Farr AL, Randall RJ (1951) Protein measurement with the folin phenol reagent. J Biol Chem 193: 265-275
22. Maggi CA, Manzini S, Meli A (1984) Evidence that $GABA_A$ receptors mediate relaxation of rat duodenum by activating nonadrenergic noncholonergic neurones. J Auton Pharmacol 4: 77-85
23. Maggi CA, Santicoli P, Meli A (1985) Dual effect of GABA on the contractile activity of the guinea pig isolated urinary bladder. J Auton Pharmacol 5: 131-141
24. McCulloch J, Kelly PAT, Grome JJ (1981) GABAergic influence upon cerebral metabolism and perfusion. In: Cervos-Navarro J, Fritschka E (eds) Cerebral microcirculation and metabolism. Raven, New York
25. Napoleone P, Bronzetti E, Cavallotti C, Amenta F (1990) Predominant epithelial localization of $GABA_A$ receptor sites within rat seminal vesicles and prostate gland (to be published)
26. Napoleone P, De Vincentis G, Amenta F (1990) Autoradiographic localization of the $GABA_A$ receptor agonist ^{3}H-muscimol in the rat intestinal musculature. Pharmacology (to be published)
27. Napoleone P, Erdö SL, Amenta F (1987) Autoradiographic localization of the $GABA_A$ receptor agonist [^{3}H]muscimol in rat cerebral vessels. Brain Res 423: 109-115
28. Okada Y, Taniguchi H, Baba S (1982) High concentration of GABA in the pancreatic islets with special emphasis of B-cells. In: Okada Y, Roberts E (eds) Problems in GABA research from brain to bacteria. Excerpta Medica, Amsterdam
29. Roberts E, Chase TN, Tower DB (1976) GABA in nervous system function. Raven, New York
30. Sangiah S, Borowitz JL, Yim GKW (1974) Actions of GABA, picrotoxin and bicuculline on adrenal medulla. Eur J Pharmacol 27: 130-155
31. Scatton B, Duverger D, L'Heureux R, Serrano A, Fage D Nowicki JP, McKenzie ET (1985) Neurochemical studies on the existence, origin and characteristics of the serotonergic innervation of small pial vessels. Brain Res 345: 219-229

32. Weiner N (1980) Drugs that inhibit adrenergic nerves and block adrenergic receptors. In: Goodman-Gilman A, Goodman LS, Gilman A (eds) The pharmacological basic of therapeutics. McMillan, New York
33. Young WS, Kuhar MJ (1979) A new method for receptor autoradiography: ^{3}H-Opioid receptors in rat brain. Brain Res 179: 255-270

The $GABA_B$ Receptor Profile of a Series of Phosphinic Acids: Agonist and Antagonist Activity in a Range of Peripheral Tissues

J.M. Hills and W. Howson

In the CNS, baclofen, a GABA analogue now known to interact with GABA receptors, was shown early on to have distinctly different effects from GABA (Fig. 1) [7]. These differences were highlighted by the work of Bowery and collegues, who showed that the baclofen induced repression of transmitter release from CNS tissue in vitro could not be blocked [11], and that a proportion of ^{3}H-GABA binding to brain membranes was not displaced [18] by the GABA antagonist bicuculline. It was proposed that a second class of GABA receptors, selectively activated by baclofen and insensitive to antagonism by bicuculline, may exist in the mammalian brain. The classical GABA receptor became known as the $GABA_A$ receptor, whereas the new bicuculline insensitve receptor was termed the $GABA_B$ receptor.

Since that time, much work confirming the early experimentation has been reported. The binding of ^{3}H-GABA, in the presence of a specific $GABA_A$ ligand, to the central $GABA_B$ receptor is readily displaced by (-) baclofen (IC_{50}=65 nM). Functionally, the effects of GABA in the presence of a $GABA_A$ antagonist are comparable to those of baclofen. The affinity of baclofen for the $GABA_B$ receptor is similar to that for the natural ligand, i.e. GABA (IC_{50}=40 nM), and this similarity in potency is mirrored in $GABA_B$ functional pharmacological tests [7, 10].

The existence of $GABA_B$ receptors on autonomic terminals, first reported by Bowery and colleagues [9], has proved to be a relatively widespread phenomenon, occurring often in the absence of an established GABAergic innervation. Early experiments showed that baclofen, or GABA in the presence of bicuculline, would reduce the effects of electrical field stimulation, an effect to be presumed to be via a prejunctional inhibitory mechanism diminishing the amount of excitatory transmitter released. Such an effect was reported in rat atrium and mouse vas deferens (where the excitatory transmitter is norepinephrine or norepinephrine with ATP respectivly [34]) and in the guinea-pig ileum (where the transmitter is ACh) [9, 21, 31]. Similar observations where also made in the rat anococcygeus [29]. The pattern was clear. Addition of baclofen, or GABA in the presence of bicuculline, reduced the postsynaptic effect of transmural electrical stimulation by approximately 30%, but did not reduce the postsynaptic response to, for example, norepinephrine or acetylcholine. As no specific $GABA_B$ antagonist was available, desensitisation techniques were often used to test receptor specifity [15, 19]. These peripheral tissues therefore provided a relativly straightforward way of investigating the effects of new putative $GABA_B$ receptor agonists or antagonists. Early on, compounds such δ-aminovaleric acid (DAVA) and homotaurine (3-aminopropanesulphonic acid, 3-APS) (Fig.1) were claimed to be $GABA_B$ antagonists. 3-APS was shown to be an effective antagonist in the guinea-pig ileum against (-)-baclofen, with a reported pA_2 of 4.3 [16]. Aware of the interaction of 3-APS with $GABA_A$ receptors, the experiments were repeated in the presence of picrotoxin and the $GABA_B$ antagonism still observed [16]. δ-AVA, a compound which also reduces the effects of baclofen in the guinea-pig ileum [28], has been shown by others to mediate these effects largely as a result of its ability to stimulate $GABA_A$ receptors, which mediate excitatory effects in this tissue [1]. In the anococcygeus, by contrast, the antagonism of baclofen by δ-AVA is considered to be via $GABA_B$ receptor antagonism, since there is no evidence for the existence of $GABA_A$ receptors in this tissue [1].

Significant advances did occur, however, in the search for more selective $GABA_B$ antagonists, with the discovery of baclofen analogues with alternative acidic moieties. These antagonists include phaclofen, saclofen and 2-OH saclofen (Fig.1). Although parallel experimentation took place in CNS tissues, definitive pharmacological data were first available from peripheral assay systems.

The phosphonic acid analogue of baclofen, phaclofen, was first described by Kerr and colleagues [27] as an antagonist of GABA and baclofen in the guinea-pig ileum and vas deferens, where its estimated pA_2 value was 4.0 [22,26]. This low degree of antagonism is not surprising in light of relatively poor binding affinity of phaclofen to $GABA_B$ receptors in a rat bain membrane preparation (IC_{50}=100 μM)[8].

The progress made with these compounds in the peripheral assay systems studied shows them to be more reliable as $GABA_B$ antagonists than previous compounds. This is principally due to their increased selectivity

Agonists

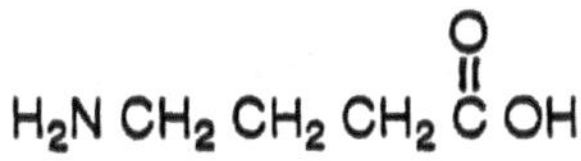

GABA
(γ-aminobutyric acid)

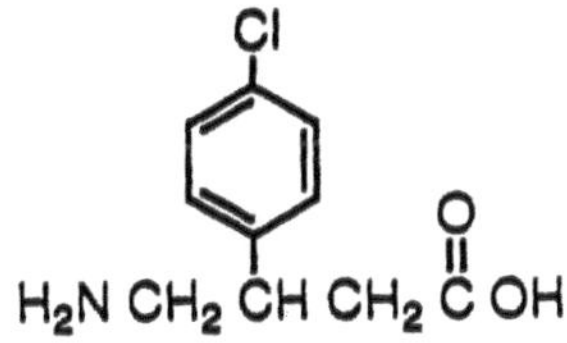

Baclofen
(4-amino-3-(4-chlorophenyl) butyric acid)

Antagonists

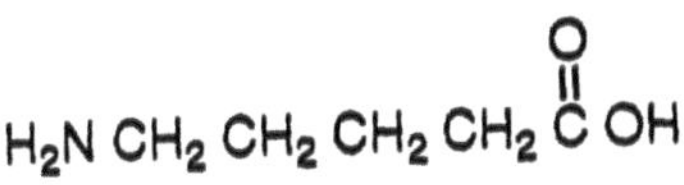

DAVA
(δ-aminovaleric acid)

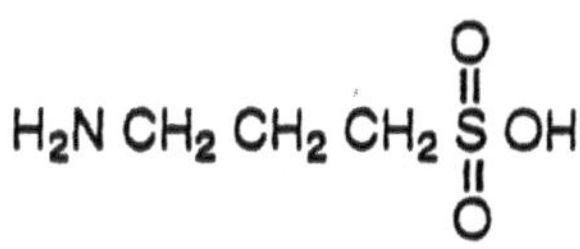

3-APS
(3-aminopropanesulphonic acid)

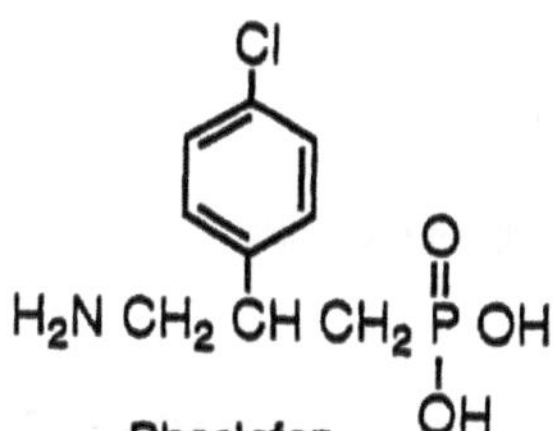

Phaclofen
(3-amino-2-(4-chlorophenyl)propyl phosphonic acid)

Fig. 1. Non-phosphinic acid analogues

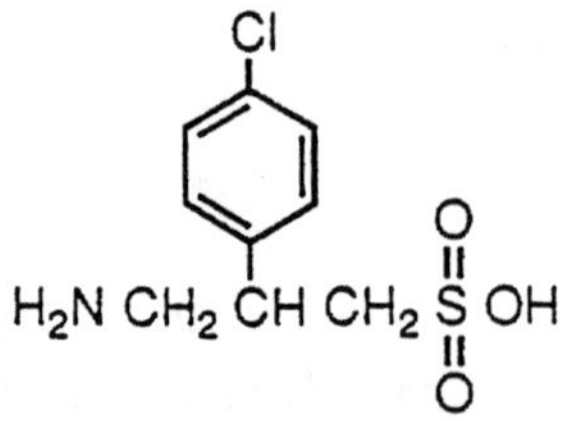

Saclofen
(3-amino-2-(4-chlorophenyl)
propylsulphonic acid)

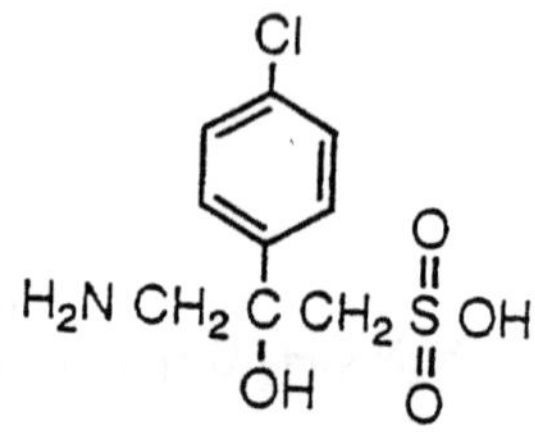

2-OH Saclofen
(3-amino-2-(4-chloro-
phenyl)-2-hydroxypropyl-
sulphonic acid)

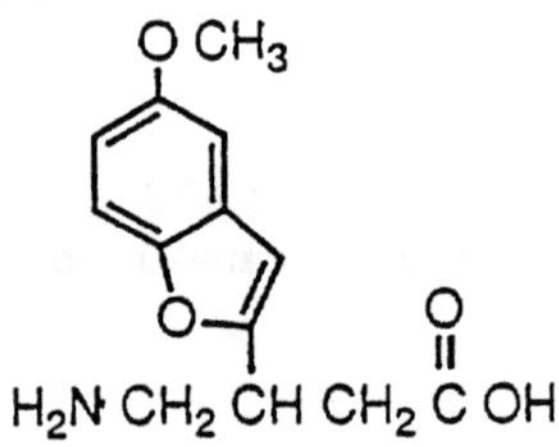

9 h
(4-amino-3-(5-methoxy-
benzo[b]furan-2-yl)butyric acid

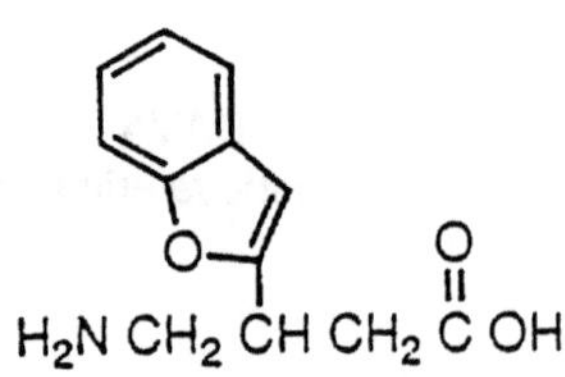

9 g
(4-amino-3-(5-benzo[b]
furan-2-yl)butyric acid)

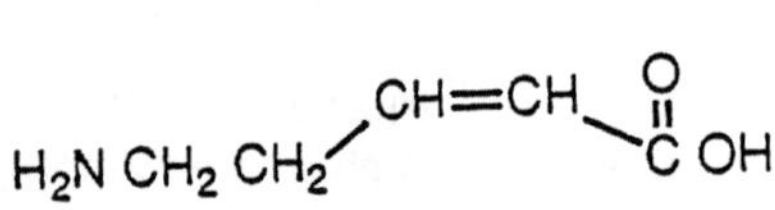

(z)-5-aminopent-2-enoic acid

Fig. 1. Non-phosphinic acid analogues (Cont.)

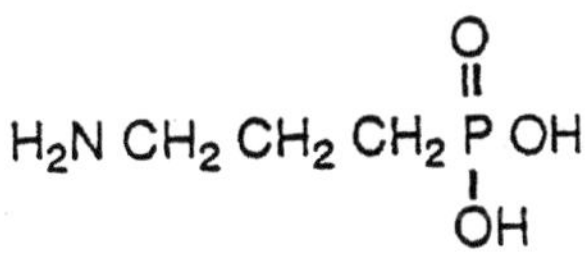

3-APPA
(3-aminopropylphosphonic
acid)

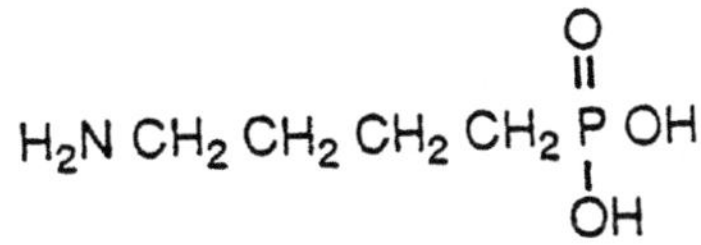

4-ABPA
(4-aminobutylphosphonic
acid)

Fig. 1. Non-phosphinic acid analogues (Cont.)

Agonists

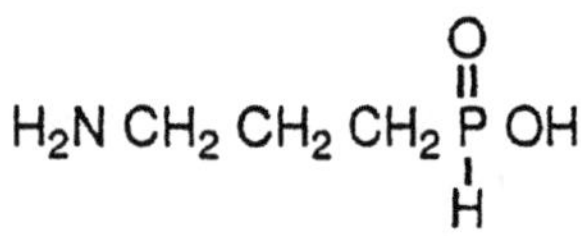

CGP 27492
(3-aminopropyl-
phosphinic acid)

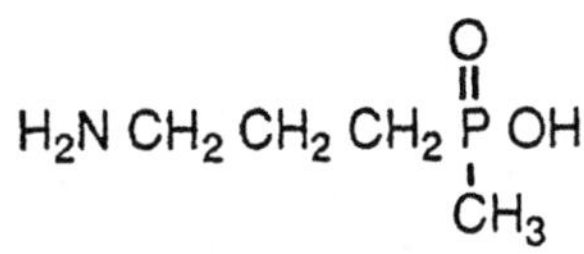

SK&F 97541
(3-aminopropyl(methyl)
phosohinic acid)

Antagonists

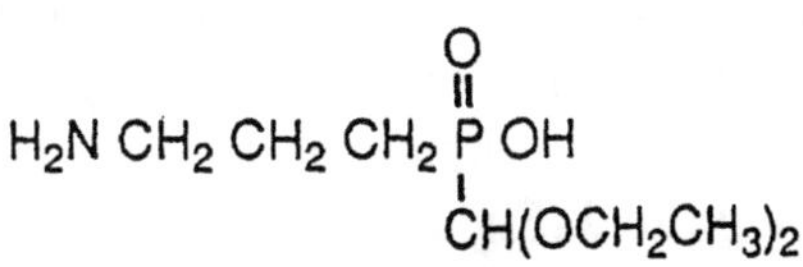

CGP 35348
(3-aminopropyl(diethoxy-
methyl)phosphinic acid)

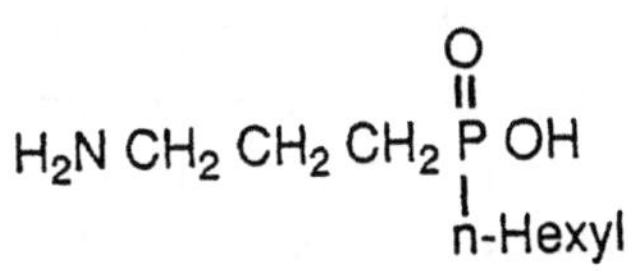

3-APHPA
(3-aminopropyl(hexyl)
phosphinic acid)

Fig. 2. Phosphinic acid analogues

(phaclofen, saclofen and 2-OH saclofen appear to be devoid of $GABA_A$ activity) and improved potency at $GABA_B$ receptor.

Two recently described baclofen analogues, in which the carboxylic acid moiety is retained but the 4-chlorophenyl group has been replaced by a benzofuran, are 4-amino-3-(5-methoxybenzo[b]furan-2-yl)butyric acid (9h) and the unsubstituted benzofuran (9g) (Fig.1). These have been claimed to be $GABA_B$ antagonists in the guinea-pig ileum, with estimated pA_2 values of 4.1 and 3.9, respectively [24]. In rat whole brain synaptic membranes 9h and 9g displace [^{3}H]-baclofen with IC_{50}s of 5.6 μM and 18 μM, respectively, while baclofen had an IC_{50} of 0.2 μM in this assay [4]. This binding data seems rather surprising in light of the poor antagonist activity described above and reflects our findings for 9h, which had an IC_{50} of approximately 100 μM in the $GABA_B$ binding assay (N. G. Bowery, personal communication) and an estimated pA_2 value of 3.4 in the rat vas deferens (J.M. Hills and M.M. Larkin, personal communication).

The above compounds are all analogues of baclofen rather than of GABA and, perhaps because of this feature, a potency barrier exists which is not readily overcome. Phosphonic analogues of GABA such as 4-aminobutylphosphonic acid (3-ABPA) and 3-aminopropylphosphonic acid (3-APPA) (Fig. 1) were shown to have a weak antagonist and mixed antagonist / agonist properties respectively at the $GABA_B$ receptor in the guinea-pig ileum and rat neocortex [27]. The apparent pA_2 value for 3-APPA in the guinea-pig ileum was reported to be 4.0 while at millimolar concentrations agonist activity was observed [27]. The higher homologue, 4-ABPA was reported in similar studies to have an estimated pA_2 of 4.0 in the guinea-pig ileum and vas deferens.

In our hands, a slightly higher pA_2 value (4.6) for 4-ABPA was obtained in the rat vas deferens. Neither compound has since been used as a pharmacological tool in further investigations. A conformationally restricted analogue of δ-AVA, Z-5-aminopent-2-enoic acid (Fig.1), has recently been reported to possess $GABA_B$ antagonist activity against baclofen in the guinea-pig ileum and vas deferens (pA_2=4.9 in the vas deferens), although this compound, like δ-AVA, is also a weak $GABA_A$ receptor agonist [12].

The most significant advance to date in the $GABA_B$ receptor field has been the discovery of pharmacological activity in a new class of $GABA_B$ ligands, the phosphinic acids. The first phosphinic acid of GABA, 3-aminopropyl phosphinic acid (CGP 27492) (Fig. 2), was synthesized by Dingwall and colleagues [13], tested in periperal systems and found to be a potent and selective $GABA_B$ receptor agonist [19, 14] (Table 1). We studied this compound in a range of tissues and found it to be some 8-24 times more potent than baclofen, depending on the tissue studied. The potency of the compound also varied between tissues; for example, the rat anococcygeus appears to be more sensitive than either the vas deferens or the guinea-pig ileum, a pattern also found for baclofen [19] (Table 1). The displacement of [^{3}H]GABA or [^{3}H]CGP 27492 from GABA$_B$ receptors by CGP 27492 has been reported by two groups to be 1 n*M* [6, 32], representing an increase in affinity over baclofen of almost 100-fold (see [8]). The increase in binding affinity of CGP 27492 over baclofen is thus greater than the increase in functional activity of this compound in the pharmacological assays. This may be partly as a result, due to technical reasons, of the lack of direct binding data on the nervous elements in the peripheral tissues studied. Alternatively, baclofen and CGP 27492 may be partial agonists at the GABA$_B$ receptor. Pratt et al. [32] have shown this to be the case for CGP 27492 in the norepinephrine stimulated cyclase assay. Recent data from experiments at SmithKline Beecham, using electrophysiology in rat brain slices, have shown that CGP 27492 both inhibits excitatory synaptic potentials in striatum and hyperpolarises substantia nigra neurones, but with a lower maximum response than baclofen [33]. Such observations may indicate that CGP 27492 is a partial agonist in certain tissues.

Table 1. The biological activity of two phosphinic analogues of GABA at GABA$_B$ receptors compared to (±)-baclofen

	IC$_{50}$ (*μM*)		
	(±)-baclofen	CGP 27492	SKF 97541
GABA$_B$ receptor binding[a]	0.1	0.001	0.0003
GABA$_A$ receptor binding[b]	> 100	0.42	23
Guinea-pig ileum,[c] inhibition of twitch ($n = 6$)	12.5±1.8	1.84±0.23	2.14± 0.37
Rat anococcygeus,[c] inhibition of twitch ($n = 12$)	1.9±0.3	0.09±0.01	0.20± 0.04
Rat vas deferens,[c] inhibition of twitch ($n = 8$)	5.9±1.4	0.80±0.20	0.40± 0.10
Guinea-pig colon, inhibition of motility ($n = 4$)	6.4±2.8	0.27±0.10	0.22± 0.02

[a] GABA$_B$ binding in rat brain cortical membranes using [^{3}H]GABA as ligand in the presence of isoguvacine (40 *μM*) to block GABA$_A$ sites. Non-specific binding determined in the presence of (±)-baclofen (100 *μM*). [b] GABA$_A$ binding as above but in the presence of (±)-baclofen (100 *μM*) to block GABA$_B$ receptors and non-specific binding determined in the presence of isoguvacine (100 *μM*). [c] Electrical field stimulation induced twitch responses induced by passing square wave pulses from Grass SD 11 stimulators via platinum electrodes. Stimulus parameters were: vas deferens, 1 ms duration, 50 Hz for 100 ms every 10 s; rat anococcygeus, 1 ms pulse duration, 10 Hz for 1 s every 10 s; guinea-pig ileum, 0.5 ms pulse duration, 0.1 Hz, all at supramaximal voltage.

A second phosphinic acid analogue of GABA, 3-aminopropyl(methyl)phosphinic acid (SKF 97541) (Fig. 2), has proved to be an equally useful tool for the study of $GABA_B$ receptor function in central and peripheral tissues. The profile of activity for SKF 97541 is similar to that for CGP 27942 (Table 1), there being little or no difference between the two compounds in agonist potency in the peripheral tissues studied. The measures binding affinity in rat brain for SKF 97541 is slightly higher than that for CGP 27942, and the compound is also more selective for the $GABA_B$ receptor over the $GABA_A$ receptor than is CGP 27492 (N.G. Bowery, personal communication; [20]). Furthermore, electrophysiological studies in rat brain slices indicate that SKF 97541 is ten times more potent than both SGP 27492 and baclofen [33], and, on the basis of these experiments, $GABA_B$ receptor heterogeneity remains a possibility.

Studies at SmithKline Beecham have shown the major difference between these two phosphinic acid $GABA_B$ receptor agonists to be their differential ability to penetrate the CNS. Hypothermia in the mouse is known to be a feature of the central action of baclofen, while $GABA_A$ agonists such as muscimol mediate hyperthermia [17]. In studies comparing the hypothermic effects of CGP 27942 and SKF 97541 with baclofen, it was found that SKF 97541 (0.1-1 mg/kg i.p.) and baclofen (1-10 mg/kg i.p.) caused a marked and dose related decrease in body temperature (A. J. Hunter, personal communication). CGP 27942, studied at doses up to 5 mg/kg (i.p.), was without effect on mouse body temperature.

Similarly, the work of Andrews and colleagues [2, 35] showed baclofen to stimulate gastric motility and acid secretion in anesthetised rats by a central mechanism. We have compared the effects of CGP 27492 and SKF 97541 with baclofen in their ability to stimulate gastric acid secretion in the lumen in perfused anesthetised rat. Baclofen (4 mg/kg i.v.) caused a stimulation of acid secretion which was closely mimicked by SKF 97541 (0.2 mg/kg i.v.) (J. M. Hills and P. A. Sullivan, personal communication). The effect of both secretagogues was markedly reduced by vagotomy. CGP 27482 (20 mg/kg i.v.), by contrast, was ineffective at causing acid secretion both before and after vagotomy, suggesting lack of CNS penetration. An additional study was carried out using radiolabelled SKF 97541, infused intravenously at 0.5 and 1 mg/kg into conscious rats to reach a steady state plasma level of 1-2 μmol equivalents per litre. The mean whole brain:blood ratio was similar in both studies and a value of 0.013 ($n = 8$) was obtained (R. Griffith and V. A. Lewis, personal communication). This value represents a relatively poor access of SKF 97541 into the whole brain, however, access in certain areas such as the thalamus and hypothalamus was as much as 0.04.

In a recent patent a series of phosphinic acids with a bulkier alkyl substituent were claimed as $GABA_B$ antagonists [3]. Published work to date with these compounds has focussed on CGP 35348 (3-aminopropyl-(diethoxymethyl)phosphinic acid) (Fig.2) and involved the use of CNS assay techniques, such as $GABA_B$ binding to rat brain membranes (IC_{50}=35 μM using ^{3}H-CGP 27492 as ligand), and electrophysiology [5]. We have investigated the antagonist potency of two of these compounds, CPG 35348 and 3-aminopropyl(*n*-hexyl)phosphinic acid (3-APHPA) (Fig. 2), in the rat anococcygeus, rat vas deferens and guinea-pig ileum. Their activity has been compared with the antagonist potency of 2-OH saclofen (Table 2).

In the rat anococcygeus, using CGP 27492 as the $GABA_B$ agonist, pA_2 values obtained for CGP 35348 and 3-APHPA were 5.4 and 4.9, respectively. This compares with a pA_2 value for 2-OH saclofen obtained in the same tissue of 4.5. As may be seen from Table 2, although CGP 35348 is the most potent antagonist that we have tested to date, the confidence limits are wide and the slope of the Schild plot is significantly less than one.Such observations suggest that either CGP 35348 is not a truly competitive antagonist or that it may not be specific for the $GABA_B$ receptor. The possibility of interaction with an as yet unidentified subclass of $GABA_B$ receptors cannot be ruled out. In our hands, therefore, while 3-APHPA appears to be less potent than CGP 35348, examination of the Schild data would suggest it to be a more useful $GABA_B$ antagonist. Similar findings were obtained in the rat vas deferens with these compounds (Table 2). In the guinea-pig ileum, we were unable to obtain significant antagonism of CGP 27492 concentration resopnse curves with 2-OH saclofen at 100 μM. CGP 35348 and 3-APHPA, unlike 2-OH saclofen,were weak but effective antagonists in the preparation (Table 2).

Studies at SmithKline Beecham, using these compounds in slice preparations of rat striatum and substantia nigra to antagonise the effects of SKF 97541, have found similar effects to those described above. For example, the depression of striatal excitatory synaptic potentials caused by SKF 97541 is antagonised by CGP 35348 and 3-APHPA with *K* values of 11.2 ± 1.7 μM and 13.3 ± 0.4 μM, respectively [33]. These *K* values approximate pA_2 values of 5.0 and are therefore in close agreement with the pA_2 values obtained in the experiments on the peripheral tissues.

Although these results confirm that CGP 35348 and 3-APHPA are $GABA_B$ receptor antagonists, they fail to show any significant improvement over compounds such as 2-OH saclofen. The relatively poor pA_2 values obtained in peripheral and central assay systems are not associated with a lack of activity in vivo [3, 5]. This in vivo activity, together with

Table 2. The $GABA_B$ antagonist activity of two phosphinic acids compared to 2-OH saclofen

	2-OH saclofen	CGP 35348	3-APHPA
Rat anococcygeus			
pA_2 (95% CI)	4.5 (4.2, 4.8)	5.4 (4.4)	4.9 (4.7, 5.1)
slope (95% CI)	1.0 (0.7, 1.3)	0.7 (-0.1, 1.4)	1.0 (0.8, 1.1)
Rat vas deferens			
pA_2 (95% CI)	4.7[a]	5.0 (4.6, 6.0)	4.6 (4.3, 5.0)
slope (95% CI)		0.6 (0.4, 0.9)[*]	1.0 (0.7, 1.4)
Guinea-pig ileum			
estimated pA_2	Inactive	4.3	4.5

Stimulus parameters are the same as for Table 1.
[a] Estimated pA_2. [*]Significantly different from 1; CI, confidence interval

the CNS penetrating properties of these compounds [35], may well turn out to be the feature which makes them better pharmacological tools than those previously available.

In conclusion, we have studied in some detail the $GABA_B$ agonist and antagonist activity of phosphinic acid analogues of GABA in a range of peripheral tissues. The rat anococcygeus preparation would appear to be the most reliable tissue to use in such studies, since it gives a greater maximum response to $GABA_B$ agonists, does not show marked $GABA_B$ receptor tachyphylaxis compared to other peripheral preparations and consistent agonist responses may be obtained over the entire duration of the experiment. The range of pharmacological tools currently available, although not ideal in terms of potency, does provide a basis for further $GABA_B$ receptor investigations, particularly the physiological role of $GABA_B$ receptors and possible receptor heterogeneity in central and peripheral systems.

Acknowledgements. The authors wish to thank Professor N. Bowery for helpful guidance during this work and his staff for assistance with experimentation. The enthusiasm and intellectual contribution made by Dr M. E. Parsons and others at SmithKline Beecham is also gratefully acknowledged.

References

1. Allan RD, Dickenson HW (1986) Evidence that antagonism by δ-aminovaleric acid of $GABA_B$ receptors in the guinea-pig ileum may be due to an interaction between $GABA_A$ and $GABA_B$ receptors. Eur J Pharmacol 120: 119-122
2. Andrews PLR, Wood KL (1986) Systemic baclofen stimulates gastric motility and secretion via a central mechanism in the rat. Br J Pharmacol 89: 461-467
3. Baylis EK, Bittiger H, Frostl W, Hall RG, Maier L, Mikel SJ, Olpe H (1989) Substituted propane phosphinic acid compounds. European Patent No. 319479
4. Berthelot P, Vaccher C, Musadad A, Flouquet N, Debaert, Luyckx M (1987) Synthesis and pharmacological evaluation of γ-aminobutyric acid analogues. New ligands for $GABA_B$ sites. J Med Chem 30: 743-746
5. Bittiger H, Reyman N, Hall RG, Steinmann MW, Van Reizen H (1990) Biochemistry, electrophysiology and pharmacology of a new $GABA_B$ antagonist: CPG 35348. In: Bowery NG, Bittiger H, Olpe H-R (eds) $GABA_B$ Receptors in mammalian function. Wiley, Chichester
6. Bittiger H, Reyman N, Hall R, Kane P, (1988) Eur Neurosci Assoc Proc 16: 10 (abstr)
7. Bowery NG (1982) Baclofen: 10 years on. Trends Pharmacol Sci 3: 400-403
8. Bowery NG (1990) $GABA_B$ receptors and their significance in mammalian pharmacology. Trends Pharmacol Sci 10: 401-407
9. Bowery NG, Doble A, Hill DR, Hudson AL, Shaw JS, Turnbull MJ,Warrington R (1981) Bicuculline-insensitive GABA receptors on peripheral autonomic nerve terminals. Eur J Pharmacol 71: 53-70
10. Bowery NG, Hill DR (1983) Characteristics of $GABA_B$ receptor binding sites in rat whole brain synaptic membranes. Br J Pharmacol 78: 191-206
11. Bowery NG, Hill DR, Hudson AL, Middlemiss DN, Shaw J, Turnbull MJ (1980) (-) Baclofen decreases neurotransmitter release in the mammalian CNS by an action at a novel GABA receptor. Nature 283: 92-94
12. Dickenson HW, Allan RD, Ong J, Johnston GAR (1988) $GABA_B$ receptor antagonist and $GABA_A$ agonist properties of a δ-aminovaleric acid derivative, Z-5-aminopent-2-enoic acid. Neurosci Lett 86: 351-355
13. Dingwall JG, Ehrenfreund J, Hall RG, Jack J (1987) Substituted propanephosphonous acids using hypophosphorous acid synthons. Phosphorus Sulfur 30: 571-575
14. Dingwall JG, Ehrenfreund J, Hall RG, Jack J (1987) Substituted propane-phosphonous acid compounds. Chem Abstr 106: 18814 (United States Patent No 4656298)
15. Giotti A, Luzzi S, Maggi CA, Spagnesi S, Zilletti L (1985) Modulatory activity of $GABA_B$ receptors on cholinergic tone in the guinea-pig distal colon. Br J Pharmacol 84: 883-895
16. Giotti A, Luzzi S, Spagnesi S, Zilletti L (1983) $GABA_A$ and $GABA_B$ receptor-mediated effects in the guinea-pig ileum. Br J Pharmacol 78: 469-478
17. Gray JA, Goodwin GM, Heal DJ, Green AR (1987) Hypothermia induced by baclofen, a possible index of $GABA_B$ receptor function in mice, is enhanced by antidepressant drugs and ECS. Br J Pharmacol 92: 863-870
18. Hill DR, Bowery NG (1981) ^{3}H-Baclofen and ^{3}H-GABA bind to bicuculline-insensitive $GABA_B$ sites in rat brain. Nature 290: 149-152

19. Hills JM, Dingsdale RA, Parsons ME, Dolle RE, Howson W (1989) 3-Aminopropyl-phosphinic acid - a potent, selective $GABA_B$ receptor agonist in the guinea-pig ileum and rat anacoccygeous muscle. Br J Pharmacol 97: 1292-1296
20. Hills JM, Howson W (1990) 3-Aminopropyl-methylphosphinic acid as a therapeutic agent and intermediates in its synthesis. European Patent No 0356128
21. Kaplita PV, Walters DH, Triggle DJ (1982) γ-Aminobutyric acid action on guinea-pig ileal myenteric plexus. Eur J Pharmacol 79: 43-51
22. Kerr DIB, Ong J, Johnston GAR, Abbenante J , Prager RH (1988) 2-Hydroxy-saclofen: an improved antagonist at central and peripheral $GABA_B$ receptors. Neurosci Lett 92: 92-96
23. Kerr DIB, Ong J, Johnston GAR, Abbenante J , Prager RH (1989) Antagonism at $GABA_B$ receptors by saclofen and related sulphonic analogues of baclofen and GABA. Neurosci Lett 107: 239-244
24. Kerr DIB, Ong J, Johnston GAR, Berthelot P, Debaert M, Vaccher C (1989) Benzo-furan analogues of baclofen: a new class of central and peripheral $GABA_B$-receptor antagonists. Eur J Pharmacol 164: 361-364
25. Kerr DIB, Ong J, Prager RH (1988) 2-OH-saclofen, a new $GABA_B$ receptor antagonist. Neurosci Lett Suppl 30: S85
26. Kerr DIB, Ong J, Johnston GAR, Prager RH (1990) Antagnosim of $GABA_B$-receptor-mediated responses in the guinea-pig isolated ileum and vas deferens by phosphono-analogues of GABA. Br J Pharmacol 99: 422-426
27. Kerr DIB, Ong J, Prager RH, Gynther BD, Curtis DR (1987) Phaclofen: a peripheral and central baclofen antagonist. Brain Res 405: 150-154
28. Luzzi S, Maggi CA, Spagnesi S, Santicioli P, Zilletti L (1985) 5-Aminovaleric acid interactions with $GABA_A$ and $GABA_B$ receptors in the guinea-pig ileum. J Auton Pharmacol 5: 65-69
29. Muhyaddin M, Roberts PJ, Woodruff GN (1982) Presynaptic γ-aminobutyric acid receptors in the rat anococcygeus muscle and their antagonism by 5-aminovaleric acid. Br J Pharmacol 77: 163-168
30. Olpe H-R, Karlsson G, Schmutz M, Klebs K, Bittiger H (1990) $GABA_B$ receptors and experimental models of epilepsy. In: Bowery NG, Bittiger H, Olpe H-R (eds) $GABA_B$ receptors in mammalian function. Wiley Chichester
31. Ong J, Kerr DIB (1983) $GABA_A$ and $GABA_B$ receptor-mediated modification of intestinal motility. Eur J Pharmacol 86: 9-17
32. Pratt GD, Knott C, Davey R, Bowery NG (1989) Characterisation of 3-aminopropyl phosphinic acid (3-APPA) as a $GABA_B$ agonist in rat brain tissue. Br J Pharmacol 96: 141P
33. Seabrook GR, Howson W, Lacey MG (1990) Electrophysiological characterisation of potent agonists and antagonists at pre- and postsynaptic $GABA_B$ receptor on neurones in rat brain slices. Br J Pharmacol (to be published).
34. Westfall DP, Stitzel RE, Rowe JN (1978) Postjunctional effects and neural release of purine compounds in guinea-pig vas deferens. Eur J Pharmacol 50: 27-38
35. Wood KL, Addae JI, Andrews PLR, Stone TW (1987) Injection of baclofen into the ventromedial hypothalamus stimulates gastric motility in the rat. Neuropharmacology 26: 1191-1194

GABAergic Mechanisms and Cardiovascular Function

K. Taniyama, N. Saito, S. Matsuyama, K. Takeda, and C. Tanaka

Introduction

It has been reported that GABA affects the regulation of cardiovascular function through both the peripheral GABAergic system and central GABAergic mechanisms. Intracerebral or intraventricular administration of GABA agonists or GABA transaminase inhibitors has been shown to lower blood pressure and heart rate via central GABAergic mechanisms [1, 4, 5, 16, 19, 25]. Intravenous administration of GABA was found to lower blood pressure and to induce bradycardia in dogs [6]. In this case, it is most likely that peripheral GABA actions were responsible for changes in cardiovascular function, since GABA is unable to cross the blood-brain barrier [11]. In in vitro experiments, GABA inhibited the stimulation-evoked release of norepinephrine from the rat atria [2]. It has been proposed that GABA induces a release of noncholinergic and nonadrenergic substance [13]. However, evidence of the presence of GABAergic neurons in the heart has not been reported. The role of central GABAergic mechanisms in cardiovascular function has been reviewed in detail and is not addressed here. Of relevance to this chapter are those studies that implicate a peripheral cardiovascular function for GABA and a possible neurotransmitter role for GABA in the sinus node of the heart.

In Vivo Experiments

When GABA is administered through intravenous, subcutaneous, or oral routes, it may act through peripheral GABAergic mechanisms, since, as noted above, it is unable to cross the blood-brain barrier [11].

Rabbit

Effects of GABA on the cardiovascular system were first noted in a study by Takahashi et al. [21], in which GABA (0.07-12 mg/kg) injected intravenously into urethane-anesthetized rabbits decreased both blood pressure and heart rate for up to 10 min. In the unanesthetized rabbit, GABA produced a transient decrease in the blood pressure followed by an increase which lasted up to 20 min. Subsequently, Elliott and Hobbiger [6] examined the effect of GABA on cardiovascular functions in rabbits, dogs, and cats. In the urethane-anesthetized rabbits, the intravenous administration of GABA (0.07-10 mg/kg) decreased blood pressure but did not affect heart rate [6], findings differing from those of Takahashi et al. [21]. In rabbits anesthetized with thiopental or chloralose, similar effects of GABA were also noted [6].

Dog

In anesthetized dogs, the intravenous administration of GABA (0.01-10 mg/kg) decreased both blood pressure and heart rate [6,20]. The hypotensive effect of GABA is atropine-resistant and vagotomy-resistant, while the bradycardiac effect is less marked in vagotomized dogs [6]. A similar result was noted in vagotomized dogs: GABA (0.03-3 mg/kg) injected intravenously produced a decrease in the blood pressure with no effect on the heart rate [10]. Low doses of GABA (0.004-0.032 mg/kg) injected intravenously produced hypertension and bradycardia (lasting 10-20 s) followed by hypotension (lasting for 30-120 s). Lightly anesthetized dogs were more susceptible to the action of GABA [20].

In vagotomized dogs, GABA (0.001-0.3 mg/kg) exerted biphasic effects on the tachycardia induced by stimulation of the preganglionic sympathetic nerve; low doses of GABA (0.001 and 0.003 mg/kg) produced augmentation of stimulation-induced tachycardia; higher doses of GABA (0.01 mg/kg) produced augmentation or depression; and large doses of GABA

(0.03-0.3 mg/kg) produced depression. However, GABA did not affect the tachycardia induced by stimulation of postganglionic sympathetic nerves [10]. Thus, GABA seems to have dual effects on ganglionic transmission, i.e., facilitation in small doses and depression in large doses. The augmentation of stimulation-induced tachycardia by GABA was picrotoxin-resistant and the depression was picrotoxin-sensitive. GABA may produce depression by inhibiting the activity of preganglionic sympathetic neurons, through the $GABA_A$ receptor.

GABA (0.001-1 mg/ml) administered through the sinus node artery did not affect either the intrinsic sinus rate or the tachycardia induced by stimulation of the stellate nerve, yet attenuated the bradycardia induced by stimulation of the vagus nerve [15]. The GABA effect was antagonized by picrotoxin, thereby suggesting that GABA acts on the $GABA_A$ receptor. However, the sinus node response to norepinephrine or acetylcholine was not affected by GABA; hence, GABA probably inhibits the activity of the parasympathetic neuron, an event which leads to a positive chronotropism.

Cat

In chloralose-anesthetized cats, GABA (0.1-100 mg/kg) given intravenously produced variable effects on blood pressure. A hypotensive or biphasic response (the hypotensive effect is often preceded by a hypertensive effect) with no effect on the heart rate was observed. In pentobarbital-anesthetized cats, GABA produced only a decrease in blood pressure [6].

Rat

In urethane-anesthetized rats, intravenous GABA (1-10 mg/kg) produced hypotension and bradycardia. In hexamethonium-treated rats, in which hypotension and bradycardia occurred, GABA did not produce any further depression [14]. GABA inhibits the bradycardia produced by norepinephrine or by electrical stimulation of the vagus nerve. From these findings it is suggested that GABA modulates baroreceptor reflexes by acting at a peripheral site.

In Vitro Experiments

In the Langendorff preparation, GABA (0.1 mg and higher) slightly enhanced the force of contractions but had no effect on the heart rate in rabbits, although it did affect the heart rate in rats and guinea pigs [6].

GABA has been found to inhibit the release of norepinephrine evoked either by electrical stimulation or high K^+ depolarization from the isolated preparation of rat atria [2]. A larger inhibition of the evoked release occurred in the presence of an α_2 blocker. The inhibitory effect of GABA was mimicked by baclofen and was not antagonized by bicuculline, therefore GABA may act on the bicuculline-insensitive $GABA_B$ receptor located on the adrenergic neuron. Such an effect of GABA may lead to negative inotropism of the atrium.

In contrast, GABA has been reported to exert positive chronotropic and inotropic effects on the atria freed of influences of adrenergic and cholinergic neurons [13]. The effect of GABA on the inotropism of the electrically driven left atria of guinea pig and on the chronotropism and inotropism of spontaneously beating right atria and whole heart, in the presence of blockers for the muscarinic receptor and α- and β-adrenoceptors, has been examined. GABA (30 μM - 1 mM) enhanced the electrically stimulated positive inotropism of the left atria. GABA (1 mM) also produced transient increases in the frequency and contractile force of the spontaneously beating right atrium and whole heart. These effects of GABA were mimicked by muscimol but not by baclofen and were antagonized by bicuculline; thus, GABA may act via the $GABA_A$ receptor. On the basis of findings that capsaicin produced positive chronotropic and inotropic effects and capsaicin desensitization prevented the GABA effect, it has been proposed that $GABA_A$ receptors are located on the preterminal region of capsaicin-sensitive sensory nerves in the atrium.

Possible Presence of GABAergic Neurons in the Sinus Node

GABA and its synthetizing enzyme (glutamic acid decarboxylase) and catabolizing enzyme (GABA transaminase) are present in the mammalian heart [7, 22]; however, no other information is available concerning the possible neurotransmitter role of GABA in the heart. We measured the amount of GABA in seven regions of the guinea pig heart, the right and left auricles, sinus nodes, atrial bodies, right and left ventricles, and interventricular septums (Fig. 1). The GABA content was highest in the sinus node (70 nmol/g tissue) and was much the same in other regions (10 nmol/g tissue), thereby suggesting the presence of GABA-containing cells in the sinus node.

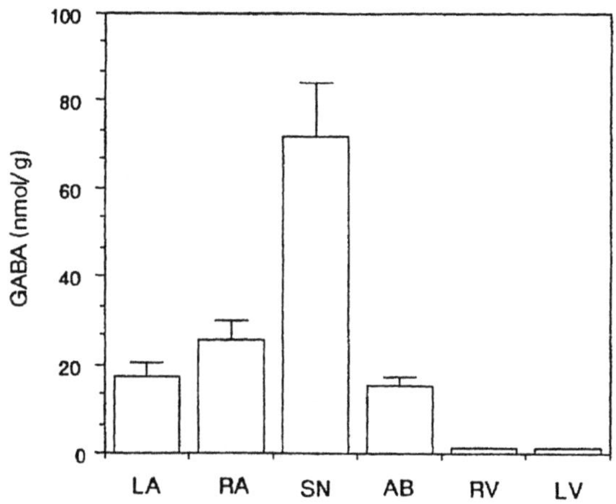

Fig.1. GABA contents in the guinea pig heart. GABA was assayed enzymatically. *LA*, left auricle; *RA*, right auricle; *SN*, sinus node; *AB*, atrical body; *RV*, right ventricle; *LV*, left ventricle

One criterion used to establish a substance as a neurotransmitter is the demonstration of its release when evoked by appropriate nerve stimulation. Electrical transmural stimulation evoked the release of [^{3}H]GABA from the sinus node preloaded with [^{3}H]GABA, in the presence of β-alanine, in a frequency-dependent manner (3-20 Hz) (Fig. 2). Such a frequency range is similar to that found in the intestine [8,9,23,24], gallbladder [17], and urinary bladder [12].

The stimulation-evoked release of [^{3}H]GABA from the sinus node was Ca^{2+}-dependent and tetrodotoxin-sensitive. The neuronal and glial release mechanisms for GABA possess strikingly different chemical sensitivities. High K^+-evoked release of [^{3}H]GABA from isolated glial cells is Ca^{2+}-independent, but [^{3}H]GABA release from synaptosomes and neuronal perikarya is Ca^{2+}-dependent [18]. In the rat superior cervical ganglia, electrical stimulation-evoked release of [^{3}H]GABA is Ca^{2+}-dependent but tetrodotoxin-resistant [3]. Thus, [^{3}H]GABA taken up mostly by the neuron is released by impulses propagated to nerve terminals from the depolarized soma-dendritic regions of the neuron.

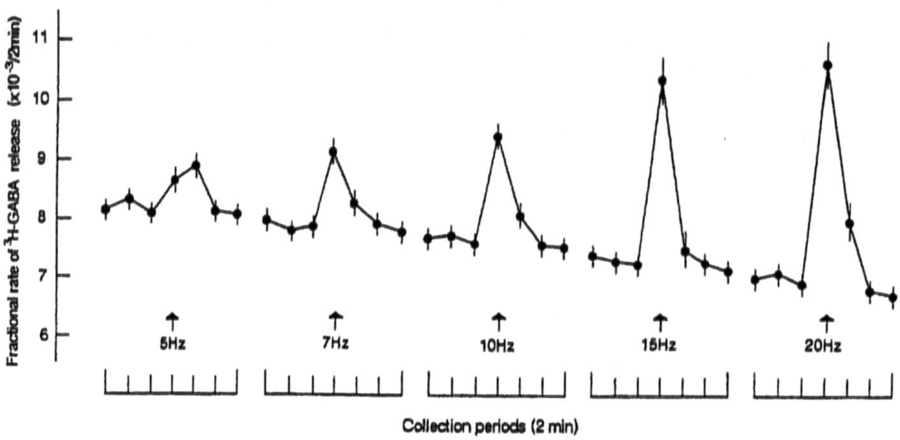

Fig.2. Electrical stimulation-evoked release of [^{3}H]GABA from the sinus node preloaded with [^{3}H]GABA. The release of [^{3}H]GABA is represented as the fractional rate obtained by dividing the amount of [^{3}H]GABA in the superfusate by the respective amount of [^{3}H]GABA in the tissue. Electrical field stimulation (3 mA, 3 ms) was applied for 1 min

Conclusion

In vivo experiments showed the response of the whole animal body to GABA, but the site of action of GABA was not elucidated. In vitro experiments revealed the property and possible localization of the GABA receptor, but GABAergic neurons at the possible site of GABA action were not identified. We presented part of the criteria required to establish GABA as a neurotransmitter in the sinus node of guinea pig heart: (a) the occurrence of GABA in the sinus node and (b) the neuronal release of GABA. There are species differences in the effect of intravenously administered GABA on blood pressure and heart rate. In vivo experiments were carried out in anesthetized animals, therefore it is possible that anesthesia masks some stimulant actions of GABA. The GABAergic neuron in the sinus node is considered to be involved in the control of the heart rate. The clarification of the physiological relevance of GABAergic neurons in the heart, the identification of GABA as a neurotransmitter in the sinus node, and the responses to GABA measured in both unanesthetized animals and the isolated preparation are the subject of ongoing studies.

References

1. Bousquet P, Feldman J, Bloch R, Schwartz J (1981) The central action of baclofen in the anesthetized cat. Eur J Pharmacol 76: 193-201.
2. Bowery NG, Doble A, Hill DR, Hudson AL, Shaw JS, Turnbull MJ, Warrington R (1981) Bicuculline-insensitive GABA receptors on peripheral autonomic nerve terminals. Eur J Pharmacol 71: 53-70
3. Bowery NG, Brown DA, Marsh S (1979) γ-Aminobutyric acid efflux from sympathetic glial cells: effect of depolarizing agents. J Physiol 293: 75-101
4. DeFeudis FV (1981) GABA and "neurocardiovascular" mechanisms. Neurochem Int 3: 113-122
5. DiMicco JA, Gale K, Hamilton B, Gillis RA (1979) GABA receptor control of parasympathetic outflow to heart: characterization and brainstem localization. Science 204: 1106-1109
6. Elliott KAC, Hobbiger F (1959) Gamma aminobutyric acid: circulatory and respiratory effects in different species: reinvestigation of the anti-strychnine action in mice. J Physiol 146: 70-84
7. Erdö SL, Kiss B (1986) Presence of GABA, glutamate decarboxylase, and GABA transaminase in peripheral tissues: collection of quantitative data. In: Erdo SL, Bowery NG (eds) GABAergic mechanisms in the mammalian periphery. Raven, New York, pp 5-17
8. Jessen KR, Hills JM, Dennison ME, Mirsky R (1983) γ-Aminobutyrate as an autonomic neurotransmitter: release and uptake of [^{3}H]-γ-aminobutyrate in guinea pig large intestine and cultured enteric neurons using physiological methods and electron microscopic autoradiography. Neuroscience 10: 1427-1442
9. Kerr DIB, Krantis A (1983) Uptake and stimulus-evoked release of [^{3}H]-γ-aminobutyric acid by myenteric nerves of guinea pig intestine. Br J Pharmacol 78: 271-276
10. Kimura T, Imamura H, Hashimoto K (1977) Facilitatory and inhibitory effects of γ-aminobutyric acid on ganglionic transmission in the sympathetic cardiac nerves of the dog. J Pharmacol Exp Ther 202: 397-403
11. Kuriyama K, Sze PY (1971) Blood brain barrier to tritiated GABA treated animals. Neuropharmacol 10: 103-108
12. Kusunoki M, Taniyama K, Tanaka C (1984) Neuronal GABA release and GABA inhibition of ACh release in guinea pig urinary bladder. Am J Physiol 246: R502-R509
13. Maggi CA, Giuliani S, Manzini S, Meli A (1989) $GABA_A$ receptor-mediated positive inotropism in guinea pig isolated left atria: evidence for the involvement of capsaicin-sensitive nerves. Br J Pharmacol 97: 103-110

14. Maggi CA, Giuliani S, Meli A (1985) The effect of peripherally administered GABA on noradrenaline-induced reflex vagal bradycardia in urethane anasthetized rats. Gen Pharmacol 16: 579-584
15. Neely BH, Hageman GR, James TN (1983) Effects of γ-aminobutyric acid on neural regulation of the canine sinus node. Am J Physiol 244: H266-H272
16. Porsius AJ, Lambrecht G, Moser U, Mutschler E (1982) Depressor effects of isoguvacine propyl ester and isoarecaidine propyl ester due to stimulation of central GABA-receptors in the cat. Eur J Pharmacol 77: 49-52
17. Saito N, Taniyama K, Tanaka C (1985) Uptake and release of GABA in guinea pig gallbladder. Am J Physiol 249: G192-G196
18. Sellstrom A, Hamberger A (1977) Potassium-stimulated γ-aminobutyric acid release from neurons and glia. Brain Res 119: 189-198
19. Snyder DW, Antonaccio MJ (1980) Central sites involved in hypotensive effects of muscimol. Brain Res Bull 5 (suppl 2): 317-323
20. Stanton HC, Woodhouse GB (1960) The effect of gamma-amino-*N*-butyric acid and some related compounds on the cardiovascular system of anesthezized dogs. J Pharmacol Exp Ther 128: 233-242
21. Takahashi H, Tiba M, Iino M, Takayasu T (1955) The effect of γ-aminobutyric acid on blood pressure. Jpn J Physiol 5: 334-341
22. Tanaka C (1985) γ-Aminobutyric acid in peripheral tissues. Life Sci 37: 2221-2235
23. Taniyama K, Miki Y, Kusunoki M, Saito N, Tanaka C (1983) Release of endogenous and labeled GABA from isolated guinea pig ileum, Am J Physiol 245: G717-G721
24. Taniyama K, Kusunoki M, Saito N, Tanaka C (1982) Release of γ-aminobutyric acid from cat colon. Science 217: 1038-1040
25. Williford D, Hamilton BL, Souza JD, Williams TP, DiMicco JA, Gillis RA (1980) Central nervous system mechanisms involving GABA influence arterial pressure and heart rate in the cat. Circ Res 47: 80-88

GABAergic Mechanisms in the Pulmonary System

R.W. Chapman, J.A. Hey, D.C. Bolser, C.A. Rizzo, R.W. Egan, and W. Kreutner

Introduction

GABA is a well known inhibitory neurotransmitter in the central nervous system (CNS) that has recently been detected in a variety of peripheral organs [25]. In the lung, GABA has also been implicated as an inhibitory neurotransmitter. It attenuates the contractile response of the isolated guinea-pig trachea to electrical field stimulation [8, 22, 23], inhibits the release of acetylcholine from guinea-pig lungs [23], and reduces bronchoconstrictor responses to nonadrenergic, noncholinergic stimulation in guinea-pig airways [2]. However, there is no consensus as to whether the effects of GABA are mediated through an interaction with the well characterized $GABA_A$ receptor or rather with the $GABA_B$ receptor whose function is not as well understood [4]. In this chapter, we will review the status of GABA as an inhibitory neurotransmitter in the pulmonary system and provide support for $GABA_B$ receptors as functionally important for the inhibitory effects of GABA.

Distribution

GABA is found in the lungs of a variety of species [10, 12, 23, 26]. In the guinea-pig, there is a regional distribution of GABA in the lung with the highest concentration in the proximal and middle lung and the lowest concentration in the lung periphery (Fig. 1). This distribution correlates with the density of vagal innervation of the tracheobronchial tree, which is highest in the central conducting airways and least in the lung parenchyma. Indeed, recent studies have clearly defined a role for GABA as an inhibitory neuromodulator of airway nerves [2, 6, 8, 9, 22, 24].

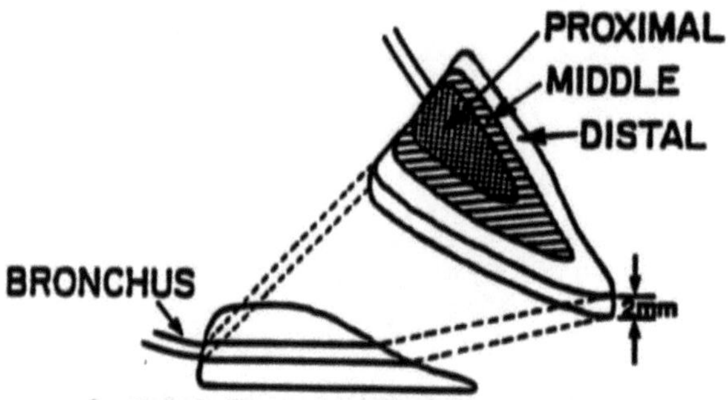

Regions	GABA Contents (nmol/g wet wt.)
Distal	8.6 ± 1.9 [7]
Middle	21.8 ± 2.5 [7]
Proximal	16.7 ± 3.9 [7]

Fig. 1. Preparation of isolated strips from guinea-pig lung. Frontal sections of lung tissue, about 2 mm thick, were excised from both lower lobes then cut into three thin strips distal, middle, and proximal, with the longitudinal axis of the strip cut vertical to the bronchus (From [23])

In the CNS, $GABA_A$ receptors are coupled to a benzodiazepine-GABA receptor complex. Peripheral type benzodiazepine receptors are also found in the lung [17, 27] where they are particularly dense over submucosal glands in the intrapulmonary bronchi [17] and in the airway epithelium and smooth muscle [17]. Peripheral type benzodiazepine receptors may interact with $GABA_A$ receptors in pulmonary tissue [23], but the physiological significance of these receptors in controlling airway function has yet to be determined.

In Vitro Pulmonary Effects of GABA

GABA inhibits the contractile response of the isolated guinea-pig trachea that is induced by electrical field stimulation (EFS) [8, 22, 24]. This tracheal contraction to EFS is blocked by tetrodotoxin and atropine, but not by hexamethonium, indicating that the contractions are mediated by the neuronal release of acetylcholine from postganglionic nerves [8, 22]. GABA has no effect on airway smooth muscle contractions induced by exogenously administered cholinergic agonists [8]. Thus, it effects tracheal EFS contractions by activating prejunctional inhibitory sites on postganglionic parasympathetic nerve terminals [8, 22, 24].

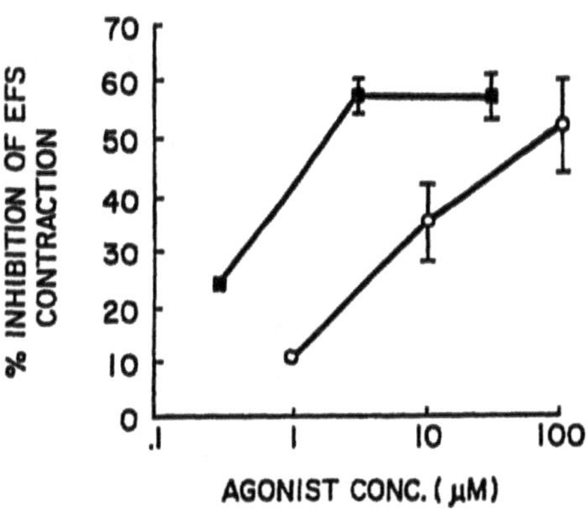

Fig. 2. Cumulative dose responses to baclofen (○) and 3-aminopropyl-phosphinic acid (■) on the contractile response of guinea-pig trachea to EFS. Each point is the mean ± SE of 4 tissues

Although GABA clearly inhibits the contractile responses of the guinea-pig trachea to EFS, there are conflicting views on the GABA-receptor subtype involved. Tamaoki et al. [24] found that GABA and baclofen reduced EFS contraction and that the effects of these GABA agonists were inhibited by bicuculline. These results suggest that $GABA_A$ receptors were involved. However, in that study, the $GABA_A$ agonist, muscimol, was inactive compared to the effects seen with GABA and baclofen. Unfortunately, the authors did not use $GABA_B$ antagonists to fully characterize their system. Results from our laboratory strongly suggest that the inhibitory effects of GABA against EFS contraction are mediated by $GABA_B$ receptors [8, 22]. We have found that the selective $GABA_B$ agonists baclofen [5] and 3-aminopropylphosphinic acid [14] duplicate the effects of GABA and partially inhibit EFS-induced tracheal contraction (Fig. 2), whereas muscimol has little activity compared with GABA, baclofen, or 3-aminopropylphosphinic acid [8, 22].

We have also found that a variety of $GABA_B$ antagonists [4] inhibit the reduction of EFS contraction by baclofen (Table 1), whereas antagonists acting at $GABA_A$ receptors such as bicuculline are inactive in this system [8, 22].

Table 1. Effect of $GABA_B$ antagonists on the inhibitory responses to baclofen against EFS-induced tracheal contraction

Compound[a]	Concentration (μM)	Antagonism of Response to baclofen[b] (mean ± SEM)
3-Aminopropylphosphinic Acid	100	27 ± 11
	300	47 ± 9*
2-Hydroxysaclofen	100	23 ± 8
	300	53 ± 6*
CGP 35348	10	5 ± 9
	30	48 ± 3*
	100	59 ± 6*

* $p < 0.05$ compared to placebo.
a Each compound was given 10 min before baclofen. b Baclofen at 30 μM.

These findings indicate that GABA inhibits neuronally induced cholinergic tracheal contractions by acting at prejunctional $GABA_B$ receptors. These results are consistent with a $GABA_B$ receptor-mediated inhibition of acetylcholine release from the postganglionic parasympathetic neurons that innervate airway smooth muscle [23].

GABA also promotes the release of acetylcholine from lungs by stimulating $GABA_A$ receptors [23], but neither GABA nor muscimol cause cholinergic contraction of tracheal smooth muscle [8, 22]. The physiological significance of this GABA-induced acetylcholine release is unclear at the present time.

In Vivo Pulmonary Effects of GABA

Cholinergic Bronchospasm

GABA inhibits a variety of neuronally induced cholinergic bronchoconstrictor responses [6, 9, 19], such as electrical stimulation of the peripheral cut ends of the vagus nerves [6] and electrical stimulation of the dorsal medulla oblongata [9]. In each of these studies, the cholinergic component of the bronchospasm was confirmed by blocking the response with anticholinergic drugs or by vagotomy [7, 13, 18]. Since GABA has no effect on bronchospasm that is induced by exogenously administered cholinergic agonists [6, 8, 9], its effect against neuronally induced cholinergic bronchospasm must be due to an effect on prejunctional or presynaptic inhibitory GABA receptors on parasympathetic nerves.

The inhibitory effect of GABA against neuronally induced cholinergic bronchospasm is by $GABA_B$ receptor activation. In each of these studies, baclofen duplicated the effect of GABA to inhibit bronchospasm, and the effects of baclofen were blocked by the $GABA_B$ antagonist 3-aminopropylphosphonic acid (3-APPA) [6, 8, 9, 19]. Compounds acting at $GABA_A$ receptors were inactive in these studies. We have expanded our studies in the vagally stimulated cholinergic bronchospasm model and now show that a variety of other $GABA_B$ antagonists, such as 2-hydroxysaclofen and CGP 35348, also antagonize the inhibitory effects of baclofen against vagal bronchospasm (Table 2). These results provide additional strong evidence that the inhibitory effect of GABA on neuronally induced cholinergic airway contractions in vivo are also mediated at $GABA_B$ receptors.

GABA and baclofen have also been studied in conscious guinea-pigs for their effects on bronchoconstrictor responses to anaphylaxis, histamine, and prostaglandin $F_{2\alpha}$ [15, 16]. Both GABA and baclofen inhibit the bronchoconstrictor responses to these spasmogens. This effect may be due to activation of prejunctional $GABA_B$ receptors on airway parasympathetic nerves because vagal cholinergic reflexes have been implicated in the bronchoconstrictor responses to these spasmogens in conscious guinea-pigs [3, 11, 20].

NANC Bronchospasm

GABA and baclofen also inhibit the nonadrenergic, noncholinergic (NANC) bronchospasm that is induced in guinea-pigs by antidromic vagal nerve stimulation [2] or by intravenous nicotine administration [9]. We

Table 2. Effect of $GABA_B$ antagonists on the inhibitory responses to baclofen against electrical vagal nerve stimulation

Compound[a]	Doses tested (mg/kg)	Potency[b] (ED_{40} mg/kg)
3-APPA	1 - 5	3
2-Hydroxysaclofen	10	10
CGP 35348	1 - 10	3

[a] Compounds given i.v. 5 min before 3 mg/kg of baclofen.
[b] Dose of compound inhibiting baclofen response by 40%.

have also found that the inhibitory effect of baclofen against nicotine-induced bronchospasm in guinea-pigs is antagonized by 3-APPA [9] providing further evidence that the GABA receptor subtype is $GABA_B$.

NANC bronchospasms are probably due to the release of tachykinins such as substance P and neurokinin A from C-fiber sensory nerve endings [1]. Neither GABA nor baclofen inhibit the bronchospasm due to exogenously administered substance P [2] or neurokinin A [9]; thus their effects against NANC bronchospasm probably reflects a prejunctional mechanism, such as inhibition of tachykinin release from C-fibers. Indeed, both GABA and baclofen inhibit the release of substance P from capsaicin sensitive nerves in the airways [21].

NANC bronchospasms appear to be also modulated by $GABA_A$ receptors. For example, the $GABA_A$ receptor antagonist bicuculline was found to potentiate the inhibitory effects of baclofen on vagally stimulated NANC bronchospasm in guinea-pigs [2]. Moreover, bicuculline by itself inhibited this NANC bronchospasm [2]. Additional studies are required to clarify the functional role of $GABA_A$ receptors on NANC contractions of airway smooth muscle.

Conclusion

GABA is found in the lungs and has been implicated as an inhibitory neurotransmitter in the pulmonary system. GABA inhibits neuronally induced cholinergic contractions in airway smooth muscle and attenuates bronchoconstrictor responses to stimulation of NANC nerves. These effects are probably due to stimulation of prejunctional $GABA_B$ receptors on cholinergic and sensory nerves to inhibit the release of acetylcholine and tachykinins, respectively. Since these neuronal mechanisms are important components of certain respiratory diseases such as asthma [1], a selective GABAergic agent may have therapeutic potential for the treatment of these respiratory disorders.

Acknowledgement. The authors thank Ms. Lisa Ramirez for the preparation of this manuscript.

References

1. Barnes PJ (1986) Neural control of human airways in health and disease. Am Rev Respir Dis 134: 1289-1314
2. Belvisi MG, Ichinose M, Barnes PJ (1989) Modulation of non-adrenergic, non-cholinergic neural bronchoconstriction in guinea-pig airways via $GABA_B$ receptors. Br J Pharmacol 97: 1225-1231
3. Bergren DR, Gustafson JM, Myers DL (1984) Effect of prostaglandin $F_{2\alpha}$ on pulmonary rapidly adapting receptors in the guinea-pig. Prostaglandins 3: 391-405
4. Bowery NG (1989) $GABA_B$ receptors and their significance in mammalian pharmacology. Trends Pharmacol Sci 10: 401-407
5. Bowery NG, Doble A, Hill DR, Hudson AL, Shaw JS, Turnbull MJ, Warrington R (1981) Bicuculline-insensitive GABA receptors on peripheral autonomic nerve terminals. Eur J Pharmacol 71: 53-70
6. Chapman RW, Danko G (1989) Inhibitory effect of $GABA_B$ receptors on vagally mediated bronchoconstriction in guinea-pigs. FASEB J 3: A440
7. Chapman RW, Danko G (1990) Role of cholinergic, vagal reflexes on the bronchoconstrictor responses to histamine during carbon dioxide inhalation in conscious guinea-pigs. Pharmacol Res 22: 133-141
8. Chapman RW, Danko G, Rizzo C, Egan RW, Mauser PJ, Kreutner W (1990) Prejunctional $GABA_B$ inhibition of cholinergic, neurally mediated airway contractions in guinea-pigs (to be published)
9. Chapman RW, Kreutner W, del Prado M, Hey JA (1990) Role of $GABA_B$ receptors on cholinergic and peptidergic bronchoconstriction in guinea-pigs. International Symposium on Peripheral GABAergic Mechanisms, 19-21 July 1990
10. Del Rio RM, Caballero AL (1980) Presence of γ-aminobutyric acid in rat ovary. J Neurochem 34: 1584-1586
11. Drazen JM, Austen KF (1975) Atropine modification of pulmonary effects of chemical mediators in the guinea-pig. J Appl Physiol 38: 834-838
12. Gerber JC, Hare TA (1979) Gamma-aminobutyric acid in peripheral tissue, with emphasis on the endocrine pancreas. Diabetes 28: 1073-1076

13. Hey JA, del Prado M, Chapman RW (1990) Evidence for a novel medullary bronchomotor pathway that modulates bronchial smooth muscle tone in guinea-pigs. Pulmon Pharmacol 3: 53-54
14. Hills JM, Dingsdale RA, Parsons ME, Dolle RE, Howson W (1989) 3-Aminopropylphosphinic acid, a potent, selective $GABA_B$ receptor agonist in the guinea-pig ileum and rat anococcygeus muscle. Br J Pharmacol 97: 1292-1296
15. Luzzi S, Franchi-Micheli S, Ciuffi M, Zilletti L (1986) Effects of GABA agonists on Herxheimer microshock in guinea-pigs. Agents Actions 1/2: 245-247
16. Luzzi S, Franchi-Micheli S, Folco G, Rossoni G, Ciuffi M, Zilletti L (1987) Effect of baclofen on different models of bronchial hyperreactivity in the guinea-pig. Agents Actions 3/4: 307-309
17. Mak JC, Barnes PJ (1989) Peripheral type benzodiazepine receptors in human and guinea-pig lung: characterization and autoradiographic mapping. J Pharmacol Exp Ther 252: 880-885
18. Mauser PJ, Edelman NH, Chapman RW (1988) Central nervous system control of airway tone in guinea-pigs. J Appl Physiol 65: 2024-2029
19. Mauser PJ (1989) Inhibitory effect of GABA on the bronchoconstriction due to intracerebroventricular histamine in guinea-pigs. Am Rev Respir Dis 139: A107
20. Michoud M-C, Fraser RS, Pare PD, Hogg JC (1976) Effect of indomethacin and atropine in experimental asthma in conscious guinea-pigs. J Appl Physiol 40: 889-894
21. Ray NJ, Jones AJ, Keen P (1989) Evidence that $GABA_B$ receptors can modulate release of substance P from capsaicin-sensitive neurons in rat trachea. Br J Pharmacol 97: 562P
22. Rizzo C, Kreutner W (1989) Prejunctional $GABA_B$ inhibition of electrically stimulated contraction in isolated guinea-pig trachea. FASEB J 3: A440
23. Shirakawa J, Taniyama K, Tanaka C (1987) γ-Aminobutyric acid-induced modulation of acetylcholine release from the guinea-pig lung. J Pharmacol Exp Ther 243: 364-369
24. Tamaoki J, Graf PD, Nadel JA (1987) Effect of γ-Aminobutyric acid on neuronally mediated contraction of guinea-pig trachealis smooth muscle. J Pharmacol Exp Ther 243: 86-90
25. Tanaka C (1985) γ-Aminobutyric acid in peripheral tissues. Life Sci 37: 2221-2235
26. Tsuji M, Nakajima T (1978) Studies on the formation of γ-aminobutyric acid from putrescine in rat organs and purification of its synthetic enzyme from rat intestine. J Biochem 83: 1407-1412
27. Verma A, Snyder SH (1989) Peripheral type benzodiazepine receptors. Ann Rev Pharmacol Toxicol 29: 307-322

GABAergic Mechanisms and Modulation of the Anaphylactic Response[1]

L. Zilletti, M. Ciufi, S. Franchi-Micheli, and G. Gentilini

Introduction

It is well known than the anaphylactic reaction can be affected by various neurotransmitters. Furthermore, exocytic processes following challenge with antigen can be modulated due to the presence of different neurotransmitter receptors located on the membranes of involved sensitized cells [23, 22]. GABA, through $GABA_A$ and $GABA_B$ receptors, appears to have a peculiar role as a neuromodulator in the neuro-vegetative system, since it can inhibit or facilitate the release of classical neurotransmitters such as acetylcholine (ACh), norepinephrine (NE) or peptides (for comprehensive and recent reviews see [5, 7]). Thus, the possibility that GABA could interfere with anaphylaxis arose, and this has been the object of our study, performed in in vivo and in vitro models. We have shown that GABAergic drugs can exert a protective effect against allergic bronchospasm in actively sensitized guinea-pigs [20] and in the Schultz-Dale reaction in isolated tissues [33].

[1] This work was supported by a grant from the University of Florence.

GABAergic Mechanisms and the Anaphylactic Reaction in Longitudinal Muscle Strips of the Guinea-Pig Ileum

Longitudinal muscle strips containing the myenteric plexus, isolated from guinea-pig preterminal ileum, were initially selected [33] for two different reasons. First, they represent a suitable model to study the Schultz-Dale reaction [4], since its occurrence is characterized by a sharp contraction due to release of mediators mostly represented by histamine. The preparation is rich in mast cells located along the neural fibers of the myenteric plexus. Secondly, in the myenteric plexus, in which a complex nerve network functions through neurotransmitters activating specific receptors [32], GABA receptors have been well characterized [15, 8, 25].

Effect of GABA on the Schultz-Dale Reaction

Contractions evoked by ovalbumin (OA) in cumulative doses (10^{-10}-10^{-3} g/ml) are strongly inhibited by 10^{-5} *M* GABA [33]. This effect appears to be specific for GABA as it is not mimicked by 2-aminoesthanesulfonic acid (taurine), an amino acid unrelated to GABA neurotransmission. The integrity of neuronal mechanisms is required as the inhibition can mostly be prevented by 3×10^{-7} *M* tetrodotoxin (TTX) [21].

OA, at a maximal concentration (10^{-3}g/ml), elicited a release of histamine that was 43% of the total amine present in the tissue. GABA decreases the anaphylactic histamine release in a time- and dose- dependent manner, having its maximal effect at 10^{-5} *M* (Figs. 1 and 2).

To obtain the maximum effect it is necessary to maintain perfusion with the drug for 10 min before antigen challenge [33]. Also, this inhibition is not mimicked by taurine and is mostly prevented by TTX, while TTX by itself does not significantly decrease histamine release, thus proving the neurogenic nature of the GABA inhibition (Fig. 3) [21].

As the effect is time-dependent, the question arises whether GABA inhibition is linked to receptor desensitization rather than to receptor activation. In fact, in the myenteric plexus, desensitization of $GABA_A$ and $GABA_B$ receptors occurs very quickly (in 2 min), although it could be reversed after washing for 20 min. However, the use of GABA antagonists (see below) contradicts this view.

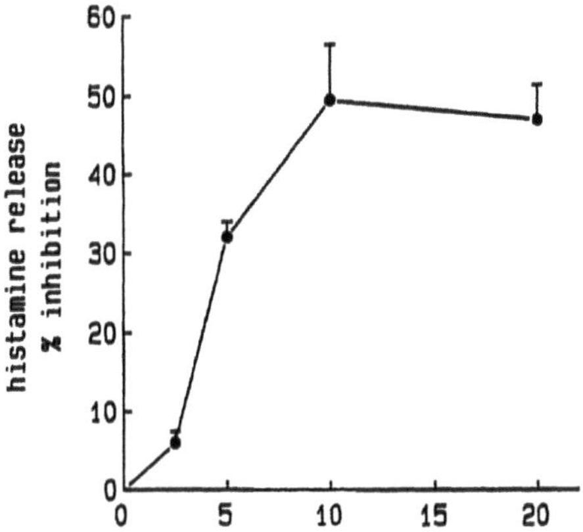

Fig. 1. Time course of GABA inhibition of antigen-evoked histamine release in ileum longitudinal muscle. Ovalbumin (0.1 mg/ml) was administered following 2.5, 5, 10, and 20 min of contact with 10 μM GABA. Values are means ± SEM ($n \geq 5$). (Modified from [21])

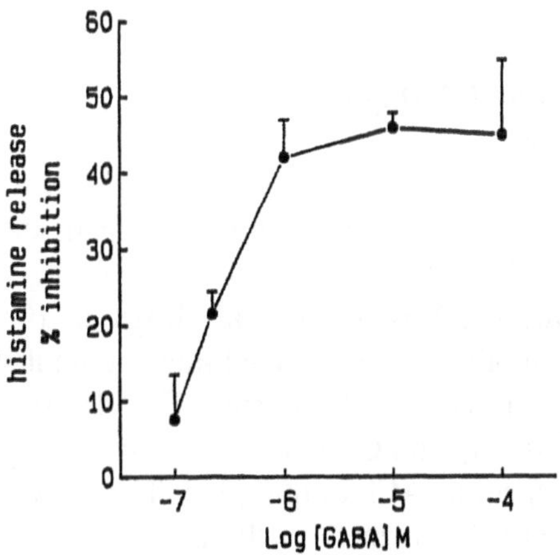

Fig. 2. Inhibition by GABA of anaphylactic histamine release from ileum longitudinal muscle. GABA was added to the perfusion medium 20 min before antigen challenge. GABA effect is expressed as percent inhibition of histamine release evoked by ovalbumin from control preparations. Values are means ± SEM ($n \geq 9$). (Modified from [21])

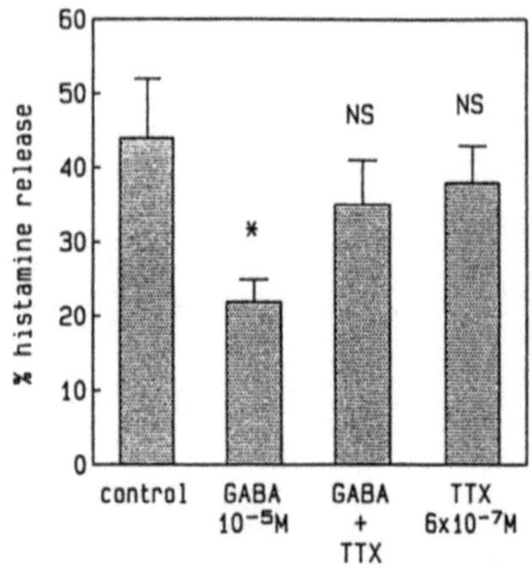

Fig. 3. Prevention by TTX of GABA-induced inhibition of anaphylactic histamine release from ileum longitudinal muscle. TTX and/or GABA were added to the perfusion medium 20 min before antigen challenge. Values are means ± SEM ($n \geq 4$); * $p < 0.025$. (Modified from [21])

Effect of GABA Receptor Agonists and Antagonists on Analphylactic Histamine Release

The inhibitory effect of GABA on anaphylactic histamine release in the longitudinal muscle of the guinea-pig ileum is shared by the $GABA_A$ agonist isoguvacine and the $GABA_B$ agonist (-)-baclofen. Both drugs, at 10^{-5} *M*, decrease histamine release by about 50%, while (+)-baclofen is completely ineffective, thus proving the stereospecifity of the drug (Fig. 4) [19]. It is remarkable that, in this organ, $GABA_A$ and $GABA_B$ activation usually leads to opposing responeses with respect to ACh output [16]. The activation of both types of GABA receptors is confirmed by the use of antagonists. Bicuculline methiodide ($3x10^{-5}M$) and 5-aminovaleric acid (5-AVA) ($3x10^{-3}M$) are effective in reversing GABA inhibition (Fig. 5). Eventually they show a tendency towards potentiation of histamine release, although it is not significant [25].

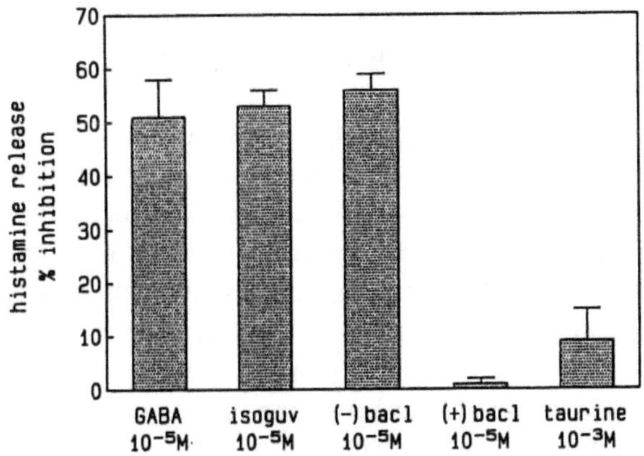

Fig. 4. Effect of various GABA-related drugs on anaphylactic histamine release from ileum longitudinal muscle. Values are means ± SEM ($n \geq 6$). (Modified from [20])

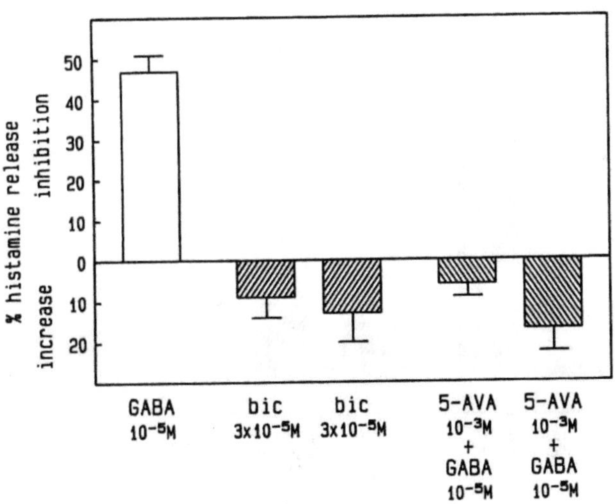

Fig. 5. Effect of receptor antagonists on GABA inhibition of anaphylactic histamine release from ileum longitudinal muscle. Bic, bicuculline; 5-AVA, 5-aminovaleric acid. Values are means ± SEM ($n \geq 6$). (Modified from [20])

Effect of Various Pharmacological Antagonists and Capsaicin on GABA Inhibition of Anaphylactic Histamine Release

Atropine ($10^{-6}M$) and phentolamine plus propranolol ($3x10^{-6}M$) were found not to affect GABA inhibitory action when they were coperfused with GABA for 20 min (the drugs by themselves were almost ineffective) (Fig. 6) [19]. This evidence would exclude the involvement of cholinergic or adrenergic mechanisms in the GABA effect, and the involvement of sensory neuropeptides appears to be uncertain. In fact, when the preparations were perfused with $10^{-5}M$ capsaicin, which depletes sensory nerve terminals of peptide mediators, anaphylactic histamine release was inhibited by 33%. When GABA perfusion was performed in the presence of capsaicin it caused a further 17% decrease in histamine release (not statistically significant), while GABA alone caused a 32% decrease. Therefore, it appears that the effects of GABA and capsaicin are only in part additive (Fig. 7).

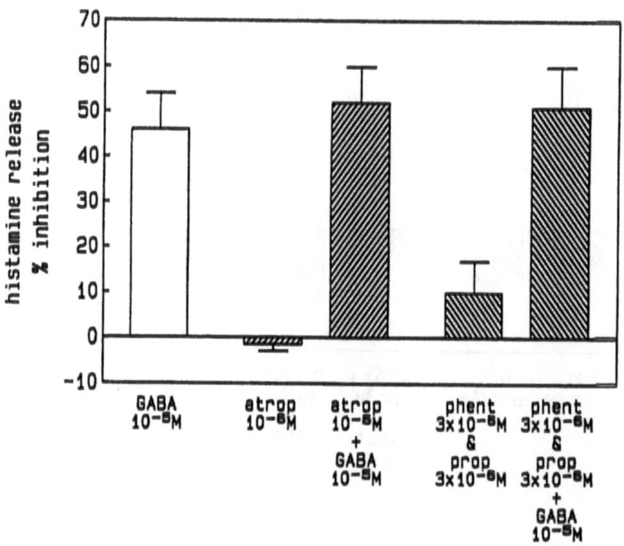

Fig. 6. Effect of atropine (*atrop*) and phentolamine (*phent*) plus propranolol (prop) on GABA inhibition of anaphylactic histamine release from ileum longitudinal muscle. Values are means ± SEM ($n \geq 6$). (Modified from [20])

Effect of GABA Agonists on Anaphylaxis in the Respiratory Tract of the Guinea-Pig

Herxheimer Microshock

Bronchospasm evoked in actively sensitized guinea-pigs by antigen aerosol, according to Herxheimer [12], was selected in view of the significant role of the neurovegetative system in this model and of some similarities to human anaphylactic bronchospasm.

GABA, 3 mmol/kg i.p., injected 20 min before challenge with OA is able to partially protect the animals against anaphylactic bronchospasm. This effect appears to be related to the activation of $GABA_B$ receptors since it is obtained both by GABA and by baclofen in a stereospecific manner; the $GABA_A$ agonist muscimol was ineffective (Fig. 8) [20].

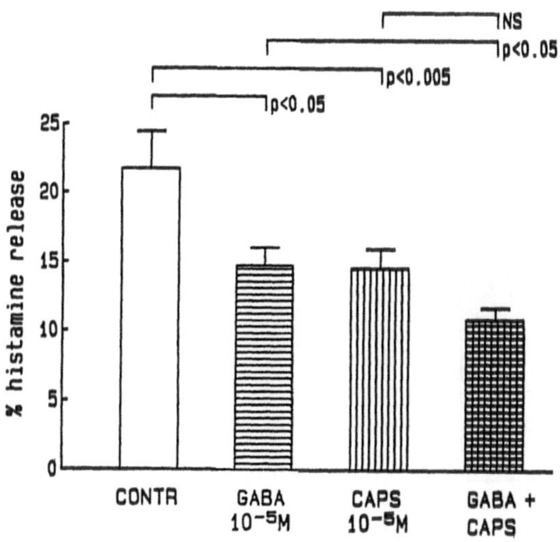

Fig. 7. Influence of GABA and capsaicin (*caps*), single and pooled, on anaphylactic histamine release from ileum longitudinal muscle. Values are means ± SEM ($n \geq 11$). Statistical analysis was performed by Student's t test for paired data

A strong increase in the preconvulsion lag time of severe dyspnea after antigen aerosol, as indicated by the drug dose ratios, shows that (-)-baclofen was about 20 times more effective than GABA (Fig. 9).

This phenomenon may be due to two features of baclofen: its high affinity for the $GABA_B$ receptor [13] and its different pharmacokinetics. In fact, GABA does not cross the blood-brain barrier but exerts its action peripherally [17], while baclofen does [3] and could be active both in the CNS and at the periphery.

Interestingly, (-)-baclofen is able to partially protect unsensitized guinea-pigs against some kinds of nonallergic bronchospasms. At 5 μmol/kg i.p., the drug significantly protected the animals from dyspnea induced by histamine and prostaglandin $(PG)F_{2\alpha}$, but not from that caused by ACh [18]. On the contrary, (-)-baclofen does not influence contractions evoked by histamine, ACh, and $PGF_{2\alpha}$ in isolated tracheal preparations, thus excluding a relaxant effect on smooth muscle.

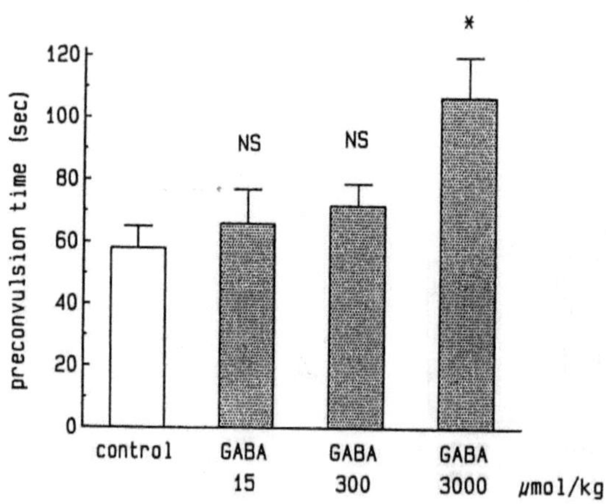

Fig. 8. Effect of GABA, administered i.p. 20 min before ovalbumin aerosol, on preconvulsion time. Values are means ± SEM ($n \geq 5$); * $p < 0.05$. (Modified from [20])

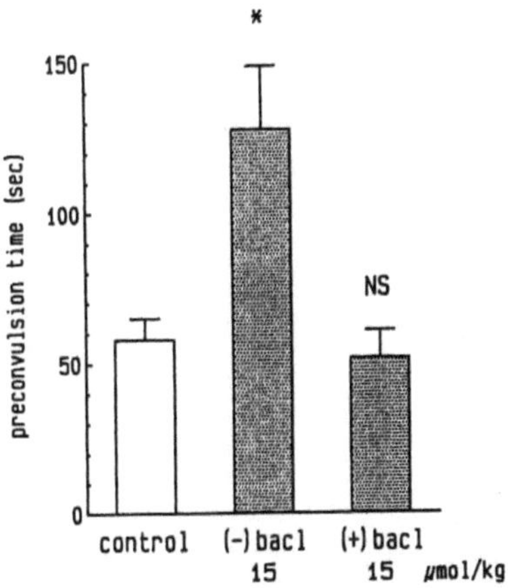

Fig. 9. Effect of (+)- and (-)-baclofen, administered i. p. 20 min before ovalbumin aerosol, on preconvulsion time. Values are means ± SEM ($n \geq 5$); * $p<0.05$. (Modified from [20])

Effect of GABA on the Analphylactic Reaction in In Vitro Preparations

In isolated and perfused lung from sensitized guinea-pigs, (-)-baclofen decreases the marked output of thromboxanes following a bolus injection of OA [18]. This observation appears to be of great interest when considering the spasmogenic and proinflammatory properties of thromboxanes, especially at the respiratory level. The relase of spasmogenic substances (evaluated by their ability to contract guinea-pig ileum smooth muscle) in the presence of $10^{-6}M$ atropine) elicited by antigen from isolated hemitracheas, is inhibited by GABA (Fig. 10) [6]. A similar effect is found in preparations desensitized in vitro to capsaicin. Experiments with capsaicin were carried out in view of the modulatory role of peptidergic sensory fibers in inflammatory processes [28]. When the analphylactic reaction is induced in the presence of both GABA and capsaicin, the release of spas-

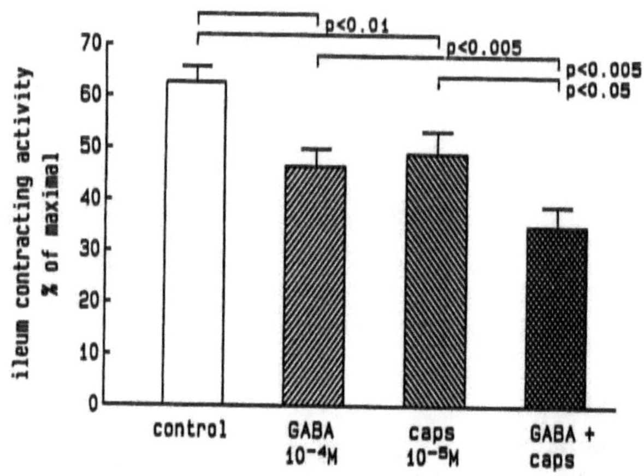

Fig. 10. Influence of GABA and capsaicin (*caps*), single and pooled, on anaphylactic histamine release from isolated guinea-pig trachea. Values are means ± SEM (n = 8). Statistical analysis was performed by Student's t test for paired data. (Modified from [6])

mogenic substances is decreased further. This suggest at least partially distinct mechanisms for the effect of the two drugs [28]. However, in other experimental conditions, a relationship between GABA and peptidergic neurons has been evidenced in peripheral districts and, in particular, in the airways, where GABA appears to modulate the release of other neurotransmitters [29, 27].

The possibility that GABA could influence the anaphylactic reaction through factors other than neurogenic ones arises from evidence that mechanisms modulating respiratory smooth muscle responsiveness are also located in epithelial cells [10, 14]. Mechanical removal of epithelium changes both the reactivity of the tracheal smooth muscle to bronchoreactive agents such as histamine [10, 14] and the reactivity to antigen of tracheal smooth muscle isolated from sensitized guinea-pigs [11]. The latter effect depends on the method of sensitization applied since it is evident in IgG but not in IgE models [2].

The release of smooth muscle relaxing factors of a prostanoid [30,31] or non-prostanoid [9] nature from epithelial cells is involved in the modulatory action of epithelium on smooth muscle responsiveness. Finally, the integrity of epithelium can affect mediator release from the tracheas of sensitized guinea-pigs [31].

The integrity of epithelium appears to be necessary to achieve GABA protection of tracheal preparations during the anaphylactic reaction. GABA (10^{-4}*M*) is able to inhibit the contraction of smooth muscle elicited in hemitracheas in a cumulative dose-dependent manner in the presence of epithelium; after its mechanical removal the amino acid is ineffective (Figs. 11-12). GABA (10^{-4}*M*) is also ineffective when cyclooxygenase activity is blocked by indomethacin (1 μg/ml) (Fig 13.).

In these experimental conditions the contractions evoked by antigen are potentiated by indomethacin when compared to controls. Moreover, the GABA effect is only slightly affected by 3×10^{-7} *M* TTX.

We can conclude that, at least in in vitro tracheal preparations, the integrity of the epithelium and the efficiency of cyclooxygenase enzymatic activity are necessary to evidence the inhibiting effect of GABA. Nonetheless, its slight TTX sensitivity conforms the prevalence of a non-neurogenic mechanism of GABA inhibition. However, the autonomy and efficiency of the tracheal strip innervation should be considered and the efficacy of the TTX effect on the various components of innervation needs to be confirmed.

Effect of GABA on Inflammatory Cells

To evaluate whether the protective effect of GABA against bronchospasm evoked by antigen aerosol in sensitized guinea-pigs could involve inflammatory cells at the pulmonary level, the effect of GABAergic drugs on the responsiveness of guinea-pig alveolar macrophages was studied. Preliminary data obtained from nonadherent alveolar macrophages show that superoxide anion formation, measured as nanomoles of reduced cytochrome in response to f-met-leu-phe-activation, can be influenced by 10^{-4}*M* GABA, (-)-baclofen, and isoguvacine; this results in a decrease in superoxide anion production. Hoewever, the extent of this effect is not great, with average values of inhibition at 10%, 21%, and 22% for GABA, (-)-baclofen, and isoguvacine, respectively (Brunelleschi et al., personal communication). Further investigation must be performed using different

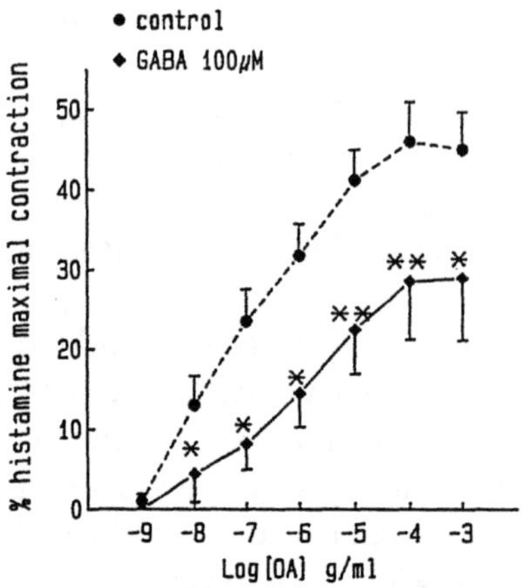

Fig. 11. GABA inhibition of antigen-induced contraction in intact tracheal strips. Values are means ± SEM ($n = 6$); * $p < 0.025$; ** $p < 0.01$. Statistical analysis was performed by Student's t test for paired data

cell populations and cell stimulations in order to establish whether any influence by GABAergic drugs on macrophage functions occurs.

Cells implicated in anaphylactic phenomena appear not to be highly responsive to GABAergic agonists. In fact, GABA and (-)-baclofen do not affect histamine release caused by 48/80 in rat serosal mast cells (unpublished data).

The little or no influence of GABAergic drugs on cells involved in inflammation may be due to the absence of specific receptors on the cell membrane. So far, except for platelets, no information is available on the presence of GABA receptors on blood cells or their responsiveness to GABAergic drugs [26].

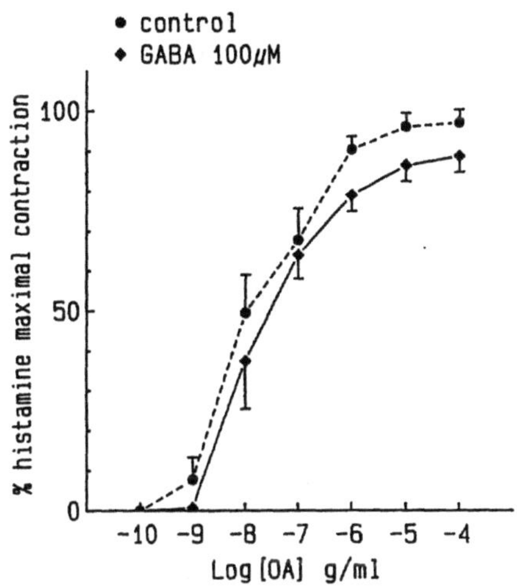

Fig. 12. GABA effect on antigen-induced contraction in epithelium-free tracheal strips. Values are means ± SEM (n = 7)

Conclusions and Discussion

GABAergic drugs appear to be effective in inhibiting anaphylactic reaction both in in vivo and in vitro models. In longitudinal muscle myenteric plexus preparations from sensitized guinea-pigs, the anaphylactic reaction is clearly inhibited by GABA, resulting in a decrease in contraction and release of histamine during challenge with antigen. The observation that the inhibitory effect of GABA is shared by baclofen in a stereospecific manner and by isoguvacine makes it difficult to interpret which type of GABA receptor is involved in the drug's effect. Time dependence suggests that desensitization rather than activation might occur. The $GABA_A$ antagonist bicuculline methiodide and the $GABA_B$ antagonist 5-AVA were

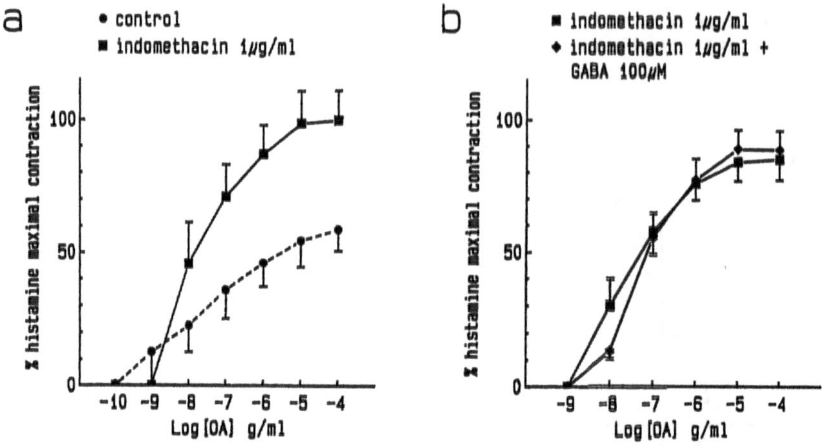

Fig. 13 a, b. Lack of GABA effect in the presence of indomethacin on antigen-induced contraction in intact tracheal strips. Values are means ± SEM.
a Increased amplitude of contractions elicited by cumulative doses of antigen in the presence of indomethacin (*n* = 4). **b** Lack of GABA inhibition (*n* = 5)

able to revert the inhibitory effect of GABA, indicating that GABA receptors have been activated. There is no answer to the question of which type of receptor is involved, at least when considering classical $GABA_A$ and $GABA_B$ receptors. Beyond the hypothesis of the presence of a type of receptor other than $GABA_A$ and $GABA_B$, it is possible to argue that prolonged contact with GABA after desensitization facilitates the emergence of a subtype of classical receptors or reveals a third type. Recent evidence of numerous variants of $GABA_A$ receptor subunits, characterized at a molecular level [24], suggests intriguing possibilities regarding the existence of receptors working in a manner not yet established.

The TTX sensitivity of the inhibitory effect of GABA on the anaphylactic response in the guinea-pig ileum suggests that this phenomenon is neurogenic in nature, although it is not yet known which neuronal system(s) is involved. In fact, cholinergic and sympathetic receptor blockade does not affect GABA inhibition. It is not clear whether capsaicin-sensitive sensory neurons play a role, since simultaneous treatment with GABA and capsaicin hardly increases the inhibitory effect of capsaicin alone.

In the respiratory tract as well as at the intestinal level GABAergic drugs show an inhibitory effect on the anaphylactic response. Evidence is provided both in in vivo and in vitro models. $GABA_B$ receptor activation appears to protect animals partially against anaphylactic bronchospasm while $GABA_A$ receptors seem not to play a role. The inhibition of mediator release during the anaphylactic reaction may explain our evidence. In isolated lungs of sensitized guinea-pigs, (-)-baclofen was found to reduce the efflux of thromboxanes elicited by antigen. Moreover, the release of smooth muscle spasmogenic substances during anaphylaxis was decreased in hemitracheas perfused with GABA. The amino acid was also able to lower the cumulative dose-response curve of antigen in tracheal segments.

The slight TTX sensitivity of the effect suggests that neuronal mechanisms might not be important, at least for the tracheal segments, but this hypothesis warrants further evaluation.

The persistent inhibition of spasmogenic substance release by GABA in the presence of capsaicin suggests that peptidergic sensory terminals are only partially involved. Therefore we can argue that, in our model of anaphylaxis in the isolated traches, mechanisms intervene which are different from those postulated by other authors in nonanaphylactic models. Nonadrenergic noncholinergic neurotransmission has been found to be affected by $GABA_B$ agonists in an in vivo model of neurogenic bronchospasm in the guinea-pig [33] and K^+-evoked release of substance P can clearly be reduced through $GABA_B$ receptor activation in isolated rat tracheal segments [23].

The integrity of tracheal epithelium and efficiency of cyclooxygenase activity appear to play an important role in the effect of GABA. It is reduced after mechanical removal of the epithelium and does not occur in the presence of indomethacin. The role of an epithelial prostanoid relaxing factor can be hypothesized, although direct evidence that GABA facilitates the release of such a substance from epithelium is lacking. The influence of GABA on the responsiveness of inflammatory cells does not seem to have an important role.

In conclusion, GABA and GABAergic drugs can influence the anaphylactic response through both neurogenic and non-neurogenic mechanisms. Whether the latter play a role in the modulatory effect of GABA in processes other than anaphylaxis is matter for further study.

Acknowledgements. We thank Prof. Alberto Giotti for valuable discussion and Mrs. Daniela Bindi for technical assistance.

References

1. Belvisi MG, Ichinose M, Barnes PJ (1989) Modulation of non-adrenegic non-cholinergic neural bronchoconstriction in guinea-pig airways via $GABA_B$ receptors. Br J Pharmacol 97: 1225-1231
2. Bertrand C, Tschirhart E, Landry Y (1989) Nedocromil sodium inhibits IgE- and IgG-related antigen-induced contraction in guinea-pig trachea. Int Arch Allergy Appl Immunol 88: 439-446
3. Brogden RN, Speight TM, Avery GS (1974) Baclofen: a preliminary report of its pharmacological properties and therapeutic efficacy in spasticity. Drugs 8: 1-14
4. Dale MN, Zilletti L (1970) The anaphylactic reaction in the longitudinal muscle strips of guinea-pig ileum. Br J Pharmacol 39: 542-550
5. Erdö SL, Bowery NG (1986) GABAergic mechanisms in the mammalian periphery. Raven, New York
6. Gentilini G, Franchi-Micheli S, Bindi D, Ziletti L (1990) GABA and neuropeptides affect anaphylaxis in guinea-pig airways. Pharmacol Res (to be published)
7. Giotti A, Bartolni A, Failli P, Gentilini G, Malcangio M, Zilletti L (1990) Review of peripheral $GABA_B$ effects. In: Bowery NG, Bittiger H, Olpe HR (eds) $GABA_B$ receptors in mammalian function. Wiley, Chichester (to be published)
8. Giotti A, Luzzi S, Spagnesi S, Zilletti L (1983) $GABA_A$ and $GABA_B$ receptor-mediated effects in guinea-pig ileum. Br J Pharmacol 78: 469-478
9. Goldie RG, Fernandes LB, Farmer SG, Hay DW (1990) Airway epithelium-derived inhibitory factor. TIPS 11: 67-69
10. Hay DWP, Farmer SG, Racburn D, Robinson VY, Fleming WW, Fedan J (1986) Airway epithelium modulates the reactivity of guinea-pig respiratory smooth muscle. Eur J Pharmacol 129: 11-18
11. Hay DWP, Racburn D, Farmer SG, Fleming WW, Fedan JS (1986) Epithelium modulates the reactivity of ovalbumin-sensitized guinea-pig airway smooth muscle. Life Sci 38: 2461-2468
12. Herxheimer H (1952) Repeatable "microshocks" of constant strength in guinea-pig anaphylaxis. J Physiol (Lond) 117: 251-255
13. Hill DR, Bowery NB (1981) ^{3}H-baclofen and 3H-GABA bind to bicuculline-insensitive $GABA_B$ sites in rat brain. Nature 290: 149-152
14. Holroyde MC (1986) The influence of epithelium on the responsiveness of guinea-pig isolated trachea. Br J Pharmacol 87: 501-507
15. Kaplita PV, Waters DH, Triggle DJ (1982) γ-Aminobutyric acid in guinea-pig ileal myenteric plexus. Eur J Pharmacol 79: 43-51
16. Kleinrock A, Kilbinger H (1983) γ-Aminobutyric acid and cholinergic transmission in the guinea-pig ileum. Naunyn Schmiedeberg's Arch Pharmacol 322: 216-220
17. Kuriyama K, Sze PY (1971) Blood brain barrier to tritiated GABA treated animals. Neuropharmacology 10: 103-108

18. Luzzi S, Franchi-Micheli S, Folco G, Rossoni G, Ciuffi M, Zilletti L (1987) Effect of baclofen on different models of bronchial hyperreactivity in the guinea-pig. Agents Actions 20: 307-309
19. Luzzi S, Franchi-Micheli S, Ciuffi M, Rosi E, Zilletti L (1987) Effect of various GABA receptor agonists and antagonists on anaphylactic histamine release in the guinea-pig ileum. Agents Actions 20: 181-184
20. Luzzi S, Franchi-Micheli S, Ciuffi M, Zilletti L (1986) Effects of GABA agonists on Herxheimer microshock in guinea-pigs. Agents Actions 18: 245-247
21. Luzzi S, Franchi-Micheli S, Spagnesi S, Rosi E, Ciuffi M, Zilletti L (1985) Inhibition of anaphylactic histamine release in vitro by GABA. Agents Actions 16: 160-162
22. Masini E, Fantozzi R, Blandina P, Brunelleschi S, Mannaioni PF (1985) The riddle of cholinergic histamine release from mast cells. In: Ellis GP, West GB (eds) Progress in medicinal chemistry, vol 22. Elsevier, Amsterdam, pp 267-291
23. Masini E, Fantozzi R, Blandina P, Galli A, Bani-Sacchi T, Giotti, Zilletti L, Mannanioni PF (1982) Presence of functionally active ß-adrenoceptors in rat mast cells. Correlation between (-) [^{3}H]-dihydroalprenolol binding and inhibition of histamine release. Naunyn Schmiedeberg's Arch Pharmacol 321: 171-176
24. Olsen RW, Tobin AJ (1990) Molecular biology of $GABA_A$ receptors. FASEB J 4: 1469-1480
25. Ong J, Kerr DIB (1983) $GABA_A$ and $GABA_B$ receptor-mediated modification of intestinal motility. Eur J Pharmacol 86: 9-17
26. Oset-Gasque MH, Launay JM, Gonzalez MP (1986) GABAergic mechanisms in blood cells: their possible role. In: Erdö SL, Bowery NG (eds) GABAergic mechanisms in the mammalian periphery. Raven, New York, pp 305-324
27. Ray NJ, Jones AJ, Keen P (1989) Evidence that $GABA_B$ receptors can modulate release of substance P from capsaicin-sensitive neurons in rat trachea. Br J Pharmacol 97: 562P
28. Saria A, Lundberg JM (1985) Neurogenic inflammation. In: Higgs GA, Williams TJ (eds) Inflammatory mediatros. VCH, Weinheim, pp 77-85
29. Shirakawa J, Taniyama K, Tanaka C (1987) γ-Aminobutyric acid-induced modulation of acetylcholine release from the guinea-pig lung. J Pharmacol Exp Ther 243: 364-369
30. Tschirhart E, Frossard N, Bertrand C, Landry Y (1987) Arachidonic acid metabolites and airway epithelium-dependent relaxant factor. J Pharmacol Exp Ther 243: 310-316
31. Undem BJ, Raible DG, Adkinson NFJr, Adams GKW (1988) Effect of removal of epithelium on antigen-induced smooth muscle contraction and mediator release from guinea-pig isolated trachea. J Pharmacol Exp Ther 244: 659-665
32. Wood JD (1987) Physiology of the enteric nervous system. In: Johnson LR (ed) Physiology of the gastrointestinal tract, vol 1. Raven, New York, pp 67-109
33. Zilletti L, Luzzi S, Franchi-Micheli S, Rizzotti M, Rosi E, Spagnesi S (1984) Influence of GABA on anaphylactic histamine release in vitro. Agents Actions 14: 478-480

Subject Index

MIX
Papier aus verantwortungsvollen Quellen
Paper from responsible sources
FSC® C105338

If you have any concerns about our products,
you can contact us on
ProductSafety@springernature.com

In case Publisher is established outside the EU,
the EU authorized representative is:
Springer Nature Customer Service Center GmbH
Europaplatz 3, 69115 Heidelberg, Germany

Printed by Libri Plureos GmbH
in Hamburg, Germany